AF576046

P38 Signaling Pathway

P38 Signaling Pathway

Editors

Ana Cuenda

Juan José Sanz-Ezquerro

MDPI • Basel • Beijing • Wuhan • Barcelona • Belgrade • Manchester • Tokyo • Cluj • Tianjin

Editors

Ana Cuenda
Centro Nacional de
Biotecnología (CSIC)
Spain

Juan José Sanz-Ezquerro
Centro Nacional de
Biotecnología (CSIC)
Spain

Editorial Office
MDPI
St. Alban-Anlage 66
4052 Basel, Switzerland

This is a reprint of articles from the Special Issue published online in the open access journal *International Journal of Molecular Sciences* (ISSN 1422-0067) (available at: https://www.mdpi.com/journal/ijms/special_issues/P38_Pathway).

For citation purposes, cite each article independently as indicated on the article page online and as indicated below:

LastName, A.A.; LastName, B.B.; LastName, C.C. Article Title. *Journal Name* **Year**, *Volume Number*, Page Range.

ISBN 978-3-0365-0858-0 (Hbk)
ISBN 978-3-0365-0859-7 (PDF)

Contents

About the Editors

Ana Cuenda (PhD) is Principal Investigator at Centro Nacional de Biotecnología—CSIC. She holds a BSc in biology and a PhD in Biochemistry from Universidad de Extremadura. After postdoctoral training and staff scientist positions in the MRC Phosphorylation unit at Dundee University (UK), she established her research group at the Department of Immunology and Oncology in CNB, where she has been the director since 2012. Her work is focused on studying the functions of alternative p38kinases, a field where she has made seminal contributions. Her current interest is the analysis of the role of p38MAPK in inflammation and cancer.

Juan José Sanz-Ezquerro (PhD) is Principal Investigator at Centro Nacional de Biotecnología —CSIC. He holds a BSc in Molecular Biology and a PhD in Virology from Universidad Autónoma de Madrid. After postdoctoral training in Developmental Biology in the UK, he established his group in Spain, working at CNB and Centro Nacional de Investigaciones Cardiovasculares (CNIC) on different aspects of organogenesis. His current interests include the analysis of signaling pathways during embryonic development and the interplay between inflammation and regeneration, including the role of p38 signaling in those processes.

Preface to "P38 Signaling Pathway"

p38 Mitogen activated protein kinases (p38MAPK) are a group of evolutionary conserved protein kinases which are central for cell adaptation to environmental changes as well as for immune response, inflammation, tissue regeneration and tumour formation. The interest in this group of protein kinases has grown continually since their discovery. Recent studies using new genetic and pharmacological tools are providing helpful information on the function of these stress-activated protein kinases and show that they have an acute impact on the development of prevalent diseases related to inflammation, diabetes, neurodegeneration, and cancer.

In this Special Issue we present novel advances and review the knowledge on the identification of p38MAPK substrates, functions, and regulation; mechanisms underlying the role of p38MAPK in malignant transformation and other pathologies; and therapeutic opportunities associated with regulation of p38MAPK activity.

This issue will be of interest to basic researchers working in cell signalling and immunology, to chemical biologists interested in drug discovery and also to clinicians interested in cell signalling pathways.

Ana Cuenda, Juan José Sanz-Ezquerro
Editors

International Journal of *Molecular Sciences*

ditorial

›38 Signalling Pathway

uan José Sanz-Ezquerro [1,*] and Ana Cuenda [2,*]

1 Department of Molecular and Cellular Biology, Centro Nacional de Biotecnología/CSIC (CNB-CSIC), Campus-UAM, 28049 Madrid, Spain
2 Department of Immunology and Oncology, Centro Nacional de Biotecnología/CSIC (CNB-CSIC), Campus-UAM, 28049 Madrid, Spain
* Correspondence: jjsanz@cnb.csic.es (J.J.S.-E.); acuenda@cnb.csic.es (A.C.); Tel.: +34-91-5855-395 (J.J.S.-E.); +34-91-5855-451 (A.C.)

Editorial on the Special Issue: p38 Signalling Pathway

p38 Mitogen activated protein kinases (p38MAPK) are a highly evolutionary conserved group of protein kinases, which are central for cell adaptation to environmental changes as well as for immune response, inflammation, tissue regeneration, and tumour formation. p38MAPKs are Ser/Thr kinases that catalyse the reversible phosphorylation of proteins in response to different stimuli such as cellular stress, infections or cytokines. This kinase family is composed by four members encoded by different genes: p38α (*MAPK14*), p38β (*MAPK11*), p38γ (*MAPK12*), and p38δ (*MAPK13*). p38α is the best characterized, whereas the functions of p38β, p38γ and p38δ have been less studied. p38α was the first p38MAPK identified by four different laboratories as a 38 kDa polypeptide that underwent tyrosine phosphorylation in response to endotoxin treatment, cytokines or cellular stress [1]. Some years later, other three p38MAPK isoforms were identified, p38β, p38γ (also known as ERK6 or SAPK3) and p38δ (initially called SAPK4). The four p38MAPKs are ubiquitously expressed, although the levels of expression of each isoform varies in specific tissues; for example, p38β is abundant in brain, whereas p38γ is highly expressed in skeletal muscle, and p38δ in endocrine glands, testis, pancreas, kidney and small intestine [1,2].

:itation: Sanz-Ezquerro, J.J.; :uenda, A. p38 Signalling Pathway. *nt. J. Mol. Sci.* **2021**, *22*, 1003. ıttps://doi.org/10.3390/ijms22031003

:eceived: 5 January 2021
\ccepted: 15 January 2021
'ublished: 20 January 2021

'ublisher's Note: MDPI stays neutral vith regard to jurisdictional claims in ublished maps and institutional affiliations.

The activation of p38MAPK is mediated by a cascade of kinases, which become sequentially activated in response to a wide range of stimuli. This cascade is typically organized in three tiers. p38MAPKs are activated by the MAPK kinases (MAP2K or MKK) MKK3 and MKK6 and, in the case of p38α, also by MKK4, by dual phosphorylation of the Thr-Gly-Tyr motif in the activation loop. The MAP2Ks are in turn activated by phosphorylation by various MAP2K kinases (MAP3Ks) depending on the stimulus and cell context [1,2]. It has been shown that the activation of p38α can also be regulated by MAP2K-independent mechanisms such as for example, autophosphorylation mediated by interaction with TAB1 (transforming growth factor β-activated protein kinase 1 (TAK1) binding protein 1) in cardiomyocytes, or by ZAP70-mediated phosphorylation at Tyr323 in T cells [1]. Based on previous published results showing that protein arginine methyltransferase 1 (PRMT1) promotes p38α activation in the differentiation of erythrocytes, in this special issue, Liu et al. provide new insight into the complexity of p38MAPK regulation by identifying a new molecular mechanism of PRTMT1-modulation of p38α activation [3]. They show that methylation of p38α at Arg49 and Arg149 by PRMT1, increased its interaction with the upstream activator MKK3 (but not MKK6) and with the downstream substrate MAPK activated protein kinase 2 (MAPKAPK2 or MK2). Liu et al. also show that p38α methylation positively regulated AraC-mediated erythroid differentiation and suggested that these findings extend the possible ways of p38MAPK pathway intervention, other than kinase activity inhibition [3].

p38MAPKs control multiple biological functions in the cell by phosphorylating numerous protein substrates, both in the cytoplasm and in the nucleus. p38MAPKs can translocate to the nucleus, either in resting cells or after stimulation. Maik-Rachline et al.

describe some of the mechanisms involved in the nucleo-cytoplasmic shuttling of p38α and p38β (p38α/p38β), which involve the binding to several importins (Imp7, 9 and 3), and to Nuclear Export Signal (NES)-containing substrates (such as MK2) for their export back to the cytoplasm [4]. In their article, the authors also discuss how prevention of p38α/p38β nuclear translocation could be a good tool to prevent some inflammatory diseases, or even cancer like colorectal cancer [4].

The interest in the group of p38MAPKs has grown continually since their discovery. Studies using genetic and pharmacological tools have provided information on the function of these kinases and show that they have a big impact not only on the coordination of the cellular responses to nearly all stressful conditions, but also on the development of prevalent pathologies related to inflammation, autoimmune diseases, neurodegeneration or cancer, among others. Since p38α is the most studied among the p38MAPKs, over the past two decades this kinase has been suggested as a potential therapeutic target for the treatment of different pathologies, particularly inflammatory diseases and cancer [5]. It has been shown that p38α plays opposing effects, either as tumour suppressor or promoter depending on different types of cancer or on phases of cancer development [6]. Recently, the implication of other p38MAPK (p38γ and p38δ) in tumour development has been studied; however, the role of p38β is still unclear. Roche et al., compile the current knowledge about the involvement of p38β in regulating oncoproteins that contribute to tumour initiation, progression, angiogenesis and metastases; or even in other aspects related to cancer disease such as cachexia or pain [7].

Idiopathic pulmonary fibrosis (IPF) is a lung pathology in which inflammation and formation of fibrotic foci are determinant for disease progression and patient mortality. Matsuda et al., examine the implication of p38MAPK activity in bleomycin-induced IPF by performing a comparative transcriptome analysis in alveolar epithelial type II cells from wild type mice and mice expressing either constitutive active MKK6 as a gain-of-function model or dominant negative p38α as a loss-of-function model [8]. These authors found that enhanced p38 signalling in the lungs was associated with increased transcription of genes driving p38MAPK pathway and also correlated with increased bleomycin-induced fibrosis severity, indicating a role of this pathway in the progression of pulmonary fibrosis [8].

p38MAPKs play a pivotal role in muscle physio-pathology; thus, there are numerous evidence implicating them in skeletal muscle and cardiovascular system differentiation/development, in muscle metabolism, as well as in cardiovascular mortality [1,2]. Romero-Becerra et al., revise the role of all p38MAPK isoforms in cardiomyocyte differentiation and growth, and discuss how they are involved in pathological conditions related to ischemia-reperfusion injury, heart failure or cardiac arrhythmia [9]. Bengal et al. focus on p38MAPK implication in glucose metabolic adaptation of skeletal muscle to exercise and obesity; they also discuss p38MAPK role in pathological conditions leading to type II diabetes and the possibility of targeting these kinases in therapies for diabetes treatment [10].

Other processes in which p38MAPK are central elements are those leading to different neuropathologies; this is studied and discussed in several articles of this special issue. Hu et al., report a novel mechanism of kainic acid (KA)-induced seizure that involves p38-mediated phosphorylation at Thr607 of the ion channel Kv4.2, which modulates hippocampal neuronal excitability and seizure strength. Importantly, the authors also show that pharmacological inhibition of p38α reduced neuronal excitability and diminished seizure intensity, indicating the importance of targeting this kinase to mitigate seizure severity in neurological disorders like epilepsy that affect a vast amount of the world population [11]. During the last decade different members of the p38MAPK pathways have joined the group of signalling pathways involved in the development of neurodegenerative diseases, such as Alzheimer's disease. In this context, Germann & Alam review the connection between the Ras-related protein Rab5 (an endosome-associated protein implicated in neurodegenerative disease development) and p38α [12]. Since p38α regulates RAB5 activity, the authors also discuss how brain-penetrant selective p38α inhibitors will provide significant therapeutic advances in neurogenerative disease through normalizing dysregulated Rab5 activity [12]. Falcicchia et al., summarize the role of p38MAPKs, in particular of p38α, in the regulation of synaptic plasticity and its implication in an animal model of neurodegeneration; and extensively describe the use of specific inhibitors for improving synaptic and memory deficit in Alzheimer's disease mouse models [13]. Finally, to shape the final outcome of this Special issue, El Rawas et al. nicely review and discuss the function of p38α in stress, anxiety, depression, and in the rewarding effect of drugs of abuse, particularly cocaine [14].

In this Special Issue, we wished to offer a platform for high-quality publications on the latest advances on p38MAPK pathway substrates, functions and regulation; the mechanisms underlying the role of p38MAPK in different pathologies; and therapeutic opportunities associated with modulation of p38MAPK activity. We expect that the information collected in this issue will be of interest to researchers working in signalling, and stimulate them to continue in their efforts to increase our knowledge on this exciting p38MAPK field. Finally, we would like to thank all researchers who dedicated their effort and time to write their articles for this Special Issue; and specially to the reviewers that generously gave their time and expertise in the field to improve the quality of the published articles.

Funding: This research was funded by grants from the Spanish Ministerio of Ciencia e Innovación (MICINN) (SAF2016-79792R, PID2019-108349RB-I00).

Conflicts of Interest: The authors declare no conflict of interest.

References

1. Cuenda, A.; Rousseau, S. p38 MAP-kinases pathway regulation, function and role in human diseases. *Biochim. Biophys. Acta* **2007**, *1773*, 1358–1375. [CrossRef] [PubMed]
2. Cuenda, A.; Sanz-Ezquerro, J.J. p38γ and p38δ: From Spectators to Key Physiological Players. *Trends Biochem. Sci.* **2017**, *42*, 431–442. [CrossRef] [PubMed]
3. Liu, M.Y.; Hua, W.K.; Chen, C.J.; Lin, W.J. The MKK-Dependent Phosphorylation of p38α Is Augmented by Arginine Methylation on Arg49/Arg149 during Erythroid Differentiation. *Int. J. Mol. Sci.* **2020**, *21*, 3546. [CrossRef] [PubMed]
4. Maik-Rachline, G.; Lifshits, L.; Seger, R. Nuclear P38: Roles in Physiological and Pathological Processes and Regulation of Nuclear Translocation. *Int. J. Mol. Sci.* **2020**, *21*, 6102. [CrossRef] [PubMed]
5. Gaestel, M.; Kotlyarov, A.; Kracht, M. Targeting innate immunity protein kinase signalling in inflammation. *Nat. Rev. Drug Discov.* **2009**, *8*, 480–499. [CrossRef] [PubMed]
6. Wagner, E.F.; Nebreda, A.R. Signal integration by JNK and p38 MAPK pathways in cancer development. *Nat. Rev. Cancer* **2009**, *9*, 537–549. [CrossRef] [PubMed]
7. Roche, O.; Fernández-Aroca, D.M.; Arconada-Luque, E.; García-Flores, N.; Mellor, L.F.; Ruiz-Hidalgo, M.J.; Sánchez-Prieto, R. p38β and Cancer: The Beginning of the Road. *Int. J. Mol. Sci.* **2020**, *21*, 7524. [CrossRef] [PubMed]
8. Matsuda, S.; Kim, J.D.; Sugiyama, F.; Matsuo, Y.; Ishida, J.; Murata, K.; Nakamura, K.; Namiki, K.; Sudo, T.; Kuwaki, T.; et al. Transcriptomic Evaluation of Pulmonary Fibrosis-Related Genes: Utilization of Transgenic Mice with Modifying p38 Signal in the Lungs. *Int. J. Mol. Sci.* **2020**, *21*, 6746. [CrossRef]
9. Romero-Becerra, R.; Santamans, A.M.; Folgueira, C.; Sabio, G. p38 MAPK Pathway in the Heart: New Insights in Health and Disease. *Int. J. Mol. Sci.* **2020**, *21*, 7412. [CrossRef] [PubMed]
10. Bengal, E.; Aviram, S.; Hayek, T. p38 MAPK in Glucose Metabolism of Skeletal Muscle: Beneficial or Harmful? *Int. J. Mol. Sci.* **2020**, *21*, 6480. [CrossRef] [PubMed]
11. Hu, J.H.; Malloy, C.; Hoffman, D.A. P38 Regulates Kainic Acid-Induced Seizure and Neuronal Firing via Kv4.2 Phosphorylation. *Int. J. Mol. Sci.* **2020**, *21*, 5921. [CrossRef] [PubMed]
12. Germann, U.A.; Alam, J.J. P38α MAPK Signaling-A Robust Therapeutic Target for Rab5-Mediated Neurodegenerative Disease. *Int. J. Mol. Sci.* **2020**, *21*, 5485. [CrossRef] [PubMed]
13. Falcicchia, C.; Tozzi, F.; Arancio, O.; Watterson, D.M.; Origlia, N. Involvement of p38 MAPK in Synaptic Function and Dysfunction. *Int. J. Mol. Sci.* **2020**, *21*, 5624. [CrossRef] [PubMed]
14. El Rawas, R.; Amaral, I.M.; Hofer, A. Is p38 MAPK Associated to Drugs of Abuse-Induced Abnormal Behaviors? *Int. J. Mol. Sci.* **2020**, *21*, 4833. [CrossRef] [PubMed]

Article

The MKK-Dependent Phosphorylation of p38α Is Augmented by Arginine Methylation on Arg49/Arg149 during Erythroid Differentiation

Mei-Yin Liu [1], Wei-Kai Hua [1], Chi-Ju Chen [2] and Wey-Jinq Lin [1,*]

1 Institute of Biopharmaceutical Sciences, National Yang-Ming University, Taipei 112, Taiwan; lsnancy51@gmail.com (M.-Y.L.); charleshua921@gmail.com (W.-K.H.)
2 Institute of Microbiology and Immunology, National Yang-Ming University, Taipei 112, Taiwan; cjchen@ym.edu.tw
* Correspondence: wjlin@ym.edu.tw; Tel.: +886-2-28267257

Received: 18 April 2020; Accepted: 15 May 2020; Published: 17 May 2020

Abstract: The activation of p38 mitogen-activated protein kinases (MAPKs) through a phosphorylation cascade is the canonical mode of regulation. Here, we report a novel activation mechanism for p38α. We show that Arg49 and Arg149 of p38α are methylated by protein arginine methyltransferase 1 (PRMT1). The non-methylation mutations of Lys49/Lys149 abolish the promotive effect of p38α on erythroid differentiation. MAPK kinase 3 (MKK3) is identified as the major p38α upstream kinase and MKK3-mediated activation of the R49/149K mutant p38α is greatly reduced. This is due to a profound reduction in the interaction of p38α and MKK3. PRMT1 can enhance both the methylation level of p38α and its interaction with MKK3. However, the phosphorylation of p38α by MKK3 is not a prerequisite for methylation. MAPK-activated protein kinase 2 (MAPKAPK2) is identified as a p38α downstream effector in the PRMT1-mediated promotion of erythroid differentiation. The interaction of MAPKAPK2 with p38α is also significantly reduced in the R49/149K mutant. Together, this study unveils a novel regulatory mechanism of p38α activation via protein arginine methylation on R49/R149 by PRMT1, which impacts partner interaction and thus promotes erythroid differentiation. This study provides a new insight into the complexity of the regulation of the versatile p38α signaling and suggests new directions in intervening p38α signaling.

Keywords: arginine methylation; erythroid differentiation; MKK3; phosphorylation, PRMT1; p38 MAPK

---

1. Introduction

The p38 mitogen-activated protein kinases (MAPKs) play pivotal roles in a number of cellular processes, such as inflammation, stress response, differentiation, and survival [1]. The four members (p38α, p38β, p38γ and p38δ) of the family share high sequence homology and common regulatory mechanisms. However, they also exhibit characteristic biochemical properties and unique cellular functions [2]. The dysregulation of p38α activity plays a dominant role in various pathological conditions. The in vitro and in vivo experiments demonstrate an essential role of p38α in various stages of hematopoiesis, including erythroid differentiation, megakaryocytic differentiation, and myelopoiesis and the involvement in hematological diseases such as myelodysplastic syndromes [3–6]. The concerted regulation of p38α with other signaling pathways can either promote tumorigenesis or suppress tumor progression as shown in colon and breast cancer, respectively [7,8]. p38α triggers the production of inflammatory mediators and cytokines, such as IL-1β and IL-6, upon stimulation [8,9]. The dysregulated activity is linked to a variety of inflammatory diseases, including chronic obstructive pulmonary disease (COPD), colitis, and rheumatoid arthritis [8,9]. The intervention of p38α signaling with an aim toward disease treatment has attracted considerable attention.

p38α MAPK is activated through a canonical cascade involving phosphorylation by upstream dual-specificity MAPK kinases (MKKs), particularly MKK3 and MKK6, on the sequence-specific threonine and tyrosine residues (Thr180 and Tyr182) located in the activation loop. This phosphorylation confers a conformational change that allows the binding of substrates and the accessibility of the catalytic center. Dual-specificity phosphatases responsible for de-phosphorylating Thr180/Tyr182 control the magnitude and duration of the signaling [2]. Accumulating evidence demonstrates the existence of alternative (or additional) mechanisms that regulate p38 activation. GRK2 (G-protein-coupled receptor kinase 2) phosphorylates p38α on Thr123 located at the docking groove for MKKs, which impairs the binding of MKK6 to p38α and diminishes the activation of p38 upon LPS (lipopolysaccharide) stimulation [10]. ZAP70 (zetachain-associated protein kinase 70) and TAB1 (transforming growth factor β-activated protein kinase 1 (TAK1) binding protein 1) can activate p38α in T cell receptor (TCR)-mediated signaling and myocardial injury, respectively [11,12]. ZAP70 phosphorylates p38α on Tyr 323, leading to autophosphorylation on Thr180 and the activation of the kinase [11]. TAB1 interacts with p38α to trigger a conformational change, leading to the autophosphorylation of p38 on Thr180 [12]. Acetylation of Lys53 in the ATP-binding pocket of p38α by PCAF/p300 increases its affinity for ATP and thus enhances the kinase activity during hypertrophy of cardiomyocytes [13]. Moreover, protein arginine methyltransferase 1 (PRMT1) promotes the activation of p38α during erythroid differentiation [3]. However, the detailed molecular mechanism is yet to be revealed. A full understanding of the regulatory mechanisms of p38α activity will provide alternative ways to develop strategies which differentiate p38α functions from other isozymes for the therapeutic needs.

PRMT1 is the predominant protein arginine methyltransferase regulating various cellular processes, including gene transcription, DNA repair, and signal transduction [14]. PRMT1 catalyzes the addition of mono- or di-methyl groups to arginine residues, leading to the alteration of protein/protein interaction, protein/nucleic acid interaction, enzymatic activity and other posttranslational modifications [15]. The malfunction of PRMT1 is tightly associated with many pathological conditions, including hematological malignancy [16,17]. PRMT1 plays important roles in hematopoiesis. PRMT1 is required for adult erythroid and lymphocyte differentiation, as shown by a conditional *Prmt1* knockout mouse model [18]. PRMT1 can methylate RUNX1, a transcription factor critical for hematopoiesis, resulting in the impairment of its association with co-repressor SIN3A, and thus impacts the maturation of myeloid and erythroid lineages in cell models [17]. PRMT1 promotes erythroid differentiation by enhancing the activation of p38α in response to EPO (erythropoietin) and AraC (1-beta-arabinofuranosyl) induction in hematopoietic $CD34^+$ progenitor and K562 cells, respectively [3]. Ca^{2+} influx is an essential event in various stages during erythroid differentiation [19,20]. We have shown that Ca^{2+} up-regulates the activity of PRMT1 and stimulates erythroid differentiation via the novel $Ca2^+$-PRMT1-p38α axis [20].

In this study, we identified Arg49 and Arg149 of p38α as PRMT1 methylation sites by in vitro methylation followed by mass spectrometric analysis (liquid chromatography-tandem mass spectrometry, LC-MS/MS). The non-methylation mutations of Lys49/Lys149 and Ala49/Ala149 abolished the promotive effect of p38α on erythroid differentiation. The activation phosphorylation of R49/149K mutant p38α was greatly reduced upon induced differentiation. The interaction of mutant p38α with the upstream MKK3 was significantly reduced. These results indicate that arginine methylation on R49/R149 enhances the interaction of p38α with MKK3 and thus promotes activation phosphorylation by MKK3. This study also indicates that phosphorylation by MKK3 is not a prerequisite for methylation by PRMT1 and PRMT1 acts directly on p38α. In addition, we identified that MAPKAPK2 (MAPK-activated protein kinase 2) was a p38α downstream effector involved in PRMT1-mediated promotion of erythroid differentiation. Interaction of MAPKAPK2 with p38α was also significantly reduced in the R49/149K mutant. Together, this study unveils a novel regulatory mechanism of p38α activation via protein arginine methylation on R49 and R149 by PRMT1, which impacts the interaction of p38α with its upstream kinase MKK3 and downstream substrate MAPKAPK2 and thus promotes erythroid differentiation.

2. Results

2.1. The Recombinant p38α Protein Is Methylated by PRMT1 on R49 and R149

PRMT1 promotes erythroid differentiation through enhancing the activation of p38α [3]. To investigate whether the enhanced activation of p38α is mediated by arginine methylation and the underlying mechanisms, we first performed in vitro methylation then identified methylation sites by LC-MS/MS analysis. Briefly, His- p38α was incubated with or without recombinant GST-PRMT1 (glutathione S-transferase-PRMT1) in the presence of S-adenosyl-methionine (AdoMet) as a methyl donor. The reaction mixtures were fractionated by SDS-PAGE (sodium dodecyl sulfate polyacrylamide gel electrophoresis), and gels containing p38α proteins were excised and subjected to mass spectrometric analysis. A schematic representation of the procedure is shown in Figure 1A. The sequence coverage of p38α from all identified peptides was above 70%. Peptides harboring dimethylated arginine residues on R49 and R149 in the presence of GST-PRMT1 were identified (Figure 1B) with Xcorr ≥ 2.0. These peptides were not identified when PRMT1 was absent in the methylation reactions. These results indicate that PRMT1 methylates p38α on R49 and R149. The R49 and R149 residues are located in the N lobe in the ATP-binding cleft and in the C lobe near the catalytic loop, respectively, based on the structure information of the kinase domain [21] (Figure 1C).

Methylation of p38α was further shown by using HA-PRMT1 (hemagglutinin-PRMT1) expressed in and immunoprecipitated from K562 cells. The His- p38α protein was also readily methylated by HA-PRMT1 in vitro (Figure 1D, lanes 1 and 4). The methyltransferase-deficient PRMT1G80R greatly lost the ability to methylate p38α, indicating that PRMT1 was indeed the enzyme in the immunoprecipitates responsible for the methylation of p38α (Figure 1D, lanes 4 and 5). The immunoprecipitated HA-PRMT6, which also catalyzes the formation of asymmetric di-methylarginine and shares some substrate recognition sequences with PRMT1 [14], did not methylate p38α. As reported [22], PRMT6 was readily automethylated (Supplementary Figure S1). The methyltransferase-deficient PRMT6KA mutant did not self-incorporate methyl groups (Supplementary Figure S1). These results further provide evidence for the specificity of p38α methylation by HA-PRMT1. The methylation level of p38α was significantly reduced when the arginine residues of 49 and 149 were mutated to lysine, which is not a substrate residue for PRMT1 (Figure 1E). The methyl incorporation into either R49K or R149K was reduced to around 60% compared to the wild-type p38α, indicating that both sites were methyl acceptors. The methylation on R49 and R149 may be a co-dependent event, since simultaneous mutation on R49 and R149 did not completely diminish methylation levels (Figure 1E, R49/149K). We cannot rule out the possibility that there are more methylation sites on p38α since mass spectrometric analysis does not guarantee a complete coverage of all peptides, including modified and non-modified. The structural integrity of these recombinant p38α proteins was examined by their sensitivity to protease digestion. After incubation with trypsin, all four purified proteins, WT, R49K, R149K, and R49/149K, exhibited a similar digestion pattern upon fractionation by SDS-PAGE (Figure 1F). These results indicate that R49K and R149K mutations do not cause a notable structural collapse, particularly in the surface regions.

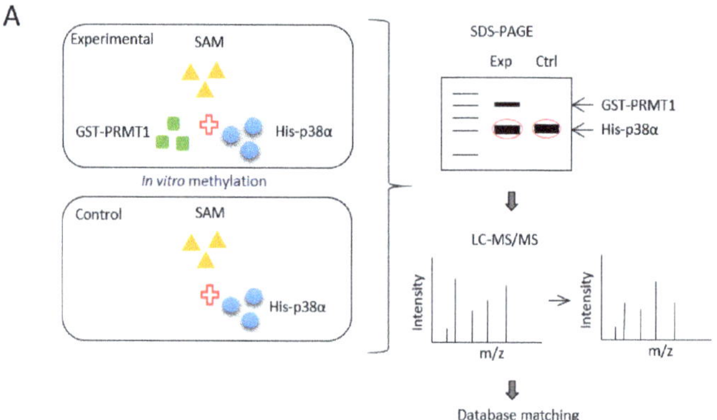

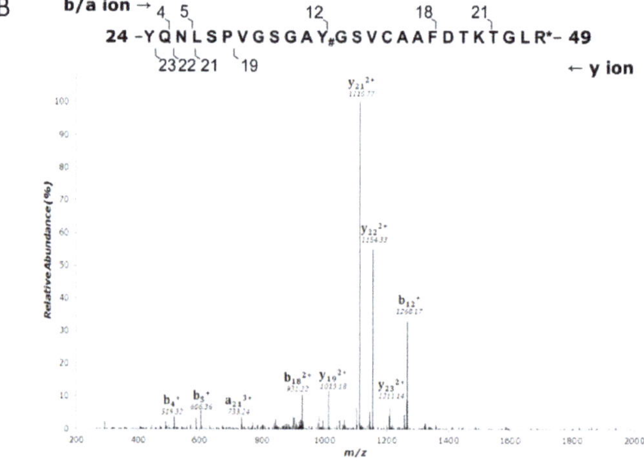

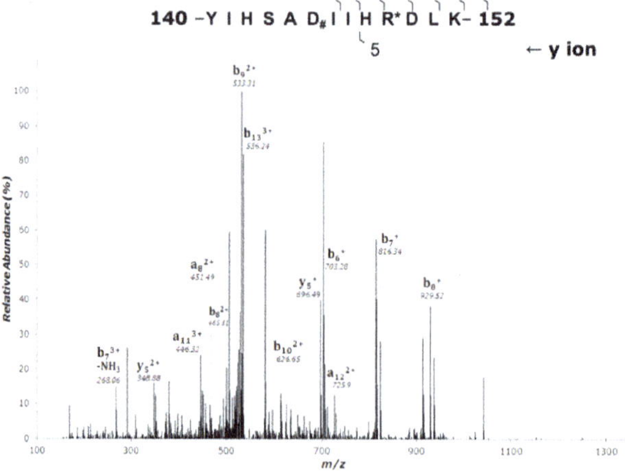

Figure 1. *Cont.*

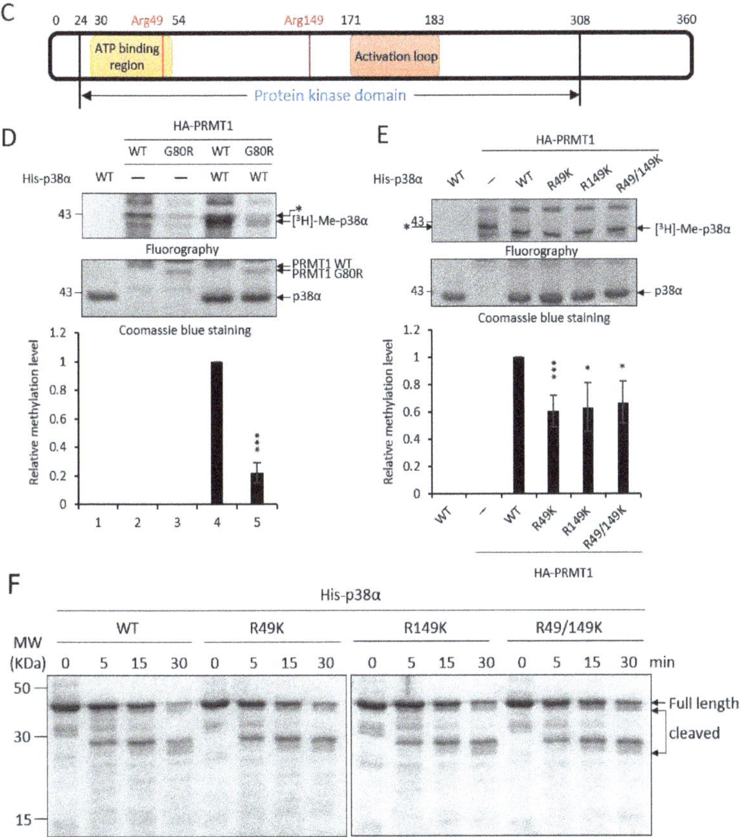

Figure 1. Recombinant p38α protein is methylated by PRMT1 (protein arginine methyltransferase 1) on R49 and R149. The methylation of His- p38α proteins was performed in the presence or absence of recombinant GST-PRMT1 (glutathione S-transferase-PRMT1) as described in the Methods section. Peptides containing di-methyl arginine were identified by mass spectrometric analysis as illustrated (**A**). The sequencing results of peptides harboring di-methylated R49 and R149 are shown (**B**). The localization of Arg49 and Arg149 is marked (**C**). The methylation of His-p38α was further carried out using HA-PRMT1 (hemagglutinin-PRMT1) immunoprecipitated from K562 cells. The methylation of p38α was greatly diminished with the methyltransferase-deficient mutant HA-PRMT1G80R (**D**). When R49 and R149 of p38α were mutated to K49 and K149, the methyl incorporation was significantly reduced as compared to the wild type (WT) (**E**). The quantification in (**D**) and (**E**) was carried out with results from three separate assays. Asterisks indicate non-specific bands. The recombinant p38α WT and mutant proteins (5 μg) were incubated with trypsin at 37 °C for various times and fractionated by SDS-PAGE to reveal their sensitivity to trypsin digestion (**F**). *** $p < 0.005$ and * $p < 0.05$ as compared with WT proteins.

2.2. The Non-Methylation R49K and R149K Mutants of p38α Lose the Ability to Stimulate Erythroid Differentiation

To investigate whether R49 and R149 mediate the promotive effect of PRMT1, we first examined how the non-methylation mutations (Arg to either Lys or Ala) might influence differentiation. The ectopic expression of wild-type p38α stimulated erythroid differentiation from 40% to 60% in p38α-knockdown K562 cells, as measured by hemoglobin accumulation (Figure 2A, Vector vs. WT). Both the R49K and R149K single mutants lost the ability to promote differentiation and so did the R49/149K double mutant (Figure 2A). Similarly, the stimulatory effect on differentiation was completely diminished in R49A and R149A mutants (Figure 2B). These indicate a crucial role of R49 and R149 in stimulating erythroid differentiation. Erythroid differentiation was further examined by the expression of key genes. GATA1 (GATA Binding Protein 1) and EKLF (erythroid Kruppel-like factor) are transcription factors critical for differentiation toward erythroid lineage and PBGD (porphobilinogen deaminase) and ALAS2 (5′-aminolevulinate synthase 2) are enzymes involved in heme synthesis [3]. These transcripts were significantly up-regulated by AraC treatment (Figure 2C, 0 h vs. 96 h); however, the extents were significantly reduced when R49 and R149 were mutated (Figure 2C, R49/149K). The R136 of p38α was also identified as a methylated arginine in our study. However, the methylation event was PRMT1 independent (unpublished results). The R136K mutant of p38α still retained the stimulatory effect on erythroid differentiation (Figure 2D), which provided supports for a selective role of R49 and R149 in modulating differentiation.

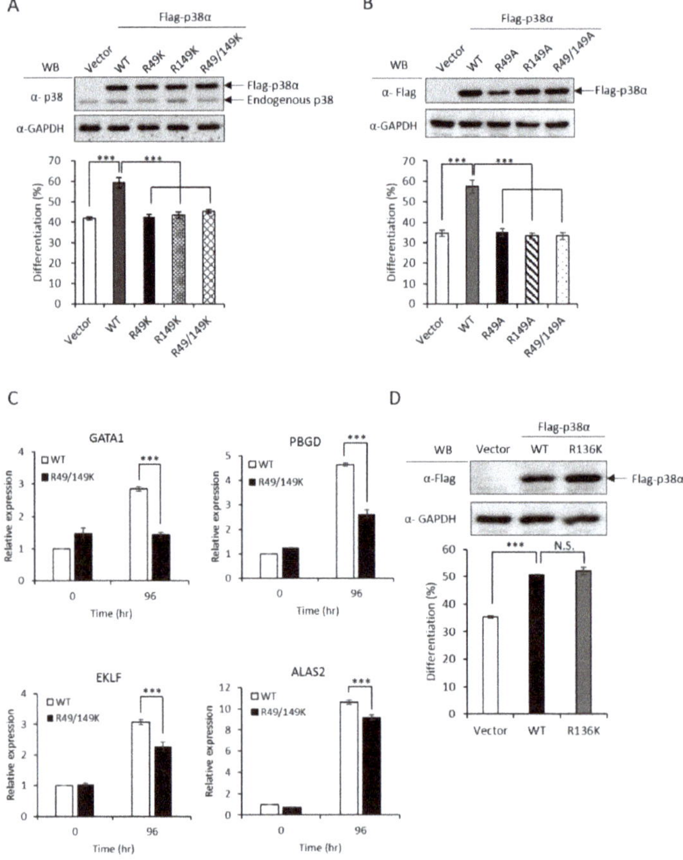

Figure 2. Non-methylation mutants of p38α lose the ability to stimulate erythroid differentiation. The wild-type and R49/R149 mutant p38α were expressed as Flag-tagged proteins in p38α-knockdown K562 cells. Erythroid differentiation was induced with AraC (1-beta-arabinofuranosyl) (1 μM) for 96 hours and analyzed by benzidine staining for the production of hemoglobin (**A** and **B**). Mutations of R49 and R149 to either lysine (R49K, R149K and R49/149K) or alanine (R49A, R149A and R49/149A)

abolished the ability to promote differentiation (**A** and **B**). Erythroid differentiation was also analyzed by RT-qPCR for the expression of key genes. These transcripts were significantly up-regulated by AraC treatment; however, the extents were significantly reduced when R49 and R149 were mutated (**C**). The mutation of R136 to K136 (R136K) did not affect the ability to promote differentiation (**D**). All results shown are representatives of three independent experiments. Erythroid differentiation is presented as the mean ± S.E. of three repeats. *** $p < 0.005$. N.S. means no significance.

2.3. The Promotive Effect of PRMT1 on Erythroid Differentiation Is Mediated by the Methylation of p38α on R49 and R149

We have demonstrated that the PRMT1 promotes erythroid differentiation in a p38α-dependent fashion in K562 cells and human primary CD34$^+$ hematopoietic progenitors [3]. To show whether the kinase activity was required, we ectopically expressed p38α in a p38α-knockdown context. Erythroid differentiation was increased from 40% to 55% (Figure 3A, Vector vs. WT). The AGF p38α mutant is deficient of kinase activity due to mutations on the Thr-Gly-Tyr motif to Ala-Gly-Phe and cannot be activated by phosphorylation via up-stream MKKs [2]. This mutant completely lost the ability to promote differentiation (Figure 3A, Vector vs. AGF), indicating a requirement for the kinase activity to promote differentiation. The catalytic activity of PRMT1 is also required to promote differentiation since the methyltransferase-deficient G80R mutant could not promote [3], indicating that PRMT1 promotes differentiation through the arginine methylation of downstream effector substrates. Notably, PRMT1 promoted differentiation only in the presence of wild-type p38α but not the p38α AGF mutant (Figure 3A, PRMT1 + WT vs. PRMT1 + AGF), indicating that the kinase activity of p38α is essential to mediate the effect of PRMT1. Together, these results provide further evidence suggesting that PRMT1 modulates the kinase activity of p38α likely via arginine methylation.

PRMT1 could not promote differentiation in a p38α-knockdown context (Figure 3B, Vector vs. PRMT1 + Vector). The co-expression of wild-type p38α greatly stimulated differentiation from around 35% to around 60% (Figure 3B, PRMT1 + Vector vs. PRMT1 + WT); however, PRMT1 was unable to stimulate when the R49/R149 of p38α were mutated to K49/K149 (Figure 3B, R49/149K vs. PRMT1 + R49/149K), indicating that the R49 and R149 of p38α mediates the stimulatory effect of PRMT1. Since the methyltransferase activity of PRMT1 is required for the promotive effect of PRMT1 [3] and PRMT1 methylated p38α on R49 and R149 (Figure 1), these results together suggest that R49 and R149 mediate the promotive effect of PRMT1 on erythroid differentiation through arginine methylation.

We further examined the phosphorylation and methylation states of p38α upon AraC stimulation. The Flag-p38α was expressed in PRMT1-knockdown cells with or without the overexpression of HA-PRMT1. Flag-p38α was immunoprecipitated from cells after AraC stimulation and examined by Western Blotting using anti-phospho-p38 or anti-methylarginine antibodies. Our results clearly show that p38α was methylated in cells and PRMT1 enhanced activation phosphorylation as well as the arginine methylation of p38α by around 1.6 folds (Figure 3C), supporting our notion that PRMT1 enhances activation of p38α by arginine methylation of the kinase. To examine whether phosphorylation was a precedent event for arginine methylation, we compared the arginine methylation status of WT and phosphorylation-deficient AGF mutant p38α. The Flag-p38α proteins were expressed in p38α-knockdown cells, immunoprecipitated after AraC stimulation and examined by Western Blotting. The WT p38α was readily phosphorylated while the AGF mutant, as expected, was not phosphorylated (Figure 3D). In spite of the dramatic difference in phosphorylation status, the arginine methylation levels were very similar in WT and AGF mutant proteins (Figure 3D), indicating that phosphorylation is not a prerequisite for arginine methylation by PRMT1.

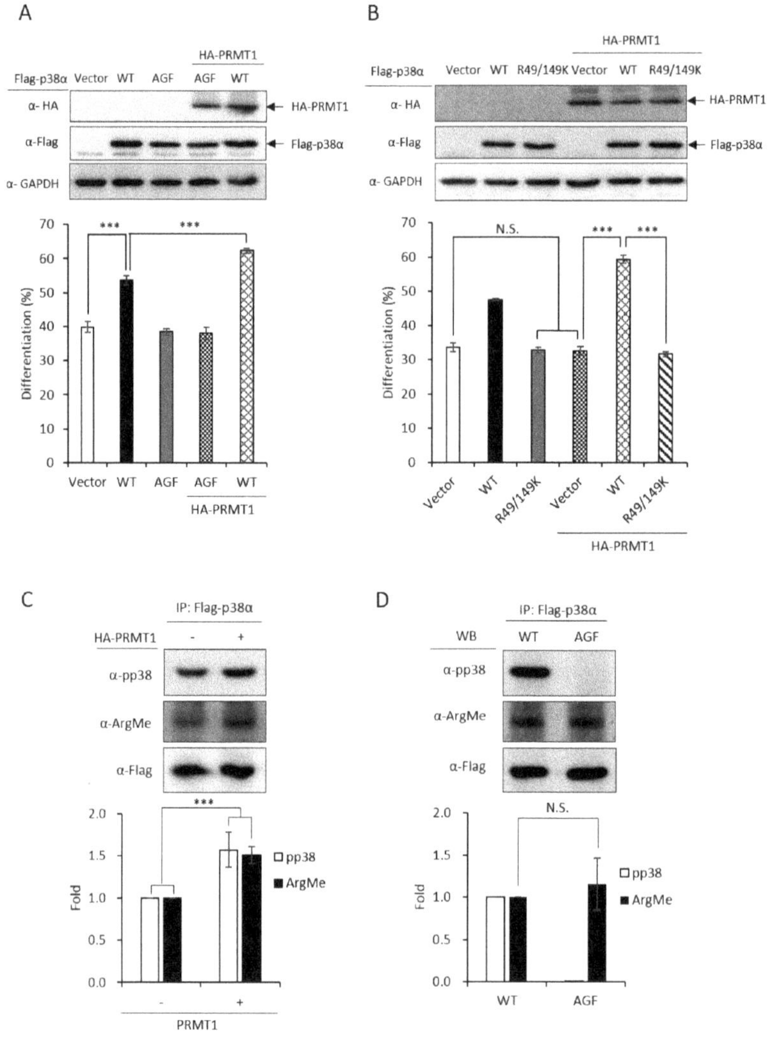

Figure 3. Promotive effect of PRMT1 on erythroid differentiation is mediated by methylation on R49 and R149 of p38α. The wild-type and AGF (Ala-Gly-Phe) mutant Flag-p38α were expressed in p38α-knockdown cells. The wild-type p38α (WT) promoted differentiation but the phosphorylation activation-deficient AGF mutant was unable to (**A**). HA-PRMT1 could further promote only in the presence of wild-type p38α but not the p38α AGF mutant (**A**). In the same p38α KD (knockdown) context, PRMT1 was unable to promote differentiation when R49 and R149 were mutated to K49 and K149 (**B**). Flag-p38α was expressed in the presence or absence of HA-PRMT1. After AraC stimulation, p38α was immunoprecipitated and examined by Western Blotting using anti-phospho-p38 or anti-methyl arginine antibodies. PRMT1 significantly enhanced p38α phosphorylation and arginine methylation (**C**). Upon AraC stimulation, both the WT and AGF mutant were methylated to a similar extent, although AGF was deficient in phosphorylation (**D**). All results shown are representatives of three independent experiments. Erythroid differentiation is presented as the mean ± S.E. of three repeats. The quantification in (**C**) and (**D**) was carried out with results from three separate experiments. *** $p < 0.005$. N.S. means no significance.

2.4. The R49 and R149 Residues Play a Crucial Role in the Activation of p38α

Upon stimulation, p38 MAPK is activated by phosphorylation on the characteristic Thr-X-Tyr motif [2]. AraC treatment stimulated phosphorylation of the wild-type p38α (Figure 4A, WT); however, the activation of the R49/149K mutant was greatly reduced (Figure 4A, R49/149K), as examined by Western Blotting analysis using anti-phospho-p38 antibodies. We further immunoprecipitated Flag-p38α proteins and examined the activation phosphorylation. Similarly, R49/149K mutation significantly reduced AraC-induced activation (Figure 4B). These results indicate that R49/R149 have a role in modulating the activation of p38α. To further assess whether R49 and R149 modulate the activation of p38α in conditions other than AraC-induced differentiation, we stimulated K562 cells with sorbitol, which is a well-known strong stimulator of p38α in hyperosmotic stress [23]. The activation of p38α wild type was stimulated by around 4.0 folds at 0.5 h; however, it was reduced to around 2 folds in R49/149K mutant (Figure 4C). As an internal control, the activation of endogenous p38α was similar, around 3.5 folds, in cells expressing either Flag-p38α WT or Flag-R49/149K mutant (Figure 4C). To examine whether PRMT1 has a role in the activation of p38α induced by sorbitol treatment, we used K562 and PRMT1-overexpressing R2-1 cells as a comparison and found that the sorbitol-stimulated activation of p38 was significantly higher when PRMT1 was overexpressed (Figure 4D, K562 vs. R2-1). The basal level of p38α phosphorylation in R2-1 cells was higher than K562 parental cells before stimulation (Figure 4D, 0 h), which is in agreement with our previous observations [3]. This is conceivable because the ectopically expressed PRMT1 is active [3,4,20]. Together, these results indicate that PRMT1 likely also plays a role in enhancing the activation of p38α via methylation on R49 and R149 upon sorbitol stimulation.

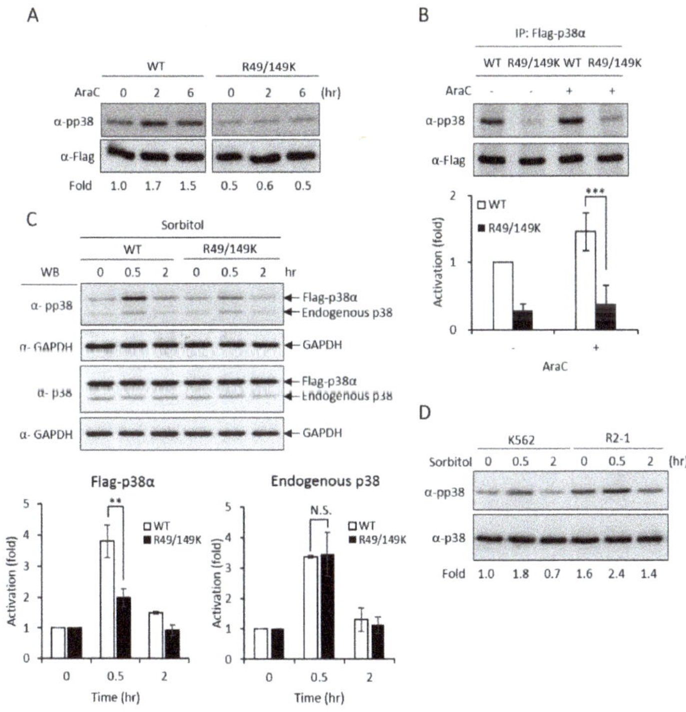

Figure 4. R49 and R149 residues play a crucial role in the activation of p38α. The Flag-p38α wild type or R49/149K mutant were expressed in p38α-knockdown cells. Cells were stimulated with AraC (1 μM)

and the activation of p38α was examined directly by Western Blotting using specific anti-phospho-p38 antibodies (**A**) or after immunoprecipitation (**B**). Phosphorylation was greatly reduced in R49/149K. Alternatively, the cells were treated with sorbitol (150 mM) and the activation of p38α was examined by Western Blotting (**C**). The overexpression of PRMT1 (R2-1) enhanced the activation of p38α upon sorbitol stimulation (**D**). The levels of pp38 and p38 were quantified by Multi-Gauge V3.0 analysis. All results shown are representatives of three independent experiments. Statistical analysis was performed with results from three separate experiments. *** $p < 0.005$ and ** $p < 0.01$ as compared with WT.

2.5. PRMT1 Acts Downstream of MKK3 to Promote Erythroid Differentiation and the R49/149K Non-Methylation Mutant p38α Exhibits a Reduced Interaction with MKK3

MKK3 and MKK6 both are upstream kinases, which phosphorylate and activate the p38 MAPK pathway [1]. To identify which MKK is responsible for activating p38α and promoting erythroid differentiation, we established MKK3 and MKK6 knockdown cell clones (Figure 5A,B, upper) and found that only MKK3 knockdown (MKK3 KD) greatly compromised erythroid differentiation (Figure 5A, lower), indicating that MKK3 is the major MAPK kinase in AraC-induced erythroid differentiation. On the contrary, MKK6 knockdown (MKK6 KD) exhibited a remarkable stimulation on erythroid differentiation (Figure 5B, lower), indicating a previously unknown role in negatively regulating erythroid differentiation. This result suggests that MKK6 is not a p38α activating MAPK kinase in erythroid differentiation. Furthermore, the AraC-induced activation of p38 was greatly reduced in MKK3 KD cell clones (Figure 5C), confirming that MKK3 plays a role in activating p38 during differentiation. This notion was further supported by the observation that overexpression of p38α in MKK3 KD cells partially rescued the differentiation but p38β, which is not involved in erythroid differentiation [3], could not (Figure 5D). Mutations of R49 and/or R149 lost the ability to compensate the inefficiency of MKK3 activity in MKK3 KD1 cells (Figure 5E), suggesting R49 and R149 are required to mediate the activation of p38α by MKK3.

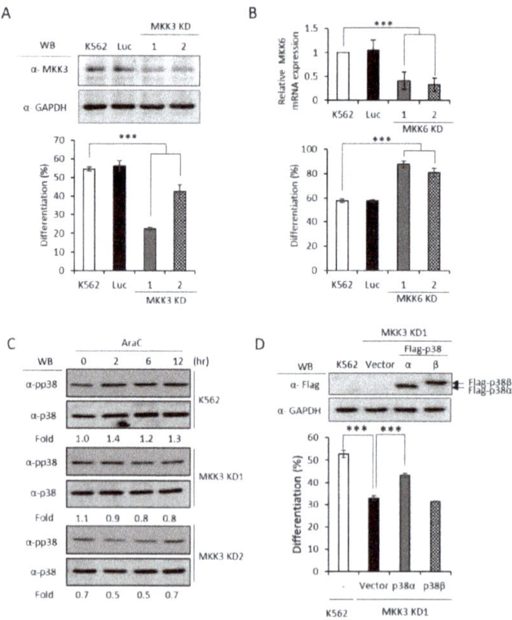

Figure 5. *Cont.*

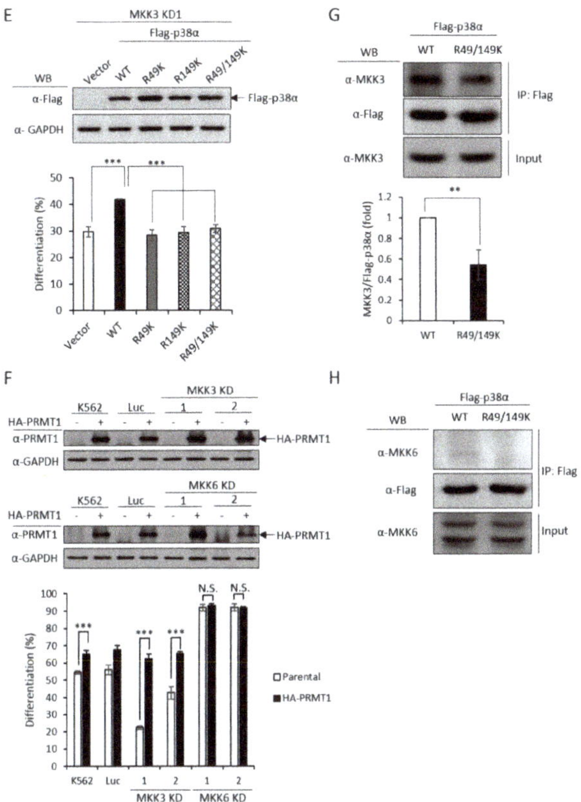

Figure 5. PRMT1 acts downstream of M KK3 (MAPK kinase 3) to promote erythroid differentiation, and the R49/149K non-methylation mutant exhibits a reduced interaction with MKK3. MKK3 and MKK6 were knocked down (**A**) and (**B**) as described in the Methods section. AraC-induced erythroid differentiation was significantly suppressed in MKK3-knockdown (KD1 and KD2) cells (**A**); however it was stimulated in MKK6-knockdown (KD1 and KD2) cells (**B**). The activation of p38 was significantly reduced in MKK3 KD1 and KD2 cells (**C**). Ectopic expression of p38α, but not p38β, in MKK3 KD1 cells could partially rescue differentiation (**D**). However, the R49K, R149K and R49/149K mutants lost the ability to promote differentiation (**E**). Ectopic expression of HA-PRMT1 promoted differentiation in parental K562, Luc and MKK3 KD1 and KD2 cells but had no effect in MKK6 KD1 and KD2 cells (**F**). Luc is the vector control cell. To examine the interaction of p38α with MKK3 and MKK6, Flag-p38α wild type and R49/149K mutant proteins were expressed in p38α KD cells. After cells were stimulated with AraC (1 μM) for 5 h, Flag-p38α proteins were immunoprecipitated and the protein levels of MKK3 (**G**) or MKK6 (**H**) in the immunoprecipitates were examined by Western Blot. The levels of MKK3, pp38 and p38 were quantified by Multi-Gauge V3.0 analysis. All results shown are representatives of three independent experiments. Statistical analysis was performed with results from three separate experiments. *** $p < 0.005$. ** $p < 0.01$.

Since PRMT1 enhanced the activation of p38α (Figure 3C), we next examined whether PRMT1 acts upstream or downstream of MKK3 to enhance the activation of p38α and to promote differentiation. We overexpressed PRMT1 in MKK3 KD cells and found that erythroid differentiation was greatly stimulated to around 65% as compared to 20–40% without PRMT1 overexpression (Figure 5F). These results indicate that higher PRMT1 levels can compensate the inefficiency of MKK3 functions and suggest that PRMT1 acts downstream of MKK3. In agreement with the notion that MKK6 was not the upstream kinase for p38α (Figure 5B), PRMT1 did not further stimulate differentiation in MKK6 KD cells (Figure 5F). Taken together, these results indicate PRMT1 acts downstream of MKK3 and likely by directly methylating p38α on 49 and R149.

The MAPKs are known to form a protein complex with upstream kinases, scaffold proteins and phosphatases to bring the players to a close proximity, which confers temporal and spatial regulation of signal transduction [1]. When immunoprecipitated from cells, the p38α protein interacted with MKK3 only but not MKK6 (Figure 5G,H, WT), supporting our notions that MKK3, but not MKK6, is the major MAPK kinase for p38α. The non-methylation p38α R49/149K mutant exhibited a significantly reduced interaction with MKK3 by around 50% (Figure 5G, R49/149K). Similar to the wild type, the R49/149K mutant did not interact with MKK6 (Figure 5H, R49/149K). The expression levels of MKK3 and MKK6 were similar in the parental and R49/149K mutant cells (Figure 5G,H, input). Together with the observation that the activation of the R49/149K mutant was significantly lower than the wild type (Figure 4A–C), our results suggest that R49 and R149 up-regulates the activation of p38α through enhancing the interaction of p38α with MKK3 via arginine methylation by PRMT1.

2.6. Identification of MAPKAPK2 as a Downstream Effector of P38α During Erythroid Differentiation and the R49/149K Non-Methylation Mutant Exhibits a Reduced Interaction with MAPKAPK2

In order to have a comprehensive understanding of the biochemical and functional role of R49 and R149 methylation, we further analyzed the protein interactors of the wild-type and mutant p38α proteins. We co-expressed either the wild-type or the R49/149K mutant with HA-PRMT1 in p38α KD cells, performed immunoprecipitation, fractionated the immunoprecipitates by SDS-PAGE and carried out mass spectrometric analysis to reveal the interacting proteins. A total of 163 proteins were found to interact with both WT and mutant p38α. We then analyzed the connection of these proteins with the p38α pathway by using Ingenuity Pathway Analysis (IPA). MAPKAPK2 and MAPKAPK3 (MAPK-activated protein kinases 2 and 3) were identified by the software within the limit of "direct interaction" and "experimentally observed". MAPKAPK2 is a known substrate of p38α [24], whereas the role of MAPKAPK3 is less described. Since, to the best of our knowledge, there was no report for the role of MAPKAPK2 in erythroid differentiation, we knocked down MAPKAPK2 and examined AraC-induced erythroid differentiation. The results showed that the knockdown of MAPKAPK2 reduced erythroid differentiation, from 50% to 35–40% (Figure 6A,B), suggesting a role of MAPKAPK2 in erythroid differentiation. The ectopic expression of p38α still promoted differentiation in the MAPKAPK-2 KD cells, likely due to the remaining low level of MAPKAPK2. However, the differentiation was significantly lower in KD-1 cells (49%) than in the K562 parental cells (61%), where the MAPKAPK2 level was normal (Figure 6B). These results suggest that p38α signals through MAPKAPK2 to promote differentiation. This notion was further supported by the immunoprecipitation experiments, showing the interaction of MAPKAPK2 with p38α (Figure 6C, WT). Notably, the interaction was significantly reduced to about 60% when R49 and R149 were mutated (Figure 6C, R49/149K). Although R49/149K mutation reduced the interaction of p38α with its upstream kinase MKK3 as well as with its downstream substrate MAPKAPK2, the interaction of p38α with protein phosphatase 2A (PP2A) was not significantly affected (Supplementary Figure S2). To examine the influence of arginine methylation on partner interaction, we immunoprecipitated Flag-p38α in the presence or absence of HA-PRMT1 upon AraC stimulation. The results show that PRMT1 enhanced the interaction of p38α with MKK3 and MAPKAPK2 by 1.7 folds and 1.2 folds, respectively (Figure 6D). Together, our results indicate that the R49 and R149 of p38α are critical for a previously unidentified role in selective partner interactions through arginine methylation by PRMT1.

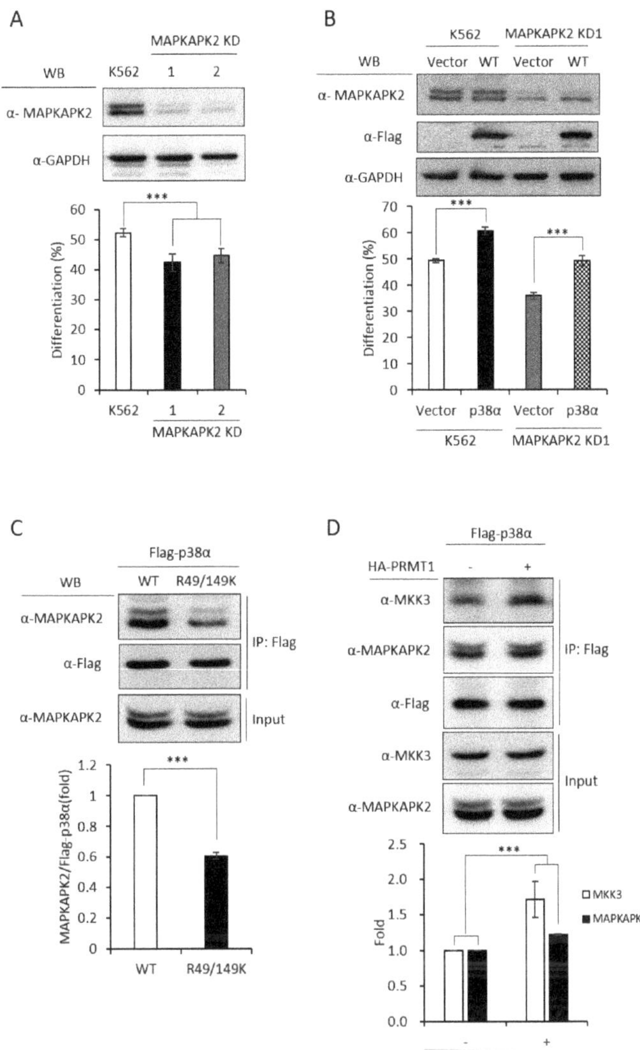

Figure 6. MAPKAPK2 is a downstream effector of p38α and the R49/149K non-methylation mutation reduces the interaction of p38α with MAPKAPK2. The wild-type and R49/149K mutant p38α proteins were expressed in p38α KD cells. Cells were stimulated with AraC and Flag-p38α was immunoprecipitated using anti-Flag antibodies. The interacting proteins were analyzed by mass spectrometric analysis. MAPKAPK2 was identified as an interactor of the p38α by Ingenuity Pathway Analysis (IPA). AraC-induced erythroid differentiation was reduced in MAPKAPK2 knockdown (KD) cells (**A**). The ectopic expression of p38α rescued differentiation of MAPKAPK2 KD1 cells (**B**). The Flag-p38α WT and R49/149K proteins were immunoprecipitated after AraC stimulation. MAPKAPK2 interacted with wild-type p38α; however, the R49/149K mutant exhibited a significantly lower interaction with MAPKAPK2, as examined by Western Blotting (**C**). PRMT1 promoted the interaction of wild-type p38α with MKK3 and MAPKAPK2 (**D**). The intensity of protein bands in Western Blots were quantified by Multi-Gauge V3.0 analysis. All results shown are representatives of three independent experiments. Statistical analysis was performed with results from three separate experiments. *** $p < 0.005$.

This study reveals a novel regulatory mechanism for p38α. Arginine methylation of R49/R149 by PRMT1 occurs upon stimulation, which does not require a prior phosphorylation on Thr180 and Tyr182. The methylation of R49/R149 facilitates the selective association of p38α with MKK3 and thus augments phosphorylation by MKK3 and enhances the activation of p38α. The methylation of R49/R149 also facilitates the association of p38α with a downstream effector MAPKAPK2, which enhances the propagation of signals to up-regulate erythroid differentiation. An illustrated model is presented in Figure 7.

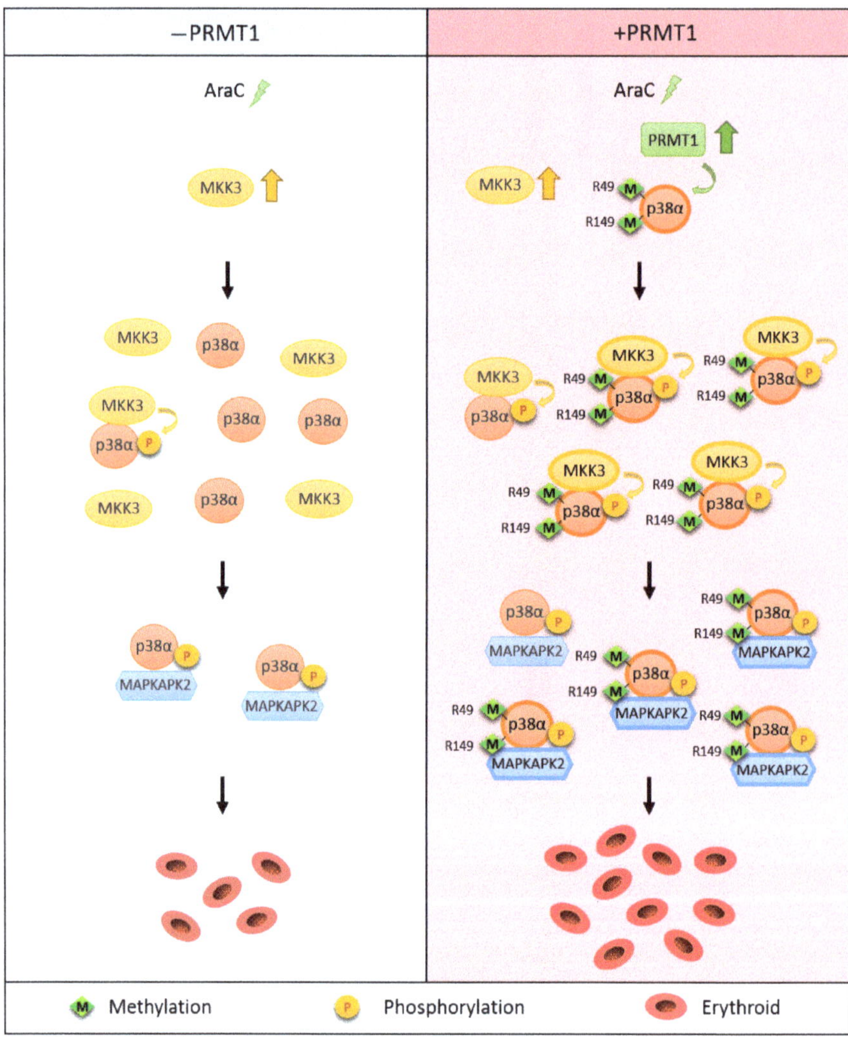

Figure 7. Illustration of the novel regulatory mechanism for p38α through arginine methylation on R49/R149 by PRMT1. AraC treatment stimulates the methyltransferase activity of PRMT1, which methylates p38α on R49 and R149. This facilitates the interaction of p38α and MKK3 and enhances the activation phosphorylation of p38α by MKK3. The interaction of p38α with downstream effector MAPKAPK2 is also increased upon R49/R149 methylation by PRMT1 methylation. Together, erythroid differentiation is promoted due to the facilitated p38α signaling. Green arrow: methylation. Yellow arrow: phosphorylation.

3. Discussion

Extensive efforts have been focused in manipulating the activity of p38 MAPK for therapeutic purposes due to its close association with various pathological conditions, including inflammatory and malignant diseases [8]. Activation by phosphorylation via the upstream MKKs is the canonical mode of kinase activation. This study identifies, for the first time, a posttranslational modification of arginine methylation on Arg49 and Arg149 of p38α by PRMT1 and demonstrates that methylation on R49/R149 modulates activation of p38α through an increased association with the upstream MKK3 and the downstream effector MAPKAPK2 and thus impacts erythroid differentiation. This study elucidates a novel regulatory mechanism for p38α activation and indicates a potential new strategy in intervening p38α signaling.

Upon stimulation, a cascade of phosphorylation events involving MEKs and MKKs result in the phosphorylation of p38 MAPKs on the conserved Thr^{180}-Gly-Tyr^{182} motif, leading to an open and extended activation loop, which allows for the binding of substrate and facilitates catalysis [21]. In addition to this canonical mode, increasing evidence has shown other modes of regulation. The phosphorylation of Thr180/Tyr182 of p38 can also be achieved by p38 auto-catalysis in response to T cell receptor (TCR)- or TNFα-mediated signaling in an MKK-independent manner [11,12]. The triggering of autophosphorylation can be induced by a precedent phosphorylation of p38 on Tyr323 by ZAP70, a T cell receptor-associated tyrosine kinase [11], or by the binding of TAB1, an adaptor protein in TNFα-mediated signaling, with p38α [12]. In the MKK-dependent mode, docking regions on p38 mediate its interactions with various partners including MKKs, substrates, and phosphatases, in a context-dependent fashion, and contributes to the selective transduction of diverse signals [25]. GRK2 (G-protein-coupled-receptor kinase 2) phosphorylates p38, in a known docking groove, compromises the binding of MKK6 and thus suppresses the activation of p38 upon LPS stimulation in macrophages [10]. In this study, we show that arginine methylation of R49 and R149 by PRMT1 enhances the activation phosphorylation of p38α upon induced erythroid differentiation (Figure 3C, Figure 4A,B). We identify that MKK3 is the responsible upstream kinase (Figure 5A,C,D and E). The enhanced activation of p38α is mediated by an increased association with the upstream kinase MKK3 via R49/R149 methylation (Figure 5F,G). MKK6 is not an upstream kinase for p38α in AraC-induced erythroid differentiation (Figure 5B,F) and is not associated with p38α no matter whether R49 and R149 are wild-type or mutated (Figure 5H). These results uncover a novel mechanism for the regulation of p38α signaling selectively through MKK3 by arginine methylation.

The correct recognition of MAPKs by the cognate interacting partners is critical for specifically transducing signals with a high fidelity. A common docking (CD) domain containing a few of conserved acidic residues is found in MAPKs that contribute mainly to the binding affinity [25,26]. The intervening sequences of these acidic residues and sequences in other regions play predominant roles in partner selectivity [25,26]. The sequence context for partner selection and binding affinity in response to different stimuli is not fully understood. Arg49 and Arg149 are not located near the conserved acidic residues Asp313, Asp315, and Asp316 in the CD domain of p38α. Arginine methylation catalyzed by PRMTs alters the hydrogen bonding capacity, hydrophobicity, and steric hindrance of the target arginine and its vicinity [27] and thus affects a wide range of protein properties. The activity of PRMT1 is up-regulated during erythroid differentiation [3], leading to the methylation of p38α and the enhanced activation of the kinase (Figure 3C). Arginine methylation on Arg49/Arg149 can potentially cause a conformational change or an increased hydrophobicity favoring partner selection and/or binding affinity. In addition, the HRD sequence (His148-Arg149-Asp150) is shown to lock p38α in an inactive conformation that can be disrupted by Tyr323 phosphorylation [28]. This observation raises the possibility that the methylation of R149 may facilitate a conformational change favorable for kinase activation. Arg49 is stereoscopically near Leu108 and Met109, which is part of a lipophilic pocket facilitating ATP binding [29]. The possibility that methylation on R49 favors the stabilization of this lipophilic pocket is worthy of further study.

p38α is a versatile MAPK participating in many important physiological and pathological conditions [1–3]. The molecular events involved in its activation and regulation have attracted considerable research attention. A number of posttranslational modifications (PTMs) in p38α have been identified via proteomics approaches [30]. However, relatively few of the identified PTMs are reported with functional impacts. In this study, we have identified several methylated arginine residues in p38α. Among those, R49 and R149 are dimethylated only in the presence of PRMT1, suggesting they are PRMT1 substrate sites (Figure 1). Non-methylation mutants of R49K and R149K lost around 40% of methyl incorporation (Figure 1), indicating the existence of other potential PRMT1 sites. In cells, the activation phosphorylation of R49/149K mutant is remarkably reduced (Figure 4), which explains the incapability of the mutant to promote erythroid differentiation (Figure 2). The lack of Thr^{180}-Gly-Tyr^{182} phosphorylation of the AGF mutant does not affect its arginine methylation level (Figure 3D), indicating that phosphorylation is not a prerequisite for methylation. Arginine methylation by PRMT1 significantly increases the association of p38α with the upstream kinase MKK3 (Figures 5G and 6D) and the downstream substrate MAPKAPK2 (Figure 6C,D). This study reveals that the arginine methylation of p38α on R49/R149 by PRMT1 renders the kinase more accessible to the enzyme (MKK3) and also facilitates its action toward the downstream effector MAPKAPK2. The importance of methylation in the MAPK signaling is also evidenced in other studies. The methylation of MAPK kinase kinase 2 (MAP3K2) on Lys260 by SMYD3 (SET and MYND domain containing 3) blocks its interaction with the negative regulator PP2A phosphatase and activate Ras-mediated MEK/ERK (extracellular signal-regulated kinase) signaling [31]. The arginine methylation of Raf by PRMT5 results in an increased degradation, a decreased Ras-Raf-Erk signaling and a reduced proliferation of PC12 cells upon EGF (epidermal growth factor) stimulation [32]. Our results demonstrate that, in addition to erythroid differentiation, the sorbitol-stimulated activation of p38α is also enhanced by PRMT1 and dependent on the methylation of R49 and R149 (Figure 4C,D). Together, methylation, in collaboration with phosphorylation, is emerging as a critical regulatory mechanism of the MAPK pathways via which extracellular cues are integrated to elicit an appropriate response in terms of signal magnitude, duration and specificity.

Although p38α has been shown to be a critical player in various stages during erythroid differentiation, its upstream activating kinase and downstream substrate had not been clearly and fully revealed. Our results decisively show that only MKK3, not MKK6, mediates the activation of p38α in AraC-induced erythroid differentiation (Figure 5A,B). The immediate downstream effectors of p38α in erythroid differentiation are much less described. We identify that MAPKAPK2 participates in erythroid differentiation via p38α signaling (Figure 6). The RNA-binding protein human antigen R (HuR) protein is an effector of MAPKAPK2 that can bind and stabilize GATA1 transcripts during embryonic erythropoiesis [33]. GATA1 is a critical transcription factor upregulated during erythroid differentiation [34] in a p38α-dependent fashion (Figure 2C). Whether MAPKAK2 promotes erythroid differentiation by stabilizing GATA1 transcripts via HuR is of interest.

Genetically modified mice have evidenced that the p38α pathway plays important roles in inflammatory responses and hematopoietic homeostasis among others [8]. A number of pharmacological inhibitors of p38 MAPK have been developed and tested in clinical trials for treating inflammatory diseases, such as rheumatoid arthritis and chronic obstructive pulmonary disease and hematopoietic diseases, such as myelodysplastic syndromes (MDS), which frequently leads to hematological malignancy [35]. Major drawbacks for p38 inhibitors are the isotype specificity and the undesired side effects often observed with kinase inhibitors. This study unveils a novel regulatory mechanism for p38α activation and signaling through arginine methylation and provides a new strategy, other than kinase inhibition, in intervening p38α signaling.

4. Materials and Methods

4.1. Materials and Plasmid

1-beta-D-arabinofuranosylcytosine (AraC) and benzidine were obtained from Sigma-Aldrich. S-adenosyl-L-[methyl-^{3}H] methionine (^{3}H-AdoMet, 0.55 mCi/ml, NET-155H) and fluorographic enhancer, EN^{3}HANCE, were from PerkinElmer. The pFlag-CMV2-p38α plasmid [3] was used as a template for PCR to generate mutations on R49 and R149. These plasmids were subsequently cloned into the pET6H vector for expression of recombinant proteins.

4.2. Cell Culture

The human chronic myelogenous leukemia (CML) K562 cells were purchased from BCRC (Bioresource Collection and Research Center, Taiwan) and cultured in RPMI1640 medium supplemented with 10% fetal bovine serum, 100 IU/mL streptomycin and 100 IU/mL penicillin as described. The pLKO.1 puro-based shRNAs, including MKK3-sh1, MKK6-sh1, MKK6-sh2, MAPKAPK2-sh1, MAPKAPK2-sh2, were purchased from National RNAi Core Facility, Taiwan. The transfection of K562 was performed by using Lipofectamine^{TM} 2000 Reagent (Invitrogen). Stable clones of gene knockdown were selected in the presence of puromycin (0.5 μg/mL). The p38α-knockdown cell clones were generated by the same procedure as described [3,4]. The protein levels of p38α in these knockdown cells were around 20% of the parental cells.

4.3. Expression and Purification of Recombinant Proteins

Recombinant His-tagged p38α WT and mutants (R49K, R149K and R49/149K) proteins were expressed in *E. coli* BL21 (DE3) pLysS by isopropyl-beta-D-thiogalactoside (IPTG) (0.4 mM) induction and immobilized on Ni $^+$ -NTA agarose (Qiagen). Proteins were eluted with elution buffer (20 mM Tris-HCl pH7.9, 0.5 M NaCl, and 0.6 M imidazole) and dialyzed to remove imidazole. The purified proteins were stored in a buffer containing 20 mM Tris-HCl pH7.9 and 50 mM NaCl. The recombinant GST-fused PRMT1 proteins were expressed in *E. coli* BL21 by IPTG (0.2 mM) induction and immobilized on glutathione Sepharose 4B (GE Healthcare). Proteins were eluted with and stored in buffer (50 mM Tris-HCl Ph 7.9 and 20 mM reduced glutathione).

4.4. Immunoprecipitation

Cells were lysed in lysis buffer (150 mM NaCl, 20 mM Tris, pH7.4, 1% Triton X-100, 1 mM EDTA, 1 mM EGTA, 1 mM PMSF, 1 μg/mL leupeptin, 1 μg/mL aprotinin, 1 μg/mL pepstatin,1 mM Na_3VO_4, and 2.5 mM β-glycerophosphate). Cell lysates (1 mg) were incubated with 1 μL α-Flag antibodies (Sigma-Aldrich, F4042) in lysis buffer containing 0.2% Triton X-100 for 3 h at 4 °C followed by incubation with 20 μL protein G agarose-beads (GE Healthcare) for 1–3 hours at 4 °C. At the end of incubation, beads were washed four times with lysis buffer without Triton X-100. The phosphorylation states of p38α and the potential associated proteins were detected by Western Blotting. Alternatively, the immunoprecipitates were fractionated by SDS-PAGE and subjected to mass spectrometric analysis to examine the interactome.

To immunoprecipitate HA-PRMTs for methylation assay, cell lysates (0.5 mg) were incubated with α-HA antibodies (0.5 μL) (BioLegend, #901501) as described above and followed by incubation with 10 μL protein G agarose-beads (GE Healthcare). At the end of incubation, beads were washed twice with lysis buffer without Triton X-100, twice with PBS contained 0.2% tween-20, then twice with 25 mM Tris-HCl pH7.9. The immunoprecipitated HA-PRMTs were used immediately as an enzyme source for in vitro methylation assay.

4.5. In Vitro Methylation Assay

For the mass spectrometric identification of methylation sites in p38α, the methylation assay was carried out in vitro using recombinant wild-type His-p38α proteins as a substrate and with or without recombinant GST-PRMT1 proteins as an enzyme. The reactions were carried out using S-adenosyl-methionine (AdoMet, PerkinElmer) as a methyl donor as described previously [20]. Reactions were terminated and subjected to SDS-PAGE fractionation. The protein band containing p38α was excised from the gel and subjected to mass spectrometric analysis to examine its posttranslational modifications.

To compare methyl incorporation into the wild-type and mutant His-p38α proteins, HA-PRMT1, HA-PRMT1G80R, HA-PRMT6, HA-PRMT6KA proteins were expressed from the pcDNA3-HA2 plasmids in K562 cells and immunoprecipitated using anti-HA antibodies. Methylation reactions were carried out using S-adenosyl-L-[methyl-^{3}H]methionine (^{3}H-AdoMet, 1.65 μCi, PerkinElmer) as a methyl donor. Methyl incorporation was visualized by fluorography as described [20].

4.6. Mass Spectrometric Analysis

The gel pieces from SDS-PAGE were minced and incubated with trypsin (Promega). Peptides were extracted from the gel pieces by sonication in 50% acetonitrile containing 0.1% formic acid. The solutions were dried. The peptides were resuspended in 0.1% formic acid and analyzed by nanoflow high-performance liquid chromatography (Agilent Technologies 1200 series) followed by a LTQ-Orbitrap Discovery hybrid mass spectrometer with a nano-eletrospray ion source (Thermoelectron). This was performed by the Proteomics Center of National Yang- Ming University. The raw data were processed by the Xcalibur 2.0 SR1 software (Thermoelectron) and were converted to DTA files. Protein identity was determined by comparing the peptide sequences against the UniProtKB protein database (http://www.uniprot.org/) with TurboSequest search server T (version 27, revision 11) [36]. A protein was identified when two peptide ions matched with an Xcorr score >2.5. For the identification of p38α methylation sites, post-translational modifications (PTMs) of the peptides were identified by the in-house PTM finder program [37]. The peptides that contained PTM were matched with an Xcorr score >2.0. For the identification of associated proteins, quantitative analysis with MS spectra counting was performed by an in-house tool within a Microsoft VBA environment. MS spectra counts were normalized to total identified spectra per sample and then normalized to p38α spectra counts.

4.7. Limited Trypsin Digestion of Recombinant p38α

The recombinant His-tagged p38α WT and mutants (R49K, R149K and R49/149K) proteins (5 μg) were incubated with 0.5 μg trypsin (Sigma) in a final volume of 20 μL for 5, 15, and 30 min at 37 °C. Samples were analyzed by SDS-PAGE.

4.8. Erythroid Differentiation

For erythroid differentiation, K562 cells were treated with 1-beta-D-arabinofuranosylcytosine (AraC, 1μM, Sigma-Aldrich) for various times as indicated. Hemoglobin production was detected by using a benzidine/hydrogen peroxide solution as described [20]. Stained cells were counted under a light microscope. Two hundred cells were examined in each assay.

4.9. Real-Time Reverse Transcription PCR

Total RNA was extracted by using an illustra RNAspin Mini Isolation Kit (GE Healthcare). Isolated RNAs (4 μg) were subjected to cDNA synthesis with a RevertAid first strand cDNA synthesis kit (Fermentas). Real-time PCR was performed on an ABI StepOne Plus Real-Time PCR System (Applied Biosystems) using SYBR-Green reagents (SensiFAST SYBR Hi-ROX mix, Bioline). The samples were examined in triplicate for each assay and analyzed by using the comparative cycle threshold C_T ($\Delta\Delta C_T$) method [3]. The expression of GAPDH (glyceraldehyde-3-phosphate dehydrogenase) was used as an internal control for cDNA contents. Gene-specific amplification was conducted with the following primer sets: for EKLF, 5′-CGGCAAGAGCTACACCAAG-3′ (sense) and 5′-CCGTGTGTTTCCGGTAGTG-3′ (antisense); for GATA1, 5′-CAGTCTTTCAGGTGTACCC -3′ (sense) and 5′-GAGTGATGAAGGCAGTGCAG-3′ (antisense); for ALAS2, 5′-GCAGCACT CAACAGCAAG-3′ (sense) and 5′-ACAGGACGGCGACAGAAA-3′ (antisense); for PBGD, 5′-CGCCTCCCTCTAGTCTCTGCTTCT-3′ (sense) and 5′- GTTGCCACCACACTGTCCGTCTG-3′ (antisense); for GAPDH, 5′-TGGTATCGTGGAAGGACTCATGAC-3′ (sense) and 5′-ATGCCA GTGAGCTTCCCGTTCAGC-3′ (antisense).

4.10. Western Blotting

Cells were lysed in RIPA buffer (150 mM NaCl, 10 mM Tris, pH7.4, 0.1% SDS, 1% Triton X-100, 5 mM EDTA, 1% Na-deoxycholate, 1 mM PMSF (phenylmethylsulfonyl fluoride), 1 μg/mL leupeptin, 1 μg/mL aprotinin, 1 μg/mL pepstatin). The primary antibodies used were α-Flag (Sigma-Aldrich, F4042, 1:2000 or Sigma-Aldrich, F7425, 1:1000), α-HA (BioLegend, #901501, 1:1000), α-p38 (Cell signaling, #9212, 1:1000), α-pp38 (Cell signaling, #9219, 1:1000), α-MKK3 (Cell signaling, #5674, 1:500), α-MKK6 (Abnova, #M02, 1:500), α-MAPKAPK2 (Cell signaling, #12155, 1:1000), α-PRMT1 (Sigma-Aldrich, #P16220, 1:1000), α-GAPDH (Cell signaling, G9545, 1:10000), α-mono- and di-methyl arginine (abcam, ab412, 1:500). The secondary antibodies used were anti-mouse (Sigma-Aldrich) or anti-rabbit (Genetex) IgG linked-horseradish peroxidase.

4.11. Statistical Analysis

All experiments were performed at least three times. The data are presented as means ± S.E.M. Statistical significance was performed by using Student's t test, and $p < 0.05$ was considered statistically significant.

Supplementary Materials: Supplementary materials can be found at http://www.mdpi.com/1422-0067/21/10/3546/s1.

Author Contributions: M.-Y.L. and W.-K.H. carried out experiments, analyzed data and prepared the figures in the paper. W.-J.L. contributed to experimental design, analysis, interpretation of data and writing of the paper. C.-J.C. contributed to insightful discussion, data interpretation and the writing of the paper. All authors have read and agreed to the published version of the manuscript.

Funding: This work was supported by the Ministry of Science and Technology (grant No: MOST 107-2320-B-010-039 to W.-J.L. and MOST 108-2320-B-010-017 to W.-J.L.)

Acknowledgments: We thank the current and former members of the group for their technical support in experiments.

Conflicts of Interest: The authors declare no conflict of interest.

References

1. Cuenda, A.; Rousseau, S. p38 MAP-kinases pathway regulation, function and role in human diseases. *Biochim. Biophys. Acta* **2007**, *1773*, 1358–1375. [CrossRef]
2. Cuadrado, A.; Nebreda, A.R. Mechanisms and functions of p38 MAPK signalling. *Biochem. J.* **2010**, *429*, 403–417. [CrossRef]
3. Hua, W.K.; Chang, Y.I.; Yao, C.L.; Hwang, S.M.; Chang, C.Y.; Lin, W.-J. Protein arginine methyltransferase 1 interacts with and activates p38alpha to facilitate erythroid differentiation. *PLoS ONE* **2013**, *8*, e56715. [CrossRef]
4. Chang, Y.I.; Hua, W.K.; Yao, C.L.; Hwang, S.M.; Hung, Y.C.; Kuan, C.-J.; Leou, J.-S.; Lin, W.-J. Protein-arginine methyltransferase 1 suppresses megakaryocytic differentiation via modulation of the p38 MAPK pathway in K562 cells. *J. Biol. Chem.* **2010**, *285*, 20595–20606. [CrossRef]

5. Stefkova, K.; Hanackova, M.; Kucera, J.; Radaszkiewicz, K.A.; Ambruzova, B.; Kubala, L.; Pacherník, J. MAPK p38alpha Kinase Influences Haematopoiesis in Embryonic Stem Cells. *Stem. Cells Int.* **2019**, *2019*, 5128135. [CrossRef]
6. Feng, Y.; Wen, J.; Chang, C.C. p38 Mitogen-activated protein kinase and hematologic malignancies. *Arch. Pathol. Lab. Med.* **2009**, *133*, 1850–1856. [CrossRef]
7. Koul, H.K.; Pal, M.; Koul, S. Role of p38 MAP Kinase Signal Transduction in Solid Tumors. *Genes. Cancer* **2013**, *4*, 342–359. [CrossRef]
8. Gupta, J.; Nebreda, A.R. Roles of p38alpha mitogen-activated protein kinase in mouse models of inflammatory diseases and cancer. *FEBS J.* **2015**, *282*, 1841–1857. [CrossRef]
9. Bohm, C.; Hayer, S.; Kilian, A.; Zaiss, M.M.; Finger, S.; Hess, A.; Engelke, K.; Kollias, G.; Krönke, G.; Zwerina, J.; et al. The alpha-isoform of p38 MAPK specifically regulates arthritic bone loss. *J. Immunol.* **2009**, *183*, 5938–5947. [CrossRef]
10. Peregrin, S.; Jurado-Pueyo, M.; Campos, P.M.; Sanz-Moreno, V.; Ruiz-Gomez, A.; Crespo, P.; Mayor, F., Jr.; Murga, C. Phosphorylation of p38 by GRK2 at the docking groove unveils a novel mechanism for inactivating p38MAPK. *Curr. Biol.* **2006**, *16*, 2042–2047. [CrossRef]
11. Mittelstadt, P.R.; Yamaguchi, H.; Appella, E.; Ashwell, J.D. T cell receptor-mediated activation of p38{alpha} by mono-phosphorylation of the activation loop results in altered substrate specificity. *J. Biol. Chem.* **2009**, *284*, 15469–15474. [CrossRef] [PubMed]
12. DeNicola, G.F.; Martin, E.D.; Chaikuad, A.; Bassi, R.; Clark, J.; Martino, L.; Verma, S.; Sicard, P.; Tata, R.; Atkinson, R.A.; et al. Mechanism and consequence of the autoactivation of p38alpha mitogen-activated protein kinase promoted by TAB1. *Nat. Struct. Mol. Biol.* **2013**, *20*, 1182–1190. [CrossRef] [PubMed]
13. Pillai, V.B.; Sundaresan, N.R.; Samant, S.A.; Wolfgeher, D.; Trivedi, C.M.; Gupta, M.P. Acetylation of a conserved lysine residue in the ATP binding pocket of p38 augments its kinase activity during hypertrophy of cardiomyocytes. *Mol. Cell. Biol.* **2011**, *31*, 2349–2363. [CrossRef] [PubMed]
14. Guccione, E.; Richard, S. The regulation, functions and clinical relevance of arginine methylation. *Nat. Rev. Mol. Cell. Biol.* **2019**, *20*, 642–657. [CrossRef] [PubMed]
15. Nicholson, T.B.; Chen, T.; Richard, S. The physiological and pathophysiological role of PRMT1-mediated protein arginine methylation. *Pharm. Res.* **2009**, *60*, 466–474. [CrossRef] [PubMed]
16. Blanc, R.S.; Richard, S. Arginine Methylation: The Coming of Age. *Mol. Cell* **2017**, *65*, 8–24. [CrossRef]
17. Greenblatt, S.M.; Liu, F.; Nimer, S.D. Arginine methyltransferases in normal and malignant hematopoiesis. *Exp. Hematol.* **2016**, *44*, 435–441. [CrossRef]
18. Zhu, L.; He, X.; Dong, H.; Sun, J.; Wang, H.; Zhu, Y.; Huang, F.; Zou, J.; Chen, Z.; Zhao, X.; et al. Protein arginine methyltransferase 1 is required for maintenance of normal adult hematopoiesis. *Int. J. Biol. Sci.* **2019**, *15*, 2763–2773. [CrossRef]
19. Wolwer, C.B.; Pase, L.B.; Russell, S.M.; Humbert, P.O. Calcium Signaling Is Required for Erythroid Enucleation. *PLoS ONE* **2016**, *11*, e0146201. [CrossRef]
20. Liu, M.Y.; Hua, W.K.; Chiou, Y.Y.; Chen, C.J.; Yao, C.L.; Lai, Y.-T.; Lin, C.-H.; Lin, W.-J. Calcium-dependent methylation by PRMT1 promotes erythroid differentiation through the p38alpha MAPK pathway. *FEBS Lett.* **2020**, *594*, 301–316. [CrossRef]
21. Kuzmanic, A.; Sutto, L.; Saladino, G.; Nebreda, A.R.; Gervasio, F.L.; Orozco, M. Changes in the free-energy landscape of p38alpha MAP kinase through its canonical activation and binding events as studied by enhanced molecular dynamics simulations. *Elife* **2017**, *6*. [CrossRef] [PubMed]
22. Singhroy, D.N.; Mesplede, T.; Sabbah, A.; Quashie, P.K.; Falgueyret, J.P.; Wainberg, M.A. Automethylation of protein arginine methyltransferase 6 (PRMT6) regulates its stability and its anti-HIV-1 activity. *Retrovirology* **2013**, *10*, 73. [CrossRef] [PubMed]
23. Godisela, K.K.; Reddy, S.S.; Reddy, P.Y.; Kumar, C.U.; Reddy, V.S.; Ayyagari, R.; Reddy, G.B. Role of sorbitol-mediated cellular stress response in obesity-associated retinal degeneration. *Arch. Biochem. Biophys.* **2020**, *679*, 108207. [CrossRef] [PubMed]
24. Soni, S.; Anand, P.; Padwad, Y.S. MAPKAPK2: The master regulator of RNA-binding proteins modulates transcript stability and tumor progression. *J. Exp. Clin. Cancer Res.* **2019**, *38*, 121. [CrossRef]
25. Tanoue, T.; Adachi, M.; Moriguchi, T.; Nishida, E. A conserved docking motif in MAP kinases common to substrates, activators and regulators. *Nat. Cell Biol.* **2000**, *2*, 110–116. [CrossRef]

26. Akella, R.; Moon, T.M.; Goldsmith, E.J. Unique MAP Kinase binding sites. *Biochim. Biophys. Acta* **2008**, *1784*, 48–55. [CrossRef]
27. Evich, M.; Stroeva, E.; Zheng, Y.G.; Germann, M.W. Effect of methylation on the side-chain pKa value of arginine. *Protein Sci.* **2016**, *25*, 479–486. [CrossRef]
28. Rothweiler, U.; Aberg, E.; Johnson, K.A.; Hansen, T.E.; Jorgensen, J.B.; Engh, R.A. p38alpha MAP kinase dimers with swapped activation segments and a novel catalytic loop conformation. *J. Mol. Biol.* **2011**, *411*, 474–485. [CrossRef]
29. Bagley, M.C.; Davis, T.; Murziani, P.G.; Widdowson, C.S.; Kipling, D. Use of p38 MAPK Inhibitors for the Treatment of Werner Syndrome. *Pharmaceuticals* **2010**, *3*, 1842–1872. [CrossRef]
30. Mertins, P.; Qiao, J.W.; Patel, J.; Udeshi, N.D.; Clauser, K.R.; Mani, D.R.; Burgess, M.W.; Gillette, M.A.; Jaffe, J.D.; Carr, S.A. Integrated proteomic analysis of post-translational modifications by serial enrichment. *Nat. Methods* **2013**, *10*, 634–637. [CrossRef]
31. Mazur, P.K.; Reynoird, N.; Khatri, P.; Jansen, P.W.; Wilkinson, A.W.; Liu, S.; Barbash, O.; Aller, G.S.V.; Huddleston, M.; Dhanak, D.; et al. SMYD3 links lysine methylation of MAP3K2 to Ras-driven cancer. *Nature* **2014**, *510*, 283–287. [CrossRef] [PubMed]
32. Andreu-Perez, P.; Esteve-Puig, R.; de Torre-Minguela, C.; Lopez-Fauqued, M.; Bech-Serra, J.J.; Tenbaum, S.; García-Trevijano, E.R.; Canals, F.; Merlino, G.; Ávila, M.A.; et al. Protein arginine methyltransferase 5 regulates ERK1/2 signal transduction amplitude and cell fate through CRAF. *Sci. Signal.* **2011**, *4*, ra58. [CrossRef] [PubMed]
33. Li, X.; Lu, Y.C.; Dai, K.; Torregroza, I.; Hla, T.; Evans, T. Elavl1a regulates zebrafish erythropoiesis via posttranscriptional control of gata1. *Blood* **2014**, *123*, 1384–1392. [CrossRef] [PubMed]
34. Dore, L.C.; Crispino, J.D. Transcription factor networks in erythroid cell and megakaryocyte development. *Blood* **2011**, *118*, 231–239. [CrossRef] [PubMed]
35. Xing, L. Clinical candidates of small molecule p38 MAPK inhibitors for inflammatory diseases. *MAP Kinase* **2015**, *4*, 24–30. [CrossRef]
36. Rajkumar, R.; Karthikeyan, K.; Archunan, G.; Huang, P.H.; Chen, Y.W.; Ng, W.V.; Liao, C.C. Using mass spectrometry to detect buffalo salivary odorant-binding protein and its post-translational modifications. *Rapid Commun. Mass Spectrom.* **2010**, *24*, 3248–3254. [CrossRef]
37. Zhang, Y.; Pan, Y.H.; Yin, Q.; Yang, T.; Dong, D.; Liao, C.-C.; Zhang, S. Critical roles of mitochondria in brain activities of torpid Myotis ricketti bats revealed by a proteomic approach. *J. Proteom.* **2014**, *105*, 266–284. [CrossRef]

Review

Nuclear P38: Roles in Physiological and Pathological Processes and Regulation of Nuclear Translocation

Galia Maik-Rachline, Lucia Lifshits and Rony Seger *

Department of Biological Regulation, Weizmann Institute of Science, Rehovot 7610001, Israel; galia.maik-rachline@weizmann.ac.il (G.M.-R.); Lucia.lifshits@weizmann.ac.il (L.L.)

* Correspondence: rony.seger@weizmann.ac.il; Tel.: +972-8-934-3602

Received: 27 July 2020; Accepted: 21 August 2020; Published: 24 August 2020

Abstract: The p38 mitogen-activated protein kinase (p38MAPK, termed here p38) cascade is a central signaling pathway that transmits stress and other signals to various intracellular targets in the cytoplasm and nucleus. More than 150 substrates of p38α/β have been identified, and this number is likely to increase. The phosphorylation of these substrates initiates or regulates a large number of cellular processes including transcription, translation, RNA processing and cell cycle progression, as well as degradation and the nuclear translocation of various proteins. Being such a central signaling cascade, its dysregulation is associated with many pathologies, particularly inflammation and cancer. One of the hallmarks of p38α/β signaling is its stimulated nuclear translocation, which occurs shortly after extracellular stimulation. Although p38α/β do not contain nuclear localization or nuclear export signals, they rapidly and robustly translocate to the nucleus, and they are exported back to the cytoplasm within minutes to hours. Here, we describe the physiological and pathological roles of p38α/β phosphorylation, concentrating mainly on the ill-reviewed regulation of p38α/β substrate degradation and nuclear translocation. In addition, we provide information on the p38α/β 's substrates, concentrating mainly on the nuclear targets and their role in p38α/β functions. Finally, we also provide information on the mechanisms of nuclear p38α/β translocation and its use as a therapeutic target for p38α/β-dependent diseases.

Keywords: p38MAPK; nuclear translocation; β-like importins; inflammation; cancer

1. Introduction

The p38 mitogen-activated protein kinase (p38MAPK, termed here p38) is a signaling protein kinase that operates within a signaling cascade to transmit extracellular signals to their intracellular targets. The p38 cascade is one of four similar cascades that are all key communication lines between the plasma membranes and the nucleus, and thereby, it is involved in fundamental cellular processes, including stress response, proliferation, differentiation and others [1–3]. The four MAPK cascades are extracellular signal-regulated kinase (ERK) 1/2 [4], c-Jun N-terminal kinase (JNK [5]), p38 [6], and ERK5 [7]. The MAPK cascades transmit signals via a sequential activation of protein kinases, which are organized in 3–5 tiers (MAP4K, MAP3K, MAPKK, MAPK, and MAPK activated protein kinases (MAPKAPKs also termed MKs)). Each of these tiers includes more than one kinase (e.g., 4 isoforms at the p38 tier), and the components involved, while the number of tiers may vary between cell lines or under different conditions (see the scheme of the MAPK cascades in ref [8]). In this review, we focus on p38 [9,10], whose cascade is composed of many kinases at the MAP4K and MAP3K levels, MKK3/6, and perhaps MKK4 at the MAPKK tier, p38α–δ at the MAPK tier, and several MKs at the next tier (MNK1/2, MSK1/2, MK2/3, and MK5). Interestingly, unlike the other MAPKs, p38 can also be activated via MKK-independent pathways, either by ZAP/LCK-mediated Tyr phosphorylation [11] or by interaction with TAB1 [12]. The downregulation/inactivation of the p38 cascade is regulated by various phosphatases, among them are several dual specificity phosphatases termed MAPK phosphatases (MKPs) that operate directly on the MAPKs [13]. As in all MAPK cascades, p38 transmits signals initiated by various agents, including cytokines and environmental queues, but it is known to operate mainly as a mediator of stress responses. Thus, the kinase is a key regulator of metabolic, oxidative, and endoplasmic reticulum (ER) stresses, but it plays an important role in other physiological processes such as cell cycle, senescence, differentiation, and several aspects of immunological processes.

Being responsible for the various distinct and even opposing fundamental cellular processes, the p38 cascade needs to be tightly regulated. Indeed, several regulatory mechanisms that determine the specificity of the cascade have been identified, including the duration and strength of the signals [13,14], which are controlled mainly by dual specificity phosphatases [15,16], scaffold proteins [17], and dynamic subcellular localization of the cascade's components [18]. Importantly, the central roles of the cascade suggest that its dysregulation may cause various diseases. Indeed, p38 was shown to participate in the induction of pathologies such as inflammation-related diseases [19], autoimmune diseases [20], some types of cancer [6], and other pathologies, as specified later in this review. Interestingly, unlike other MAPKs, p38 demonstrates distinct and even opposing effects in different cancers, as it was shown to serve either as a tumor suppressor [21] or tumor promoter [22]. It was also shown that in some cases, it can perform both activities in different stages of cancer development [23]. Although all p38 isoforms have been implicated in the processes listed above, they can be divided into two somewhat distinct subgroups: p38α and p38β (p38α/β) versus p38γ and p38δ. In this review, we focus on p38α/β, mainly discussing the physiological and pathological roles of these protein kinases, providing information on nuclear p38α/βs and their substrates as well as the importance of their phosphorylation specificity. We also describe the mechanisms involved in the nuclear translocation of p38α/β and compare it to other mechanisms of nuclear shuttling. The fact that p38α/β has so many nuclear targets indicates that the prevention of their nuclear translocation may affect their physiological and pathological functions. Indeed, we show that prevention of the nuclear translocation can be used as a tool to combat inflammation and cancer.

2. Physiological Roles of Nuclear p38α/β

The p38α/β are best known for their involvement in stress signaling, and indeed, these kinases as well as JNKs were initially termed stress-activated protein kinases (SAPKs [24]). However, it was established that their activity is not confined to stress responses, and under some conditions, the p38α/β may participate in the regulation of other processes, such as proliferation, differentiation, immune response, migration, and apoptosis. In many cases, p38α/β mediate their effects by activating and regulating transcription factors. One interesting example is the modulation of endoplasmic reticulum (ER) stress in breast cancer cells, which is mediated by the p38α/β-dependent activation of the transcription factor XBP-1 that decreases the expression of the ER protein ERp29 [25]. Another example is the transcriptional inhibition of autophagy genes downstream of p38α/β in response to oxidative stress in HeLa cells [26]. Other stresses such as UV radiation translational inhibition and others were shown to operate by p38α/β–activated transcription factors, such as ATF, MEF2, Elk1, and p53 [3]. However, p38α/β can also affect other regulators (e.g., MKs, proteasome, EGFR [27]) to coordinate their signaling. Notable transcription factor-independent targets that exert p38α/β functions are cell cycle regulators that modulate (usually inhibiting) cell cycle progression. Thus, various stresses induce the downregulation of cyclinD, thereby arresting cells at G1 [28]. In addition, p38α/β cascades were shown to induce the expression of CDK inhibitors, activate p53, or inhibit the transcription factor E2F and the G2/M regulator Cdc25B phosphatase, all leading to the inhibition of cell cycle progression [6,29]. By contrast, in some systems, p38α/β seem to enhance proliferation. For example, such effects were detected in hematopoietic cells and in some cancer cell lines [30]. These differential effects may be mediated by changes in the duration of p38α/β signals, where transient signals lead to fibroblasts' proliferation, while sustained signals induce cell cycle arrest [31]. However, the molecular mechanisms by which p38α/β are involved in proliferation have not been fully deciphered yet. Thus, the effects of p38α/β on the regulation of stress response or cell cycle progression are well-reviewed (e.g., [6,27,32,33]), and we will not elaborate on these effects. However, not less important are the roles of p38α/β in regulating protein degradation and the translocation of proteins upon stimulation. The molecular mechanisms involved are described in detail next.

2.1. 38α/β Regulation of Protein Degradation

The role of p38α/β in the regulation of protein degradation is widespread, mainly upon stress signals, and may involve several distinct mechanisms in both the cytoplasm and the nucleus. One such mechanism that mostly occurs in the nucleus involves phosphorylation of ubiquitin E3 ligases, such as Siah2, which is known to regulate PHD3 that further controls the stability of the transcription factor HIF1 α. p38α/β phosphorylate Siah2 on Ser24 and Thr29, thereby facilitating its activity towards degradation of PHD, and in turn destabilization of HIF1 α [34,35]. Similarly, p38α/β phosphorylate the E3 ligase Skp2 at Ser64, leading to enhanced degradation of the transcription factor Nkx3-1 and thereby blocking its effects on estrogen receptor-mediated gene expression [36]. Another mechanism involves the phosphorylation of the ubiquitination target which can either facilitate or inhibit the ubiquitination process. Examples for enhanced degradation are the phosphorylation of RBP-Jk at Thr339 which subsequently induces its degradation [37], or phosphorylation of p300 at Ser 1834 (together with AKT) that induces its degradation to allow DNA repair [38]. On the other hand, phosphorylation of the inflammation regulator TRIM9s at Ser76/80 stabilizes it, thereby causing a positive feedback loop for the degradation of the upstream MKK6 [39]. Interestingly, p38α/β may also regulate proteasomal activity and localization to govern protein stability in general. It was shown that osmotic stress inhibits proteasome by p38α/β-dependent phosphorylation of the proteasome subunit Rpn2 at Thr273, which is important for peptide degrading activity [40]. This inhibitory effect was supported by the finding that p38 inhibitors elevate proteasome activity under varying conditions [41]. In addition, p38α/β may regulate the subcellular localization of the proteasome by phosphorylating the proteasome-binding protein PI31. Consequently, this phosphorylation facilitates the association of the proteasome with the motor dynein complex, and regulates its transport on axons [42]. Other proteins whose stability is regulated by direct p38 phosphorylation are Cdt1, HBP1, p18Hamlet, Rb1, SRC3, CDC25A/B, CyclineD1/3, TACE, p53, Snail, Twist, Nav1.6, PGC1 α, HuR and Drosha [27,43]. Thus, p38α/β use various molecular mechanisms to regulate stimulation-dependent proteins stability.

2.2. p38α/β Regulation of Stimulated Nuclear Translocation

Another important process that is regulated by p38α/β is the dynamic change of protein localization upon stimulation. As described above regarding the regulation of protein degradation, the effect of p38α/β on nuclear translocation can be either global or specific to certain phosphorylated proteins. The global effects may be derived by p38α/β phosphorylation of either nuclear pore proteins or of karyopherins (importins/exportins). Indeed, it was shown that the nuclear pore proteins Nup62, Nup153, and Nup214 are phosphorylated by p38 (or ERK), and this phosphorylation inhibits the global nuclear protein shuttling initiated by viruses that affect the heart such as the encephalomyocarditis virus [44]. A similar effect was detected in cardiomyocytes of failing hearts in rats and humans, where p38α/β phosphorylation mediates the rearrangement of nuclear pores, leading to a decreased uptake of nuclear localization signal (NLS)-containing proteins [45]. As for karyopherins, it was shown that p38α/β regulate the expression of the beta-like importins (Imp) Imp7 and Imp8 [46], which are important for the nuclear translocation of various signaling proteins. Aside from the global changes, p38α/β is known to phosphorylate the translocating proteins themselves to mediate either nuclear accumulation or nuclear export. For example, the active SMAD3 phosphorylation by p38α/β upon TGF β stimulation reduces the rate of its nuclear translocation [47]. A similar effect was detected for FOXO3 α, which is phosphorylated by p38α/β at Ser7 to promote its nuclear localization [48]. The nuclear translocation of RhoA due to p38 phosphorylation upon LPS treatment and of actin upon

TPA stimulation [49] was reported as well [50]. On the other hand, p38α/β phosphorylation may be responsible for the nuclear export of proteins; the most famous among them are its downstream MKs. It was shown that MK2/3 contain an NLS, which directs them to the nucleus of resting cells. Following phosphorylation by p38α/β, MK2/3 are exported to the cytoplasm, due to unmasking of the C-terminal NES of the MK2/3 (reviewed in [51]). Moreover, MK5 contains an NLS as well, and can be found in the nucleus under certain conditions. However, its export after p38α/β phosphorylation seems to be mediated not only by exposure of NES but also by anchoring to ERK3/4 [52]. Other proteins whose localization is directly regulated by p38α/β phosphorylation are retinoic acid receptor-γ, [53], androgen receptor [54], estrogen receptor-α [55], 5-lipoxygenase [56], the Hippo pathway transcription factor TEAD4 [57], as well as other proteins (NFATc4, Xbp1s, Drosha, CRTC2, HuR, Rabenosyn5, Lamin-B, FGFR1, PIP4K2B, EZH2, and Tripeptidyl-Peptidase II) as specified in previous reviews [27,43]. Thus, p38α/β use several distinct mechanisms for the regulation of nuclear translocation of proteins upon various stimulations.

3. Role of Nuclear p38α/β in Pathologies

Abnormal activity and dysregulation of the p38α/β cascade are associated with a variety of diseases. Indeed, p38α/β were implicated in the induction and maintenance of several pathologies such as inflammation [19], cancer [6], and autoimmune diseases [20] mentioned above, but also Friedreich's ataxia [58], Parkinson's disease [59], Alzheimer's disease [60], cardiac hypertrophy [61], hypoxic nephropathy [62], and diabetes [63]. In many cases, the role of p38α/β is not direct, but it is mediated by p38α/β-regulated inflammation, which in turn contributes to the development of the diseases. For example, Parkinson's disease is induced in part by neuroinflammation associated with glial cells [61], and colorectal cancer often develops due to initial inflammatory disease of the colon [64]. Moreover, p38α was first identified due to its involvement in the production of pro-inflammatory cytokines upon endotoxin treatment, mainly via nuclear processes [65]. It was later found that p38α/β are involved in the production of pro-inflammatory cytokines, such as TNF-α, IL-1 β, IL-2 IL-6, IL-7, and IL-8, and also in the regulation of other inflammation mediators such as Cox2 [66,67]. Although many of these effects involve nuclear processes in some instances, it may be regulated also by translation, due to AU-rich elements (ARE) in the 3′ untranslated region of their mRNA. The presence of these elements is known to shorten the half-life of mRNA containing them or block their translation, mainly due to the phosphorylation of ARE binding proteins such as HuR [68] by MK2 downstream of p38α/β) [69]. The pro-inflammatory cytokines are known players in many inflammation-related diseases such as inflammatory bowel diseases (IBD), psoriasis asthma, rheumatoid arthritis, inflammation-induced cancer, and more [66]. However, the actual trigger for inflammation in some systems, either acute or chronic, is not known, but it still requires p38α/β for its mediation. The means by which p38α/β are involved in these processes and the upstream components involved need further investigation. Due to the involvement of p38α/β in several inflammatory diseases, it became clear that specific inhibitors of these kinases should become a beneficial therapeutic approach. Indeed, in the past decade, more than 20 specific inhibitors of p38α/β were developed and proven very effective in pre-clinical investigations, demonstrating good tolerability and efficacy in several mouse models [70,71]. However, when subjected to clinical trials, the effects were much less favorable. Apparently, except for one inhibitor, pirfenidone, which by itself demonstrated a weak and unselective effect, no durable therapeutic effects have been detected for any of the others tested. The reasons for these failures among the different drugs are numerous, and while in some cases they were toxic, the more common problem was that after an initial good response, there was a rebound effect that increased inflammation within weeks. The reason for this rebound is still not clear and is currently under investigation.

In the past decade, substantial research has been devoted to studying the role of p38α/β in cancer, confirming that these kinases can act either as tumor promotors, or more frequently, as tumor suppressors [6,30]. The tumor suppressor effects were corroborated by immortalized MKK3/6 knockout fibroblasts that were shown to have a higher tendency to develop xenografts in nude mice [72]. Moreover, it was shown that the expression of MKK3/6, or other components of the cascade, is reduced in many cancers [73]. The function of p38α/β as tumor suppressors usually affect early stages of tumor initiation, and it generally involves either an inhibition of cell cycle progression, enhanced apoptosis, senescence, or differentiation. Similar to the effects in non-transformed cells, the inhibition of cell cycle by p38α/β in tumors can be mediated by the direct or indirect phosphorylation of several nuclear substrates. Among them are CyclinD, whose inhibition may cause apoptosis in colorectal cancer cells [28], p53, that leads to the upregulation of p21Cip1/WAF1, GADD45, and 14-3-3 proteins to cause cell cycle arrest [74], RB1, which prevents the metastasis of prostate cancer [75], and others [6]. Additionally, it was shown that p38α/β facilitate the production of apoptotic cytokines such as TNF-α [76]. The terminal differentiation of cancer was detected in rhabdomyosarcoma cells overexpressing MKK3 or MKK6 [77], and premature senescence was related to p38α/β-induced phosphorylation of the transcription factor HBP1 [78]. Some other effects on tumor promotion, although less frequent, may be mediated via inflammation in relevant cancers. Other mechanisms that may initiate cancer by p38α/β are elevated migration/invasion, increased angiogenesis [6], or a direct effect on proliferation [79,80]. As mentioned above, p38α/β are central regulators of inflammation, which in many cases is involved in cancer initiation and progression [81]. Indeed, some of the specific p38α/β inhibitors that have been developed over the years, although having failed in inflammation-related clinical trials, were proven useful in treating cancers [82]. Interestingly, in some cancer cells, p38α/β may facilitate migration by several mechanisms, including an enhanced production of chemoattractants [83] or reduced expression of fibulin 3, which is a cell migration blocker [84]. Finally, the involvement of p38α/β in angiogenesis, which supplies blood vessels to the tumors and enhances their growth, was shown to occur in head and neck cancer [85]. Some reports have shown that the molecular mechanisms involved include expression of vascular endothelial growth factor (VEGFA) and hypoxia-inducible factor 1 α (HIF1 α [30]. Interestingly, in models of colorectal cancer, p38α/β may act as either tumor suppressors or promoters in different stages of cancer development [23].

4. p38α/β's Substrates and their Phosphorylation Specificity

More than 150 direct substrates of p38α/β have been identified so far [27,43,86–89]. However, since their minimal consensus phosphorylation sequence on their substrates (Ser/Thr-Pro) is so limited, and estimated to be present on the surface of approximately 50% of the cellular proteins, the actual number is likely to increase. Moreover, the effects of p38α/β are propagated by their MKs, and therefore, the number of phosphorylated proteins downstream of p38α/β may reach several thousands. Direct substrates of p38α/β were categorized into several subgroups in a previous review [43], including DNA binding proteins, RNA binding proteins, Ser/Thr kinases, regulatory proteins, as well as membranal, endosomal, and structural proteins. Additional information has been accumulated over the years on the role of p38α/β in activating their downstream protein kinases (MKs), transcription factors, and other regulatory elements [6,9]. The phosphorylation of these substrates is important for orchestrating the various cellular processes that may be, under certain circumstances, opposing signals. For example, the stress-related transcription factors ATF2, MEF2s, and CHOP have long been known to transmit p38α/β stress signals, while Elk1 and cFos may transmit its downstream mitogenic signals [3]. These distinct effects raise the question as to how the specificity of p38 signals is regulated.

As all other MAPKs, activated p38α/β execute their functions through the phosphorylation of downstream proteins. To the best of our knowledge, unlike ERK [90], no phosphorylation-independent effects of p38α/β have been identified. The full (Pro-Xaa-Ser/Thr-Pro) or minimal (Ser/Thr-Pro) consensus phosphorylation sites of all MAPKs are similar to each other [91]. Therefore, the interaction with the residues in the phosphorylation site is not sufficient to provide p38α/β specificity to its substrates. Rather, the specificity is mostly achieved by docking motifs that are localized on the substrates (D, DEF) that interact with specific docking sites on p38α/β (CD, Hydrophobic pocket).

Thus, a docking motif that is found in many substrates of p38α/β is a D domain with consensus sequence Arg/Lys$_2$-Xaa$_{2\text{-}6}$-Φaa-Xaa-Φaa (where Φaa is a hydrophobic residue). The D-domains in the substrates bind to their counterpart docking site on p38α/β termed common docking motif (CD), which is composed of three negatively charged residues and at least two hydrophobic residues [92]. The CD domain and the catalytic site of p38α/β are located in distinct regions of the kinases, which allows the phosphorylation to occur at a fixed distance from the substrates D domain [93]. Binding of the D domain to the CD occurs in many cases when p38α/β are inactive, indicating that it may be responsible for a pre-activation association, which facilitates the rates of phosphorylation. Interestingly, similar to the consensus phosphorylation sites, also the D–CD domains interactions are nearly identical among all MAPKs, and therefore, p38α/β specificity requires extra determinants. Indeed, it was shown that in some cases, p38, ERK, and JNK may interact with distinct residues within the D-domain [94], and that two hydrophobic residues in the domain may determine specificity to some extent [95].

However, the differences between the hydrophobic residues cannot fully explain the specificity of p38α/β phosphorylation. Rather, this is likely explained by a second docking site on p38α/β substrates termed DEF (docking site for ERK, also known as FXF), which consists of two Phe residues separated by one residue and is often followed by a Pro residue [96]. The DEF domain binds the hydrophobic pocket in p38α/β, which is located at several residues C-terminal of the activatory Thr-Gly-Tyr phosphorylation site of the kinases. Structural studies show that the hydrophobic pocket is formed only upon phosphorylation, and therefore, unlike the interaction of the D-domain, only active p38α/β molecules bind to the DEF motif. Finally, it was also shown that p38, but not p38α, possesses intrinsic autophosphorylation activity, which may by itself elevate the basal activity of the kinase [97]. Thus, the combination of the three substrate interaction motifs together with intrinsic kinase activity contribute to the substrate specificity and affinity required for the proper p38α/β's functions under various conditions. In addition, other regulators that affect signaling specificity may contribute to the kinase activity, including the level of substrates' expression and their stability after stimulation as well as compartmentalization, scaffold proteins, or distinct phosphatases in a given cell. These regulating elements may also lead to sustained rather than transient activation of p38α/β, which is another specificity determinant. This is because the sustained phosphorylation results in the substrate's phosphorylation at later stages after stimulation [98], as seen in cases of cell cycle facilitation versus senescence [31,99]. Taken together, these effects regulate the outcome of the p38α/β signal that are important both for the physiological and pathological fates of the cells.

5. Subcellular Localization of p38α/β and their Substrates

One of the mechanisms that determines the ability of MAPKs to phosphorylate distinct substrates upon varying conditions or cell lines is the cellular localization of the substrates' phosphorylation [100]. Indeed, the large number of p38α/β's substrates was shown to localize in various distinct compartments, including the nucleus, cytoplasm, cytoskeletal elements, and other sites [86]. In many cases, the distribution of the substrates is dynamic and changed upon stimulation, or in different cell lines. As mentioned above, some of these changes can be regulated by p38α/β, but others are regulated by other signaling proteins. The dynamic changes in localization raise the question of where the phosphorylation by p38α/β is actually taking place. Unexpectedly, the transcription factors that serve as substrates and function within the nucleus are almost always cytoplasmic in resting cells and translocate to the nucleus upon stimulation. As mentioned above, the translocations of some of these transcription factors are regulated by p38α/β. For example, the transcription factor ATF2 is phosphorylated by p38α/β to facilitate its dimerization with other AP-1 transcription factors, which enhances nuclear translocation [101]. Additionally, p38α/β phosphorylate the transcription factor Xbp1s at Thr48 and Ser61 enhancing migration to the nucleus, thus regulating glucose homeostasis in obesity [102]. The nuclear translocation of the transcription factor MEF2A is regulated by p38α/β as well [103], leading to the expression of neonatal myosin heavy chain in C2C12 myoblasts. Thus, counterintuitively, the phosphorylation of most, if not all, transcription factors by p38α/β may take place primarily in the cytoplasm, although some phosphorylations can also occur in the nucleus upon translocation. Similar effects were shown for other effectors of p38α/β, which translocate to the nucleus upon stimulation. Importantly, the localization of some substrates might be cell-type or condition-dependent [86], but others are less variable, and they are either constantly localized in the cytoplasm (e.g., keratin-8), in the nucleus (e.g., histones), or in the nucleus of resting cells followed by export upon stimulation (e.g., MKs). A list of various p38α/β substrates, which includes several transcription factors but mostly other proteins (Table 1), indicates that about 30% of the p38α/β substrates are confined to the nucleus under most/all conditions, and therefore, their phosphorylation upon stimulation should be nuclear. Almost all these proteins were shown to participate in the regulation of cancer and inflammation at least under some conditions. These effects are supported by our recent findings that the inhibition of nuclear p38α/β translocation prevents DSS-induced colitis and DSS/AOM-induced colon cancer [80], confirming the importance of nuclear p38α/β. However, it should be noted that the involvement of nuclear p38α/β in cancer and inflammation was not always confirmed, and the effects are sometimes cell line-dependent. Taken together, the information included in Table 1, as well as our recent study, clearly indicate that p38α/β are involved in either positive or negative regulations of cancer and inflammation.

The ability of p38α/β to phosphorylate substrates in the cytoplasm as well as in the nucleus and the role of nuclear p38α/β in the regulation of cancer and inflammation diverted attention toward the subcellular localization of these kinases. Thus, it was initially shown that p38α/β may be localized in the cytoplasm of resting cells, and similarly to ERK [104], it may translocate to the nucleus upon stimulation [105,106]. In that case, the kinases remain in the nucleus for minutes to hours, after which they are exported back to the cytoplasm. Surprisingly, in other systems, p38α/β were detected in the nucleus of resting cells [107] and were exported out of the nucleus shortly upon stimulation [108]. The nuclear localization in resting cells likely occurs because of continued stress signals in these cells, or due to the expression of specific nuclear anchors that attract p38α/β to the nucleus. However, this localization seems to occur in only a limited number of systems, whereas, in most cases, p38α/β are localized in the cytoplasm of resting cells. As for the mechanisms that regulate the dynamic subcellular localization, similarly to ERKs [109], it was shown that the cytoplasmic localization of p38α/β is mediated by various cytoplasmic anchoring proteins such as PTP-SL [106], keratins [110], and others [111]. Upon stimulation, the p38α/β detach from their anchors by mechanisms that may [112] or may not [113] require the activating TGY phosphorylation. Then, p38α/β translocate to the nucleus via the nuclear pore and can stay there for minutes to hours, either in the nucleoplasm, bound to chromatin [114], or nuclear proteins [108]. Interestingly, the mechanism of export either in cells with constant nuclear p38α/β, or in later stages upon stimulation-induced translocation, involves binding to the nuclear export signal (NES)-containing p38α/β's substrate MK2 but probably not to those containing MK3 and MK5 [108,112].

6. Mechanism of Nuclear p38α/β Translocation and the Effect of its Inhibition

Although it is clear that p38α/β shuttle to the nucleus, either upon or without stimulation, the mechanism that regulates this is not fully understood. Similarly to ERK and JNK, p38α/β do not contain the canonical nuclear localization signals (NLS). In addition, they do not seem to interact with the classical Impα/β [111], or use passive diffusion for their nuclear shuttling. On the other hand, our group has recently shown that the nuclear translocation is mediated by three β-like importins, Imp3, 7, and 9 [115]. Thus, we found that upon stimulation, p38α/β as well as JNK1/2 are released from their cytoplasmic anchoring proteins and interact with either Imp7 or Imp9, each one in a complex with Imp3 (see the schematic representation in Figure 1). Then, the trimers formed (Imp7/Imp3/kinase or Imp9/Imp3/kinase) are shuttled to the nuclear pore, where Imp3 remains, while Imp7 or Imp9 escort the shuttling p38α/β or JNK1/2 into the nucleus [111]. In the nucleus, the small GTPase Ran dissociates the importins from p38α/β, and the latter are freed to execute their functions. This mechanism has some similarity to the translocation of ERK, which is detached from its anchoring protein upon stimulation and interacts with Imp7 that escorts it to the nucleus [18,116,117]. The Imp7 binding site of p38α/β resides in the N terminus of the kinase, and it is composed of at least nine residues: PERYQNLSP. Indeed, deletion of this region or substitution of its residues to Ala residues prevents the interaction [80]. Interestingly, the nuclear translocation of p38α/β may require HSP70 interaction in the nucleus [118], rely on microtubules and dynein, which may indicate that the translocation is helped by a trafficking machinery [112] and might involve SUMOylation of the kinases [119]. However, the mechanisms by which these components participate in the translocation and how they are related to the importins involved are not clear.

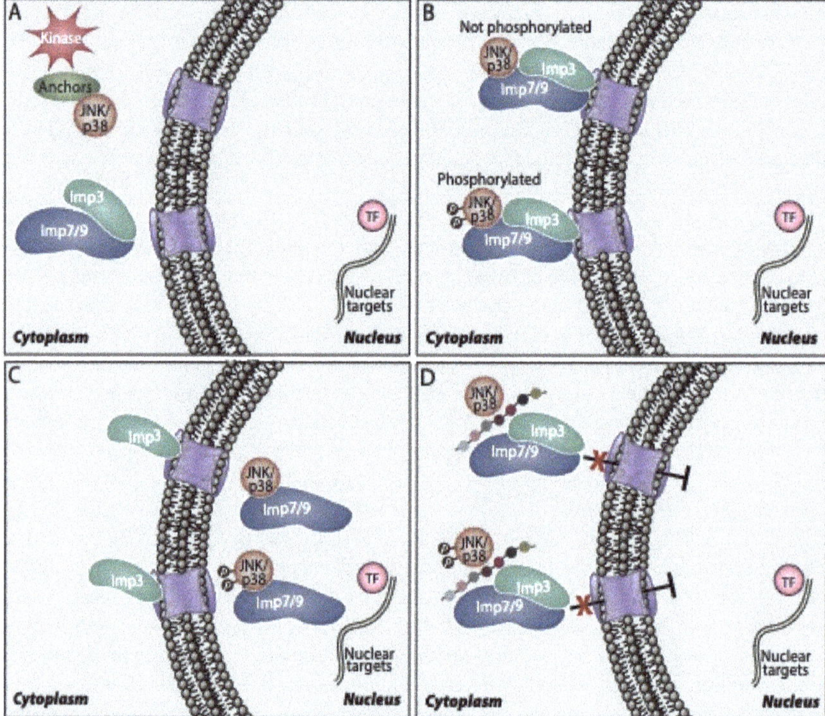

Figure 1. Scheme showing the mechanism of nuclear p38α/β translocation and its inhibition by the PERY peptide. **A**. p38α/β are localized in the cytoplasm of resting cells and some of the molecules are phosphorylated upon stimulation. **B**. Phosphorylated or non-phosphorylated p38α/β bind to a dimer of Imp7/3 or Imp9/3, which escort them to the nuclear pores. **C**. Imp3 stays outside, while Imp7 or Imp9 escort the p38α/β through the nuclear pores, to the nucleus, where they dissociate from the importins, and are free to phosphorylate their substrates. **D**. Translocation of the kinases to the nucleus can be inhibited using the inhibitory PERY peptide that was synthesized according to the sequence of the p38α–Imp binding site as described in the text.

In previous studies on the nuclear translocation of ERK, our group designed a myristoylated peptide (EPE peptide) based on the ERK interaction site with Imp7 [120,121]. When added to cells, the peptide completely abolished the interaction between ERK and Imp7, and it prevented both stimulated and non-stimulated translocation of the kinase, indicating that the great majority of the translocation is mediated by Imp7, but not by passive diffusion [122] under both conditions. The potency of the peptide varied between cell lines, completely abolishing the proliferation of melanoma and other ERK-addicted cancer cell lines, but not the growth of non-transformed cells [120,123]. The peptide also diminished the growth of melanoma xenografts, better than the Raf inhibitor vemurafenib.

Based on the successful development of the EPE peptide, we developed a myristoylated peptide targeting the binding site of p38α/β to Imp7/9, which we termed the PERY peptide. As expected [80], the peptide completely abolished the interaction of p38α/β to both Imp7 and Imp9, preventing the nuclear translocation of the kinases and the phosphorylation of nuclear targets. Since p38α/β signaling is involved in the proliferation of only limited types of cancers [6], the PERY peptide inhibited the proliferation of some breast cancer and melanoma cells but not of other cancers. In most cases, the effects were similar to those of the commercial p38 activity inhibitors, indicating that the effect on inflammation and cancer is mediated mainly by the nuclear p38α/β. Importantly, the peptide inhibited DSS-induced colon inflammation (colitis-like) and impressively, it demonstrated a significant inhibitory effect on a model of colitis-associated colon cancer [80]. Its effect was stronger when added throughout the DSS treatment than when added during the last two cycles of the treatment, and it was much stronger when compared to the effect of the commercial p38 activity inhibitor. Moreover, we have shown that the effect is mediated by macrophages [80] and not via enterocytes or other colon cells. Hence, these results clearly indicate that the nuclear translocation plays a role in the induction of inflammation and not the proliferation of colon cells. Taken together, we demonstrated that the nuclear translocation of p38α/β (and JNK) is important for the induction of inflammation, which may further lead to the development of some cancers. Additionally, it can be involved in the induction of proliferation of other cancer types such as triple negative breast cancers. Moreover, the results demonstrate that inhibiting the nuclear translocation of p38α/β may serve as a therapeutic strategy to combat various cancers as well as inflammatory diseases (Table 1).

Table 1. Nuclear substrates of p38α/β and their role in cancer and inflammation. More than 120 substrates of p38α/β were found in several reviews [27,43,86–89], and the translocation of each one of them was inspected in various databases. Substrates with constant (> 80%) nuclear localization in all cell lines described are shown under "Mostly Nuclear Proteins", while proteins that are mostly nuclear in resting cells but are exported to the cytoplasm after stimulation are shown under "Nuclear Export". The role of the p38α/β phosphorylation, as well as their general involvement in cancer or inflammation (independent of the phosphorylation in the nucleus) is described for each nuclear substrate. ND—not determined.

Localization	p38-phosphorylated Protein	Role of Phosphorylation	Involvement in Cancer	Involvement in Inflammation
Mostly Nuclear Proteins	Cyclin D3	Targets cyclin D3 for proteasomal degradation [124].	Together with CDK6 regulates cell metabolism to promote cancer [125].	Together with CDK6 phosphorylates NFκB to induce inflammatory gene expression [126].
	E47	Promotes MyoD/E47 association and muscle-gene transcription [127].	Induces EMT and therefore may facilitate tumor formation [128].	Required for the efficient recruitment of GR (anti-inflammatory) to chromatin [129].
	FBP2 (KSRP)	Controls stability of myogenic transcripts [130].	Regulates c-Fos RNA stability and therefore cancers [131].	Induce pro-inflammatory genes upon resveratrol treatment [132].
	FBP3	Controls prothrombin expression [133].	May regulate Myc expression [134].	May be involved in thrombin-induced inflammation [133].
	H2AX	Chromatin remodeling. Involved in G_2 checkpoint that protects cells from DNA breaks [135].	Phosphorylation of Ser139 by RSK (the same site phosphorylated by p38) inhibits cell transformation [136].	Colonocytes from ulcerative colitis patients showed an increase in H2AX content. Not necessarily related to phosphorylation [137].
	H3	Related to chromatin remodeling and chromosome condensation [138].	p38 phosphorylation of Ser10 causes aggressive gastric cancer [139].	p38-dependent H3 phosphorylation may mark promoters for increased NFκB recruitment and inflammation [140].
	HBP1	Stabilizes the proteins that leads to cell cycle inhibition [141].	Inhibits cell cycle and functions as a tumor suppressor [78].	Promote vascular inflammation in atherogenesis [142].
	Id2	Regulates transcription, cell cycle, and differentiation [143].	Participate in VHL inactivation in cancer [144].	Maintains regulatory T cell to suppress inflammatory diseases [145].

Table 1. *Cont.*

Localization	p38-phosphorylated Protein	Role of Phosphorylation	Involvement in Cancer	Involvement in Inflammation
	IWS1	Likely regulates RNA processing and export [89].	Regulates trimethylation of Histone H3 that may lead to cancer [146].	ND
	JDP2	Phosphorylation at Thr148 likely leads to proteasomal degradation (as with JNK [147]).	Implicated in progression and suppression of different cancers [148].	Involved in liver inflammation [149].
	MEF2d	Regulates recruitment of proteins to specific genes [150].	Enhances proliferation migration and invasion in pancreatic cancer [151].	Regulates IL-10 production in microglia to protect neuronal cells from inflammation-induced death [152].
	Mnk2b	Induces activation [153].	Mnk2b is oncogenic, by enhancing eIF4E phosphorylation [154].	MNK2 is involved in adipose tissue inflammation (possibly both isoforms) [155].
	MSK1	Induces activation [156].	Induces the transcription of immediate-early oncogenes [32].	Activation of the pro-inflammatory NF-κB signaling pathway through MSK1 in microglial cells [157].
	MSK2	Induces activation [158].	Induces the transcription of immediate-early oncogenes [32].	Plays a role in limiting Toll-like receptor-driven inflammation [159].
	P18Hamlet (Znhit1)	Stimulates p53-dependent apoptosis [160].	Regulates p53 and therefore cancer [160].	May affect p53-dependent inflammation [160,161].
	P53	Regulates apoptosis [162].	Tumor suppressor [161].	Suppressor of inflammation and autoimmunity [161].
	PGC-1α	Regulates cytokine-induced energy expenditure [163].	PGC-1α expression is altered in tumors and metastasis in relation to modifications in cellular metabolism [164].	Connects oxidative stress and mitochondrial metabolism with inflammatory response and metabolic syndrome [165].
	PPARalpha	Plays a role in cardiac metabolic stress response [166].	Modulates metabolic pathways and attenuates kidney tumor growth [167].	Exerts a major anti-inflammatory action in human liver [168].
	Ranbp2	Probably regulates SUMOylation and myotube formation [89].	Involved in inflammatory myofibroblastic tumor formation [169].	Inflammatory myofibroblastic tumor with RANBP2 and ALK gene rearrangement [169].
	Rb1	Mediates Fas-effects on inactivation of Rb1, independent of CDKs [170].	Functions as a tumor suppressor. Inactivation induces retinoblastoma and other cancers [171].	RB inactivation enhances pro-inflammatory signaling that can lead to cancer [172].
	RNF2	Modulates the expression of transcription factors and histone 2B acetylation [173].	Monoubiquitinates H2AK119 at the promoter of LTBP2, thus regulates TGFβ signaling to induce melanoma [174].	Inhibit interferon-dependent responses that may include inflammation [175].
	Rpn2	Negatively regulates proteasome activity [40].	Promotes metastasis of hepatocellular carcinoma [176].	Downregulated the inflammatory-associated JAK1/STAT3 pathway [177].
	RUNX2	Increases transcriptional activity [178].	Abnormally expressed in prostatecancerand associates with metastatic disease [179].	May have a role in the inflammatory remodeling of the collagen matrix [180].
	SPF45	Regulates alternative splicing site utilization [181], which may lead to multidrug resistance phenotypes [182].	The phosphorylation inhibits proliferation and therefore may block cancer [181].	Highly expressed in lung's inflammatory cells, which might be involved in their function [182].
	SRC3	Controls the dynamics of interactions with RARalpha to facilitate gene activation [183].	Promotes breast and prostate cancer cell proliferation and survival [184].	Regulates inflammation during wound healing [185].
Nuclear Export	AHNAK	Probably induces its differentiation-related activity [89].	Promotes metastasis through TGF-β-mediated EMT [186].	Silencing of AHNAK in dental pulp cells led to reduced inflammation-related proteins [187].
	c/EBPalpha	Inhibits enhancer activity [188].	Suppresses tumor metastasis and growth in gastric cancer [189].	Interacts with NF-κB to regulate inflammation [190].
	c/EBPbeta	Activates enhancer activity [191].	Regulates tumor progression [192].	Induces inflammation and ER stress [193].
	ERalpha	Induces activation and nuclear export [55].	Functions as an oncogene in breast cancer [194].	Abnormal ERalpha signaling leads to inflammation [195].

Table 1. *Cont.*

Localization	p38-phosphorylated Protein	Role of Phosphorylation	Involvement in Cancer	Involvement in Inflammation
	MK2	Induces activation [196].	Plays a role in the induction of lung cancer [197]. Activates cancer-related proteins (Cdc25B/C, Plk1, and TSC2) [198].	Plays a role in inflammatory pulmonary diseases [197]. Regulates inflammatory cytokines, transcript stability, and critical cellular processes [69].
	MK3	Induces activation [199].	Leads to pancreatic cancer growth [200].	Induces TNF biosynthesis and inflammation [201].
	MK5	Induces activation [202].	Induces breast cancer [203].	Phosphorylates HSP27 to induce inflammation [204].
	MRF4	Reduces transcriptional activity [205].	May regulate hairy cell leukemia (HCL) [206].	ND
	NFATc4	Activation and nuclear export [207].	Correlates with decreased proliferation and poor prognosis of ovarian cancer [208].	Involved in the secretion of inflammatory factors [209].
	NR4A	Regulates dopamine synthesis genes [210].	Has both tumor suppressor and oncogenic functions in different cells [211].	May contribute to the cellular processes that control inflammation [212].
	Pax6	Elevates transcriptional activity [213].	Induces cell proliferation in lung cancer [214].	ND

7. Concluding Remarks

The response of cells to stress and other extracellular stimuli leads to the activation of several signaling pathways, including primarily those of p38α/β and JNK. The p38α/β signaling cascade is well-known for its ability to transmit stress signals to various targets within the cells. Thus, it regulates the activity of many transcription factors involved in stress response, and it was shown to also regulate cell cycle, RNA processes, and cytoskeletal elements. Here, we discussed in detail its involvement in the degradation and nuclear translocation of many proteins, which is an effect that may be in both cases either global, by affecting executing enzymes, or individual by acting specifically on degrading/translocating molecules. Regarding protein degradation, the global effects may be mediated by regulating proteasome or ubiquitin ligases, while the phosphorylation of individual substrates may either stabilize them or enhance their degradation. As for nuclear translocation, the global effect is possibly mediated by regulating importins or nuclear pore proteins (NUPs), while the individual effects can be mediated by either reduced or enhanced binding to the translocation machinery or to anchoring proteins. Being such central signaling pathways, the dysregulation of the p38α/β cascade results in pathologies, and indeed, almost all constantly nuclear targets were shown to play a role in the regulation of cancer and inflammation.

One of the hallmarks of many stimulations is the rapid and robust nuclear translocation of p38α/β. This translocation is essential for the regulation of targets that are only localized in the nucleus, although it can enhance the phosphorylation of proteins that are phosphorylated in the cytoplasm and translocate to the nucleus upon stimulation (e.g., transcription factors). We found that the translocation is mediated by the binding of p38α/β with either Imp7 or Imp9, which further bind individually to Imp3. Then, the complex moves to the NUPs, where Imp3 stays, while Imp7 or Imp9 shuttles the p38α/β into the nucleus. In the nucleus, p38 is freed from the importins by Ran and then, it is able to execute its nuclear functions. The duration of p38α/β residence in the nucleus may vary between cells and conditions, after which the kinases are exported out of the nucleus by their NES-containing substrate, MK2. A few years ago, our group developed the PERY peptide that completely prevents the nuclear translocation of p38α/β and thereby prevents the growth of some cancer cells—particularly DSS-induced colon inflammation and inflammation-induced cancer. Thus, the nuclear translocation of p38α/β can serve as a good target for inflammation and cancer, and inhibitors of this kinase translocation should be further developed for clinical use.

Author Contributions: All three authors contributed to the preparation of this review. All authors have read and agreed to the published version of the manuscript.

Funding: This research received no external funding. Rony Seger is an incumbent of the Yale S. Lewine and Ella Miller Lewine professorial chair for cancer research.

Conflicts of Interest: The authors declare no conflict of interest.

Abbreviations

DEF	docking site for ERK
ERK	extracellular signal-regulated kinase
Imp	importin
JNK	c-Jun N-terminal kinase
MAPK	mitogen-activated protein kinase
MAPKAPK	MAPK-activated protein kinase (also known as MK)
MKK	MAPK kinase
NUP	nuclear pore protein

References

1. Peti, W.; Page, R. Molecular basis of MAP kinase regulation. *Protein Sci.* **2013**, *22*, 1698–1710. [CrossRef]
2. Keshet, Y.; Seger, R. The MAP kinase signaling cascades: A system of hundreds of components regulates a diverse array of physiological functions. *Methods Mol. Biol.* **2010**, *661*, 3–38. [CrossRef] [PubMed]
3. Cargnello, M.; Roux, P.P. Activation and function of the MAPKs and their substrates, the MAPK-activated protein kinases. *Microbiol. Mol. Biol. Rev.* **2011**, *75*, 50–83. [CrossRef] [PubMed]
4. Eblen, S.T. Extracellular-Regulated Kinases: Signaling From Ras to ERK Substrates to Control Biological Outcomes. *Adv. Cancer Res.* **2018**, *138*, 99–142. [CrossRef] [PubMed]
5. Ha, J.; Kang, E.; Seo, J.; Cho, S. Phosphorylation Dynamics of JNK Signaling: Effects of Dual-Specificity Phosphatases (DUSPs) on the JNK Pathway. *Int. J. Mol. Sci.* **2019**, *20*, 6157. [CrossRef] [PubMed]
6. Martinez-Limon, A.; Joaquin, M.; Caballero, M.; Posas, F.; de Nadal, E. The p38 Pathway: From Biology to Cancer Therapy. *Int. J. Mol. Sci.* **2020**, *21*, 1913. [CrossRef]
7. Tubita, A.; Lombardi, Z.; Tusa, I.; Dello Sbarba, P.; Rovida, E. Beyond Kinase Activity: ERK5 Nucleo-Cytoplasmic Shuttling as a Novel Target for Anticancer Therapy. *Int. J. Mol. Sci.* **2020**, *21*, 938. [CrossRef]
8. Plotnikov, A.; Zehorai, E.; Procaccia, S.; Seger, R. The MAPK cascades: Signaling components, nuclear roles and mechanisms of nuclear translocation. *Biochim. Biophys. Acta* **2011**, *1813*, 1619–1633. [CrossRef]
9. Cuadrado, A.; Nebreda, A.R. Mechanisms and functions of p38 MAPK signalling. *Biochem. J.* **2010**, *429*, 403–417. [CrossRef]
10. Shiryaev, A.; Moens, U. Mitogen-activated protein kinase p38 and MK2, MK3 and MK5: Menage a trois or menage a quatre? *Cell. Signal.* **2010**, *22*, 1185–1192. [CrossRef]
11. Salvador, J.M.; Mittelstadt, P.R.; Guszczynski, T.; Copeland, T.D.; Yamaguchi, H.; Appella, E.; Fornace, A.J., Jr.; Ashwell, J.D. Alternative p38 activation pathway mediated by T cell receptor-proximal tyrosine kinases. *Nat. Immunol.* **2005**, *6*, 390–395. [CrossRef] [PubMed]
12. Ge, B.; Gram, H.; Di Padova, F.; Huang, B.; New, L.; Ulevitch, R.J.; Luo, Y.; Han, J. MAPKK-independent activation of p38alpha mediated by TAB1-dependent autophosphorylation of p38alpha. *Science* **2002**, *295*, 1291–1294. [CrossRef] [PubMed]
13. Ferguson, B.S.; Nam, H.; Morrison, R.F. Dual-specificity phosphatases regulate mitogen-activated protein kinase signaling in adipocytes in response to inflammatory stress. *Cell. Signal.* **2019**, *53*, 234–245. [CrossRef] [PubMed]
14. Marshall, C.J. Specificity of receptor tyrosine kinase signaling: Transient versus sustained extracellular signal-regulated kinase activation. *Cell* **1995**, *80*, 179–185. [CrossRef]
15. Patterson, K.I.; Brummer, T.; O'Brien, P.M.; Daly, R.J. Dual-specificity phosphatases: Critical regulators with diverse cellular targets. *Biochem. J.* **2009**, *418*, 475–489. [CrossRef] [PubMed]
16. Yao, Z.; Seger, R. The molecular Mechanism of MAPK/ERK inactivation. *Curr. Genom.* **2004**, *5*, 385–393. [CrossRef]

17. Wang, X.D.; Zhao, C.S.; Wang, Q.L.; Zeng, Q.; Feng, X.Z.; Li, L.; Chen, Z.L.; Gong, Y.; Han, J.; Li, Y. The p38-interacting protein p38IP suppresses TCR and LPS signaling by targeting TAK1. *EMBO Rep.* **2020**. [CrossRef]
18. Flores, K.; Yadav, S.S.; Katz, A.A.; Seger, R. The Nuclear Translocation of Mitogen-Activated Protein Kinases: Molecular Mechanisms and Use as Novel Therapeutic Target. *Neuroendocrinology* **2019**, *108*, 121–131. [CrossRef]
19. Kumar, S.; Boehm, J.; Lee, J.C. p38 MAP kinases: Key signalling molecules as therapeutic targets for inflammatory diseases. *Nat. Rev. Drug Discov.* **2003**, *2*, 717–726. [CrossRef]
20. Mavropoulos, A.; Orfanidou, T.; Liaskos, C.; Smyk, D.S.; Billinis, C.; Blank, M.; Rigopoulou, E.I.; Bogdanos, D.P. p38 mitogen-activated protein kinase (p38 MAPK)-mediated autoimmunity: Lessons to learn from ANCA vasculitis and pemphigus vulgaris. *Autoimmun. Rev.* **2013**, *12*, 580–590. [CrossRef]
21. Dolado, I.; Swat, A.; Ajenjo, N.; De Vita, G.; Cuadrado, A.; Nebreda, A.R. p38alpha MAP kinase as a sensor of reactive oxygen species in tumorigenesis. *Cancer Cell* **2007**, *11*, 191–205. [CrossRef] [PubMed]
22. Tomas-Loba, A.; Manieri, E.; Gonzalez-Teran, B.; Mora, A.; Leiva-Vega, L.; Santamans, A.M.; Romero-Becerra, R.; Rodriguez, E.; Pintor-Chocano, A.; Feixas, F.; et al. p38gamma is essential for cell cycle progression and liver tumorigenesis. *Nature* **2019**, *568*, 557–560. [CrossRef]
23. Gupta, J.; del Barco Barrantes, I.; Igea, A.; Sakellariou, S.; Pateras, I.S.; Gorgoulis, V.G.; Nebreda, A.R. Dual function of p38alpha MAPK in colon cancer: Suppression of colitis-associated tumor initiation but requirement for cancer cell survival. *Cancer Cell* **2014**, *25*, 484–500. [CrossRef] [PubMed]
24. Seger, R.; Krebs, E.G. The MAPK signaling cascade. *FASEB J.* **1995**, *9*, 726–735. [CrossRef] [PubMed]
25. Bambang, I.F.; Lu, D.; Li, H.; Chiu, L.L.; Lau, Q.C.; Koay, E.; Zhang, D. Cytokeratin 19 regulates endoplasmic reticulum stress and inhibits ERp29 expression via p38 MAPK/XBP-1 signaling in breast cancer cells. *Exp. Cell Res.* **2009**, *315*, 1964–1974. [CrossRef] [PubMed]
26. Zhong, W.; Zhu, H.; Sheng, F.; Tian, Y.; Zhou, J.; Chen, Y.; Li, S.; Lin, J. Activation of the MAPK11/12/13/14 (p38 MAPK) pathway regulates the transcription of autophagy genes in response to oxidative stress induced by a novel copper complex in HeLa cells. *Autophagy* **2014**, *10*, 1285–1300. [CrossRef] [PubMed]
27. Han, J.; Wu, J.; Silke, J. An overview of mammalian p38 mitogen-activated protein kinases, central regulators of cell stress and receptor signaling. *F1000Research* **2020**, *9*. [CrossRef]
28. Thoms, H.C.; Dunlop, M.G.; Stark, L.A. p38-mediated inactivation of cyclin D1/cyclin-dependent kinase 4 stimulates nucleolar translocation of RelA and apoptosis in colorectal cancer cells. *Cancer Res.* **2007**, *67*, 1660–1669. [CrossRef]
29. Barnum, K.J.; O'Connell, M.J. Cell cycle regulation by checkpoints. *Methods Mol. Biol.* **2014**, *1170*, 29–40. [CrossRef]
30. Wagner, E.F.; Nebreda, A.R. Signal integration by JNK and p38 MAPK pathways in cancer development. *Nat. Rev. Cancer* **2009**, *9*, 537–549. [CrossRef]
31. Faust, D.; Schmitt, C.; Oesch, F.; Oesch-Bartlomowicz, B.; Schreck, I.; Weiss, C.; Dietrich, C. Differential p38-dependent signalling in response to cellular stress and mitogenic stimulation in fibroblasts. *Cell Commun. Signal.* **2012**, *10*, 6. [CrossRef] [PubMed]
32. Adewumi, I.; Lopez, C.; Davie, J.R. Mitogen and stress- activated protein kinase regulated gene expression in cancer cells. *Adv. Biol. Regul.* **2019**, *71*, 147–155. [CrossRef] [PubMed]
33. Hotamisligil, G.S.; Davis, R.J. Cell Signaling and Stress Responses. *Cold Spring Harb. Perspect. Biol.* **2016**, *8*. [CrossRef] [PubMed]
34. Khurana, A.; Nakayama, K.; Williams, S.; Davis, R.J.; Mustelin, T.; Ronai, Z. Regulation of the ring finger E3 ligase Siah2 by p38 MAPK. *J. Biol. Chem.* **2006**, *281*, 35316–35326. [CrossRef] [PubMed]
35. Hong, J.; Min, Y.; Wuest, T.; Lin, P.C. Vav1 is Essential for HIF-1alpha Activation via a Lysosomal VEGFR1-Mediated Degradation Mechanism in Endothelial Cells. *Cancers* **2020**, *12*, 1374. [CrossRef] [PubMed]
36. Bhatt, S.; Stender, J.D.; Joshi, S.; Wu, G.; Katzenellenbogen, B.S. OCT-4: A novel estrogen receptor-alpha collaborator that promotes tamoxifen resistance in breast cancer cells. *Oncogene* **2016**, *35*, 5722–5734. [CrossRef] [PubMed]
37. Kim, S.M.; Kim, M.Y.; Ann, E.J.; Mo, J.S.; Yoon, J.H.; Park, H.S. Presenilin-2 regulates the degradation of RBP-Jk protein through p38 mitogen-activated protein kinase. *J. Cell Sci.* **2012**, *125*, 1296–1308. [CrossRef]
38. Wang, Q.E.; Han, C.; Zhao, R.; Wani, G.; Zhu, Q.; Gong, L.; Battu, A.; Racoma, I.; Sharma, N.; Wani, A.A. p38 MAPK- and Akt-mediated p300 phosphorylation regulates its degradation to facilitate nucleotide excision repair. *Nucleic Acids Res.* **2013**, *41*, 1722–1733. [CrossRef]

39. Liu, K.; Zhang, C.; Li, B.; Xie, W.; Zhang, J.; Nie, X.; Tan, P.; Zheng, L.; Wu, S.; Qin, Y.; et al. Mutual Stabilization between TRIM9 Short Isoform and MKK6 Potentiates p38 Signaling to Synergistically Suppress Glioblastoma Progression. *Cell Rep.* **2018**, *23*, 838–851. [CrossRef]
40. Lee, S.H.; Park, Y.; Yoon, S.K.; Yoon, J.B. Osmotic stress inhibits proteasome by p38 MAPK-dependent phosphorylation. *J. Biol. Chem.* **2010**, *285*, 41280–41289. [CrossRef]
41. Leestemaker, Y.; de Jong, A.; Witting, K.F.; Penning, R.; Schuurman, K.; Rodenko, B.; Zaal, E.A.; van de Kooij, B.; Laufer, S.; Heck, A.J.R.; et al. Proteasome Activation by Small Molecules. *Cell Chem. Biol.* **2017**, *24*, 725–736.e7. [CrossRef] [PubMed]
42. Liu, K.; Jones, S.; Minis, A.; Rodriguez, J.; Molina, H.; Steller, H. PI31 Is an Adaptor Protein for Proteasome Transport in Axons and Required for Synaptic Development. *Dev. Cell* **2019**, *50*, 509–524. [CrossRef] [PubMed]
43. Trempolec, N.; Dave-Coll, N.; Nebreda, A.R. SnapShot: p38 MAPK substrates. *Cell* **2013**, *152*, 924. [CrossRef] [PubMed]
44. Porter, F.W.; Brown, B.; Palmenberg, A.C. Nucleoporin phosphorylation triggered by the encephalomyocarditis virus leader protein is mediated by mitogen-activated protein kinases. *J. Virol.* **2010**, *84*, 12538–12548. [CrossRef]
45. Chahine, M.N.; Mioulane, M.; Sikkel, M.B.; O'Gara, P.; Dos Remedios, C.G.; Pierce, G.N.; Lyon, A.R.; Foldes, G.; Harding, S.E. Nuclear pore rearrangements and nuclear trafficking in cardiomyocytes from rat and human failing hearts. *Cardiovasc. Res.* **2015**, *105*, 31–43. [CrossRef]
46. Jiang, Y.; Wu, C.; Boye, A.; Wu, J.; Wang, J.; Yang, X.; Yang, Y. MAPK inhibitors modulate Smad2/3/4 complex cyto-nuclear translocation in myofibroblasts via Imp7/8 mediation. *Mol. Cell. Biochem.* **2015**, *406*, 255–262. [CrossRef]
47. Hayes, S.A.; Huang, X.; Kambhampati, S.; Platanias, L.C.; Bergan, R.C. p38 MAP kinase modulates Smad-dependent changes in human prostate cell adhesion. *Oncogene* **2003**, *22*, 4841–4850. [CrossRef]
48. Ho, K.K.; McGuire, V.A.; Koo, C.Y.; Muir, K.W.; de Olano, N.; Maifoshie, E.; Kelly, D.J.; McGovern, U.B.; Monteiro, L.J.; Gomes, A.R.; et al. Phosphorylation of FOXO3a on Ser-7 by p38 promotes its nuclear localization in response to doxorubicin. *J. Biol. Chem.* **2012**, *287*, 1545–1555. [CrossRef]
49. Xu, Y.Z.; Thuraisingam, T.; Morais, D.A.; Rola-Pleszczynski, M.; Radzioch, D. Nuclear translocation of beta-actin is involved in transcriptional regulation during macrophage differentiation of HL-60 cells. *Mol. Biol. Cell* **2010**, *21*, 811–820. [CrossRef]
50. Li, Y.; Chen, Y.; Xu, J. Factors influencing RhoA protein distribution in the nucleus. *Mol. Med. Rep.* **2011**, *4*, 1115–1119. [CrossRef]
51. Gaestel, M. MAPKAP kinases-MKs-two's company, three's a crowd. *Nat. Rev. Mol. Cell Biol.* **2006**, *7*, 120–130. [CrossRef] [PubMed]
52. Seternes, O.M.; Mikalsen, T.; Johansen, B.; Michaelsen, E.; Armstrong, C.G.; Morrice, N.A.; Turgeon, B.; Meloche, S.; Moens, U.; Keyse, S.M. Activation of MK5/PRAK by the atypical MAP kinase ERK3 defines a novel signal transduction pathway. *EMBO J.* **2004**, *23*, 4780–4791. [CrossRef] [PubMed]
53. Han, Y.H.; Zhou, H.; Kim, J.H.; Yan, T.D.; Lee, K.H.; Wu, H.; Lin, F.; Lu, N.; Liu, J.; Zeng, J.Z.; et al. A unique cytoplasmic localization of retinoic acid receptor-gamma and its regulations. *J. Biol. Chem.* **2009**, *284*, 18503–18514. [CrossRef] [PubMed]
54. Gioeli, D.; Black, B.E.; Gordon, V.; Spencer, A.; Kesler, C.T.; Eblen, S.T.; Paschal, B.M.; Weber, M.J. Stress kinase signaling regulates androgen receptor phosphorylation, transcription, and localization. *Mol. Endocrinol.* **2006**, *20*, 503–515. [CrossRef]
55. Lee, H.; Bai, W. Regulation of estrogen receptor nuclear export by ligand-induced and p38-mediated receptor phosphorylation. *Mol. Cell. Biol.* **2002**, *22*, 5835–5845. [CrossRef]
56. Flamand, N.; Luo, M.; Peters-Golden, M.; Brock, T.G. Phosphorylation of serine 271 on 5-lipoxygenase and its role in nuclear export. *J. Biol. Chem.* **2009**, *284*, 306–313. [CrossRef]
57. Lin, K.C.; Moroishi, T.; Meng, Z.; Jeong, H.S.; Plouffe, S.W.; Sekido, Y.; Han, J.; Park, H.W.; Guan, K.L. Regulation of Hippo pathway transcription factor TEAD by p38 MAPK-induced cytoplasmic translocation. *Nat. Cell Biol.* **2017**, *19*, 996–1002. [CrossRef]
58. Cotticelli, M.G.; Xia, S.; Kaur, A.; Lin, D.; Wang, Y.; Ruff, E.; Tobias, J.W.; Wilson, R.B. Identification of p38 MAPK as a novel therapeutic target for Friedreich's ataxia. *Sci. Rep.* **2018**, *8*, 5007. [CrossRef]
59. Bohush, A.; Niewiadomska, G.; Filipek, A. Role of Mitogen Activated Protein Kinase Signaling in Parkinson's Disease. *Int. J. Mol. Sci.* **2018**, *19*, 2973. [CrossRef]

60. Kheiri, G.; Dolatshahi, M.; Rahmani, F.; Rezaei, N. Role of p38/MAPKs in Alzheimer's disease: Implications for amyloid beta toxicity targeted therapy. *Rev. Neurosci.* **2018**, *30*, 9–30. [CrossRef]
61. Liu, R.; Molkentin, J.D. Regulation of cardiac hypertrophy and remodeling through the dual-specificity MAPK phosphatases (DUSPs). *J. Mol. Cell. Cardiol.* **2016**, *101*, 44–49. [CrossRef] [PubMed]
62. Luo, F.; Shi, J.; Shi, Q.; Xu, X.; Xia, Y.; He, X. Mitogen-Activated Protein Kinases and Hypoxic/Ischemic Nephropathy. *Cell. Physiol. Biochem.* **2016**, *39*, 1051–1067. [CrossRef] [PubMed]
63. Nandipati, K.C.; Subramanian, S.; Agrawal, D.K. Protein kinases: Mechanisms and downstream targets in inflammation-mediated obesity and insulin resistance. *Mol. Cell. Biochem.* **2017**, *426*, 27–45. [CrossRef] [PubMed]
64. Grossi, V.; Peserico, A.; Tezil, T.; Simone, C. p38alpha MAPK pathway: A key factor in colorectal cancer therapy and chemoresistance. *World J. Gastroenterol.* **2014**, *20*, 9744–9758. [CrossRef]
65. Han, J.; Lee, J.D.; Bibbs, L.; Ulevitch, R.J. A MAP kinase targeted by endotoxin and hyperosmolarity in mammalian cells. *Science* **1994**, *265*, 808–811. [CrossRef]
66. Cuenda, A.; Rousseau, S. p38 MAP-kinases pathway regulation, function and role in human diseases. *Biochim. Biophys. Acta* **2007**, *1773*, 1358–1375. [CrossRef]
67. Yong, H.Y.; Koh, M.S.; Moon, A. The p38 MAPK inhibitors for the treatment of inflammatory diseases and cancer. *Expert Opin. Investig. Drugs* **2009**, *18*, 1893–1905. [CrossRef]
68. Tiedje, C.; Holtmann, H.; Gaestel, M. The role of mammalian MAPK signaling in regulation of cytokine mRNA stability and translation. *J. Interferon Cytokine Res.* **2014**, *34*, 220–232. [CrossRef]
69. Winzen, R.; Kracht, M.; Ritter, B.; Wilhelm, A.; Chen, C.Y.; Shyu, A.B.; Muller, M.; Gaestel, M.; Resch, K.; Holtmann, H. The p38 MAP kinase pathway signals for cytokine-induced mRNA stabilization via MAP kinase-activated protein kinase 2 and an AU-rich region-targeted mechanism. *EMBO J.* **1999**, *18*, 4969–4980. [CrossRef]
70. Haller, V.; Nahidino, P.; Forster, M.; Laufer, S.A. An updated patent review of p38 MAP kinase inhibitors (2014–2019). *Expert Opin. Ther. Pat.* **2020**, *30*, 453–466. [CrossRef]
71. Xing, L. Clinical candidates of small molecule p38 MAPK inhibitors for inflammatory diseases. *MAP Kinase* **2015**, *4*, 24–30. [CrossRef]
72. Brancho, D.; Tanaka, N.; Jaeschke, A.; Ventura, J.J.; Kelkar, N.; Tanaka, Y.; Kyuuma, M.; Takeshita, T.; Flavell, R.A.; Davis, R.J. Mechanism of p38 MAP kinase activation in vivo. *Genes Dev.* **2003**, *17*, 1969–1978. [CrossRef] [PubMed]
73. MacNeil, A.J.; Jiao, S.C.; McEachern, L.A.; Yang, Y.J.; Dennis, A.; Yu, H.; Xu, Z.; Marshall, J.S.; Lin, T.J. MAPK kinase 3 is a tumor suppressor with reduced copy number in breast cancer. *Cancer Res.* **2014**, *74*, 162–172. [CrossRef] [PubMed]
74. Kishi, H.; Nakagawa, K.; Matsumoto, M.; Suga, M.; Ando, M.; Taya, Y.; Yamaizumi, M. Osmotic shock induces G1 arrest through p53 phosphorylation at Ser33 by activated p38MAPK without phosphorylation at Ser15 and Ser20. *J. Biol. Chem.* **2001**, *276*, 39115–39122. [CrossRef] [PubMed]
75. Yu-Lee, L.Y.; Yu, G.; Lee, Y.C.; Lin, S.C.; Pan, J.; Pan, T.; Yu, K.J.; Liu, B.; Creighton, C.J.; Rodriguez-Canales, J.; et al. Osteoblast-Secreted Factors Mediate Dormancy of Metastatic Prostate Cancer in the Bone via Activation of the TGFbetaRIII-p38MAPK-pS249/T252RB Pathway. *Cancer Res.* **2018**, *78*, 2911–2924. [CrossRef]
76. Lalaoui, N.; Hanggi, K.; Brumatti, G.; Chau, D.; Nguyen, N.Y.; Vasilikos, L.; Spilgies, L.M.; Heckmann, D.A.; Ma, C.; Ghisi, M.; et al. Targeting p38 or MK2 Enhances the Anti-Leukemic Activity of Smac-Mimetics. *Cancer Cell* **2016**, *29*, 145–158. [CrossRef]
77. Puri, P.L.; Wu, Z.; Zhang, P.; Wood, L.D.; Bhakta, K.S.; Han, J.; Feramisco, J.R.; Karin, M.; Wang, J.Y. Induction of terminal differentiation by constitutive activation of p38 MAP kinase in human rhabdomyosarcoma cells. *Genes Dev.* **2000**, *14*, 574–584.
78. Bollaert, E.; de Rocca Serra, A.; Demoulin, J.B. The HMG box transcription factor HBP1: A cell cycle inhibitor at the crossroads of cancer signaling pathways. *Cell. Mol. Life Sci.* **2019**, *76*, 1529–1539. [CrossRef]
79. Wen, S.Y.; Cheng, S.Y.; Ng, S.C.; Aneja, R.; Chen, C.J.; Huang, C.Y.; Kuo, W.W. Roles of p38alpha and p38beta mitogenactivated protein kinase isoforms in human malignant melanoma A375 cells. *Int. J. Mol. Med.* **2019**, *44*, 2123–2132. [CrossRef]
80. Maik-Rachline, G.; Zehorai, E.; Hanoch, T.; Blenis, J.; Seger, R. The nuclear translocation of the kinases p38 and JNK promotes inflammation-induced cancer. *Sci. Signal.* **2018**, *11*. [CrossRef]

81. Gupta, J.; Nebreda, A.R. Roles of p38alpha mitogen-activated protein kinase in mouse models of inflammatory diseases and cancer. *FEBS J.* **2015**, *282*, 1841–1857. [CrossRef] [PubMed]
82. Vergote, I.; Heitz, F.; Buderath, P.; Powell, M.; Sehouli, J.; Lee, C.M.; Hamilton, A.; Fiorica, J.; Moore, K.N.; Teneriello, M.; et al. A randomized, double-blind, placebo-controlled phase 1b/2 study of ralimetinib, a p38 MAPK inhibitor, plus gemcitabine and carboplatin versus gemcitabine and carboplatin for women with recurrent platinum-sensitive ovarian cancer. *Gynecol. Oncol.* **2020**, *156*, 23–31. [CrossRef] [PubMed]
83. Rousseau, S.; Dolado, I.; Beardmore, V.; Shpiro, N.; Marquez, R.; Nebreda, A.R.; Arthur, J.S.; Case, L.M.; Tessier-Lavigne, M.; Gaestel, M.; et al. CXCL12 and C5a trigger cell migration via a PAK1/2-p38alpha MAPK-MAPKAP-K2-HSP27 pathway. *Cell. Signal.* **2006**, *18*, 1897–1905. [CrossRef] [PubMed]
84. Arechederra, M.; Priego, N.; Vazquez-Carballo, A.; Sequera, C.; Gutierrez-Uzquiza, A.; Cerezo-Guisado, M.I.; Ortiz-Rivero, S.; Roncero, C.; Cuenda, A.; Guerrero, C.; et al. p38 MAPK down-regulates fibulin 3 expression through methylation of gene regulatory sequences: Role in migration and invasion. *J. Biol. Chem.* **2015**, *290*, 4383–4397. [CrossRef]
85. Leelahavanichkul, K.; Amornphimoltham, P.; Molinolo, A.A.; Basile, J.R.; Koontongkaew, S.; Gutkind, J.S. A role for p38 MAPK in head and neck cancer cell growth and tumor-induced angiogenesis and lymphangiogenesis. *Mol. Oncol.* **2014**, *8*, 105–118. [CrossRef]
86. Trempolec, N.; Dave-Coll, N.; Nebreda, A.R. SnapShot: p38 MAPK signaling. *Cell* **2013**, *152*, 656. [CrossRef]
87. Roux, P.P.; Blenis, J. MAPK Signaling in Human Diseases. In *Apoptosis, Cell Signaling, and Human Diseases*; Srivastava, R., Ed.; Humana Press: Totowa, NJ, USA, 2006; pp. 135–149.
88. Iida, N.; Fujita, M.; Miyazawa, K.; Kobayashi, M.; Hattori, S. Proteomic identification of p38 MAP kinase substrates using in vitro phosphorylation. *Electrophoresis* **2014**, *35*, 554–562. [CrossRef]
89. Knight, J.D.; Tian, R.; Lee, R.E.; Wang, F.; Beauvais, A.; Zou, H.; Megeney, L.A.; Gingras, A.C.; Pawson, T.; Figeys, D.; et al. A novel whole-cell lysate kinase assay identifies substrates of the p38 MAPK in differentiating myoblasts. *Skelet. Muscle* **2012**, *2*, 5. [CrossRef]
90. Rodriguez, J.; Crespo, P. Working without kinase activity: Phosphotransfer-independent functions of extracellular signal-regulated kinases. *Sci. Signal.* **2011**, *4*, re3. [CrossRef]
91. Mayor, F., Jr.; Jurado-Pueyo, M.; Campos, P.M.; Murga, C. Interfering with MAP kinase docking interactions: Implications and perspective for the p38 route. *Cell Cycle* **2007**, *6*, 528–533. [CrossRef]
92. Tanoue, T.; Adachi, M.; Moriguchi, T.; Nishida, E. A conserved docking motif in MAP kinases common to substrates, activators and regulators. *Nat. Cell Biol.* **2000**, *2*, 110–116. [CrossRef] [PubMed]
93. Chang, C.I.; Xu, B.E.; Akella, R.; Cobb, M.H.; Goldsmith, E.J. Crystal structures of MAP kinase p38 complexed to the docking sites on its nuclear substrate MEF2A and activator MKK3b. *Mol. Cell* **2002**, *9*, 1241–1249. [CrossRef]
94. Yang, S.H.; Whitmarsh, A.J.; Davis, R.J.; Sharrocks, A.D. Differential targeting of MAP kinases to the ETS-domain transcription factor Elk-1. *EMBO J.* **1998**, *17*, 1740–1749. [CrossRef] [PubMed]
95. Bardwell, A.J.; Bardwell, L. Two hydrophobic residues can determine the specificity of mitogen-activated protein kinase docking interactions. *J. Biol. Chem.* **2015**, *290*, 26661–26674. [CrossRef] [PubMed]
96. Tzarum, N.; Komornik, N.; Ben Chetrit, D.; Engelberg, D.; Livnah, O. DEF pocket in p38alpha facilitates substrate selectivity and mediates autophosphorylation. *J. Biol. Chem.* **2013**, *288*, 19537–19547. [CrossRef]
97. Beenstock, J.; Ben-Yehuda, S.; Melamed, D.; Admon, A.; Livnah, O.; Ahn, N.G.; Engelberg, D. The p38beta mitogen-activated protein kinase possesses an intrinsic autophosphorylation activity, generated by a short region composed of the alpha-G helix and MAPK insert. *J. Biol. Chem.* **2014**, *289*, 23546–23556. [CrossRef]
98. Purvis, J.E.; Lahav, G. Encoding and decoding cellular information through signaling dynamics. *Cell* **2013**, *152*, 945–956. [CrossRef]
99. Haq, R.; Brenton, J.D.; Takahashi, M.; Finan, D.; Finkielsztein, A.; Damaraju, S.; Rottapel, R.; Zanke, B. Constitutive p38HOG mitogen-activated protein kinase activation induces permanent cell cycle arrest and senescence. *Cancer Res.* **2002**, *62*, 5076–5082.
100. Shaul, Y.D.; Seger, R. The MEK/ERK cascade: From signaling specificity to diverse functions. *Biochim. Biophys. Acta* **2007**, *1773*, 1213–1226. [CrossRef]
101. Lau, E.; Ronai, Z.A. ATF2 at the crossroad of nuclear and cytosolic functions. *J. Cell Sci.* **2012**, *125*, 2815–2824. [CrossRef]
102. Lee, J.; Sun, C.; Zhou, Y.; Lee, J.; Gokalp, D.; Herrema, H.; Park, S.W.; Davis, R.J.; Ozcan, U. p38 MAPK-mediated regulation of Xbp1s is crucial for glucose homeostasis. *Nat. Med.* **2011**, *17*, 1251–1260. [CrossRef]

103. Rauch, C.; Loughna, P.T. Static stretch promotes MEF2A nuclear translocation and expression of neonatal myosin heavy chain in C_2Cl_2 myocytes in a calcineurin- and p38-dependent manner. *Am. J. Physiol. Cell Physiol.* **2005**, *288*, C593–C605. [CrossRef] [PubMed]
104. Chen, R.H.; Sarnecki, C.; Blenis, J. Nuclear localization and regulation of erk- and rsk-encoded protein kinases. *Mol. Cell. Biol.* **1992**, *12*, 915–927. [CrossRef] [PubMed]
105. Wood, C.D.; Thornton, T.M.; Sabio, G.; Davis, R.A.; Rincon, M. Nuclear localization of p38 MAPK in response to DNA damage. *Int. J. Biol. Sci.* **2009**, *5*, 428–437. [CrossRef] [PubMed]
106. Blanco-Aparicio, C.; Torres, J.; Pulido, R. A novel regulatory mechanism of MAP kinases activation and nuclear translocation mediated by PKA and the PTP-SL tyrosine phosphatase. *J. Cell Biol.* **1999**, *147*, 1129–1136. [CrossRef]
107. Posen, Y.; Kalchenko, V.; Seger, R.; Brandis, A.; Scherz, A.; Salomon, Y. Manipulation of redox signaling in mammalian cells enabled by controlled photogeneration of reactive oxygen species. *J. Cell Sci.* **2005**, *118*, 1957–1969. [CrossRef]
108. Ben-Levy, R.; Hooper, S.; Wilson, R.; Paterson, H.F.; Marshall, C.J. Nuclear export of the stress-activated protein kinase p38 mediated by its substrate MAPKAP kinase-2. *Curr. Biol.* **1998**, *8*, 1049–1057. [CrossRef]
109. Wainstein, E.; Seger, R. The dynamic subcellular localization of ERK: Mechanisms of translocation and role in various organelles. *Curr. Opin. Cell Biol.* **2016**, *39*, 15–20. [CrossRef]
110. Lee, S.Y.; Kim, S.; Lim, Y.; Yoon, H.N.; Ku, N.O. Keratins regulate Hsp70-mediated nuclear localization of p38 mitogen-activated protein kinase. *J. Cell Sci.* **2019**, *132*. [CrossRef]
111. Morrison, D.K.; Davis, R.J. Regulation of MAP kinase signaling modules by scaffold proteins in mammals. *Annu. Rev. Cell Dev. Biol.* **2003**, *19*, 91–118. [CrossRef]
112. Gong, X.; Ming, X.; Deng, P.; Jiang, Y. Mechanisms regulating the nuclear translocation of p38 MAP kinase. *J. Cell. Biochem.* **2010**, *110*, 1420–1429. [CrossRef] [PubMed]
113. Zehorai, E.; Seger, R. Beta-Like Importins Mediate the Nuclear Translocation of MAPKs. *Cell. Physiol. Biochem.* **2019**, *52*, 802–821. [CrossRef] [PubMed]
114. Ferreiro, I.; Barragan, M.; Gubern, A.; Ballestar, E.; Joaquin, M.; Posas, F. The p38 SAPK is recruited to chromatin via its interaction with transcription factors. *J. Biol. Chem.* **2010**, *285*, 31819–31828. [CrossRef] [PubMed]
115. Flores, K.; Seger, R. Stimulated nuclear import by β-like importins. *F1000Prime Rep.* **2013**, *5*. [CrossRef]
116. Chuderland, D.; Konson, A.; Seger, R. Identification and characterization of a general nuclear translocation signal in signaling proteins. *Mol. Cell* **2008**, *31*, 850–861. [CrossRef] [PubMed]
117. Plotnikov, A.; Chuderland, D.; Karamansha, Y.; Livnah, O.; Seger, R. Nuclear ERK Translocation is Mediated by Protein Kinase CK2 and Accelerated by Autophosphorylation. *Cell. Physiol. Biochem.* **2019**, *53*, 366–387. [CrossRef]
118. Gong, X.; Luo, T.; Deng, P.; Liu, Z.; Xiu, J.; Shi, H.; Jiang, Y. Stress-induced interaction between p38 MAPK and HSP70. *Biochem. Biophys. Res. Commun.* **2012**, *425*, 357–362. [CrossRef]
119. Wang, P.Y.; Hsu, P.I.; Wu, D.C.; Chen, T.C.; Jarman, A.P.; Powell, L.M.; Chen, A. SUMOs Mediate the Nuclear Transfer of p38 and p-p38 during Helicobacter Pylori Infection. *Int. J. Mol. Sci.* **2018**, *19*, 2482. [CrossRef]
120. Plotnikov, A.; Flores, K.; Maik-Rachline, G.; Zehorai, E.; Kapri-Pardes, E.; Berti, D.A.; Hanoch, T.; Besser, M.J.; Seger, R. The nuclear translocation of ERK1/2 as an anticancer target. *Nat. Commun.* **2015**, *6*, 6685. [CrossRef]
121. Maik-Rachline, G.; Hacohen-Lev-Ran, A.; Seger, R. Nuclear ERK: Mechanism of Translocation, Substrates, and Role in Cancer. *Int. J. Mol. Sci.* **2019**, *20*, 1194. [CrossRef]
122. Jivan, A.; Ranganathan, A.; Cobb, M.H. Reconstitution of the nuclear transport of the MAP kinase ERK2. *Methods Mol. Biol.* **2010**, *661*, 273–285. [CrossRef] [PubMed]
123. Arafeh, R.; Flores, K.; Karen-Paz, A.; Maik-Rachline, G.; Gutkind, N.; Rosenberg, S.; Seger, R.; Samuels, Y. Combinede inhibition of MEK and nuclear ERK translocation has synergistic antitumor activity in mutated NRAS, BRAF and NF1 melanoma cells. *Sci. Rep.* **2017**, *7*, 16345. [CrossRef]
124. Casanovas, O.; Jaumot, M.; Paules, A.B.; Agell, N.; Bachs, O. P38SAPK2 phosphorylates cyclin D3 at Thr-283 and targets it for proteasomal degradation. *Oncogene* **2004**, *23*, 7537–7544. [CrossRef] [PubMed]
125. Wang, H.; Nicolay, B.N.; Chick, J.M.; Gao, X.; Geng, Y.; Ren, H.; Gao, H.; Yang, G.; Williams, J.A.; Suski, J.M. et al. The metabolic function of cyclin D3-CDK6 kinase in cancer cell survival. *Nature* **2017**, *546*, 426–430. [CrossRef] [PubMed]

126. Buss, H.; Handschick, K.; Jurrmann, N.; Pekkonen, P.; Beuerlein, K.; Muller, H.; Wait, R.; Saklatvala, J.; Ojala, P.M.; Schmitz, M.L.; et al. Cyclin-dependent kinase 6 phosphorylates NF-kappaB P65 at serine 536 and contributes to the regulation of inflammatory gene expression. *PLoS ONE* **2012**, *7*, e51847. [CrossRef]
127. Lluis, F.; Ballestar, E.; Suelves, M.; Esteller, M.; Munoz-Canoves, P. E47 phosphorylation by p38 MAPK promotes MyoD/E47 association and muscle-specific gene transcription. *EMBO J.* **2005**, *24*, 974–984. [CrossRef]
128. Teng, Y.; Li, X. The roles of HLH transcription factors in epithelial mesenchymal transition and multiple molecular mechanisms. *Clin. Exp. Metastasis* **2014**, *31*, 367–377. [CrossRef] [PubMed]
129. Hemmer, M.C.; Wierer, M.; Schachtrup, K.; Downes, M.; Hubner, N.; Evans, R.M.; Uhlenhaut, N.H. E47 modulates hepatic glucocorticoid action. *Nat. Commun.* **2019**, *10*, 306. [CrossRef]
130. Briata, P.; Forcales, S.V.; Ponassi, M.; Corte, G.; Chen, C.Y.; Karin, M.; Puri, P.L.; Gherzi, R. p38-dependent phosphorylation of the mRNA decay-promoting factor KSRP controls the stability of select myogenic transcripts. *Mol. Cell* **2005**, *20*, 891–903. [CrossRef] [PubMed]
131. Degese, M.S.; Tanos, T.; Naipauer, J.; Gingerich, T.; Chiappe, D.; Echeverria, P.; LaMarre, J.; Gutkind, J.S.; Coso, O.A. An interplay between the p38 MAPK pathway and AUBPs regulates c-fos mRNA stability during mitogenic stimulation. *Biochem. J.* **2015**, *467*, 77–90. [CrossRef]
132. Bollmann, F.; Art, J.; Henke, J.; Schrick, K.; Besche, V.; Bros, M.; Li, H.; Siuda, D.; Handler, N.; Bauer, F.; et al. Resveratrol post-transcriptionally regulates pro-inflammatory gene expression via regulation of KSRP RNA binding activity. *Nucleic Acids Res.* **2014**, *42*, 12555–12569. [CrossRef]
133. Danckwardt, S.; Gantzert, A.S.; Macher-Goeppinger, S.; Probst, H.C.; Gentzel, M.; Wilm, M.; Grone, H.J.; Schirmacher, P.; Hentze, M.W.; Kulozik, A.E. p38 MAPK controls prothrombin expression by regulated RNA 3′ end processing. *Mol. Cell* **2011**, *41*, 298–310. [CrossRef] [PubMed]
134. Weber, A.; Kristiansen, I.; Johannsen, M.; Oelrich, B.; Scholmann, K.; Gunia, S.; May, M.; Meyer, H.A.; Behnke, S.; Moch, H.; et al. The FUSE binding proteins FBP1 and FBP3 are potential c-myc regulators in renal, but not in prostate and bladder cancer. *BMC Cancer* **2008**, *8*, 369. [CrossRef] [PubMed]
135. Dmitrieva, N.I.; Bulavin, D.V.; Fornace, A.J., Jr.; Burg, M.B. Rapid activation of G2/M checkpoint after hypertonic stress in renal inner medullary epithelial (IME) cells is protective and requires p38 kinase. *Proc. Natl. Acad. Sci. USA* **2002**, *99*, 184–189. [CrossRef] [PubMed]
136. Zhu, F.; Zykova, T.A.; Peng, C.; Zhang, J.; Cho, Y.Y.; Zheng, D.; Yao, K.; Ma, W.Y.; Lau, A.T.; Bode, A.M.; et al. Phosphorylation of H2AX at Ser139 and a new phosphorylation site Ser16 by RSK2 decreases H2AX ubiquitination and inhibits cell transformation. *Cancer Res.* **2011**, *71*, 393–403. [CrossRef] [PubMed]
137. Risques, R.A.; Lai, L.A.; Brentnall, T.A.; Li, L.; Feng, Z.; Gallaher, J.; Mandelson, M.T.; Potter, J.D.; Bronner, M.P.; Rabinovitch, P.S. Ulcerative colitis is a disease of accelerated colon aging: Evidence from telomere attrition and DNA damage. *Gastroenterology* **2008**, *135*, 410–418. [CrossRef] [PubMed]
138. Zhong, S.P.; Ma, W.Y.; Dong, Z. ERKs and p38 kinases mediate ultraviolet B-induced phosphorylation of histone H3 at serine 10. *J. Biol. Chem.* **2000**, *275*, 20980–20984. [CrossRef] [PubMed]
139. Khan, S.A.; Amnekar, R.; Khade, B.; Barreto, S.G.; Ramadwar, M.; Shrikhande, S.V.; Gupta, S. p38-MAPK/MSK1-mediated overexpression of histone H3 serine 10 phosphorylation defines distance- dependent prognostic value of negative resection margin in gastric cancer. *Clin. Epigenetics* **2016**, *8*, 88. [CrossRef]
140. Saccani, S.; Pantano, S.; Natoli, G. p38-Dependent marking of inflammatory genes for increased NF-kappa B recruitment. *Nat. Immunol.* **2002**, *3*, 69–75. [CrossRef]
141. Xiu, M.; Kim, J.; Sampson, E.; Huang, C.Y.; Davis, R.J.; Paulson, K.E.; Yee, A.S. The transcriptional repressor HBP1 is a target of the p38 mitogen-activated protein kinase pathway in cell cycle regulation. *Mol. Cell. Biol.* **2003**, *23*, 8890–8901. [CrossRef]
142. Chen, H.; Li, X.; Liu, S.; Gu, L.; Zhou, X. MircroRNA-19a promotes vascular inflammation and foam cell formation by targeting HBP-1 in atherogenesis. *Sci. Rep.* **2017**, *7*, 12089. [CrossRef]
143. Roschger, C.; Cabrele, C. The Id-protein family in developmental and cancer-associated pathways. *Cell Commun. Signal.* **2017**, *15*, 7. [CrossRef] [PubMed]
144. Lee, S.B.; Frattini, V.; Bansal, M.; Castano, A.M.; Sherman, D.; Hutchinson, K.; Bruce, J.N.; Califano, A.; Liu, G.; Cardozo, T.; et al. An ID2-dependent mechanism for VHL inactivation in cancer. *Nature* **2016**, *529*, 172–177. [CrossRef] [PubMed]
145. Miyazaki, M.; Miyazaki, K.; Chen, S.; Itoi, M.; Miller, M.; Lu, L.F.; Varki, N.; Chang, A.N.; Broide, D.H.; Murre, C. Id2 and Id3 maintain the regulatory T cell pool to suppress inflammatory disease. *Nat. Immunol.* **2014**, *15*, 767–776. [CrossRef] [PubMed]

146. Oqani, R.K.; Lin, T.; Lee, J.E.; Kang, J.W.; Shin, H.Y.; Il Jin, D. Iws1 and Spt6 Regulate Trimethylation of Histone H3 on Lysine 36 through Akt Signaling and are Essential for Mouse Embryonic Genome Activation. *Sci. Rep.* **2019**, *9*, 3831. [CrossRef]
147. Katz, S.; Aronheim, A. Differential targeting of the stress mitogen-activated protein kinases to the c-Jun dimerization protein 2. *Biochem. J.* **2002**, *368*, 939–945. [CrossRef]
148. Tsai, M.H.; Wuputra, K.; Lin, Y.C.; Lin, C.S.; Yokoyama, K.K. Multiple functions of the histone chaperone Jun dimerization protein 2. *Gene* **2016**, *590*, 193–200. [CrossRef]
149. Bitton-Worms, K.; Pikarsky, E.; Aronheim, A. The AP-1 repressor protein, JDP2, potentiates hepatocellular carcinoma in mice. *Mol. Cancer* **2010**, *9*, 54. [CrossRef]
150. Rampalli, S.; Li, L.; Mak, E.; Ge, K.; Brand, M.; Tapscott, S.J.; Dilworth, F.J. p38 MAPK signaling regulates recruitment of Ash2L-containing methyltransferase complexes to specific genes during differentiation. *Nat. Struct. Mol. Biol.* **2007**, *14*, 1150–1156. [CrossRef]
151. Song, Z.; Feng, C.; Lu, Y.; Gao, Y.; Lin, Y.; Dong, C. Overexpression and biological function of MEF2D in human pancreatic cancer. *Am. J. Transl. Res.* **2017**, *9*, 4836–4847.
152. Yang, S.; Gao, L.; Lu, F.; Wang, B.; Gao, F.; Zhu, G.; Cai, Z.; Lai, J.; Yang, Q. Transcription factor myocyte enhancer factor 2D regulates interleukin-10 production in microglia to protect neuronal cells from inflammation-induced death. *J. Neuroinflamm.* **2015**, *12*, 33. [CrossRef] [PubMed]
153. Waskiewicz, A.J.; Flynn, A.; Proud, C.G.; Cooper, J.A. Mitogen-activated protein kinases activate the serine/threonine kinases Mnk1 and Mnk2. *EMBO J.* **1997**, *16*, 1909–1920. [CrossRef] [PubMed]
154. Maimon, A.; Mogilevsky, M.; Shilo, A.; Golan-Gerstl, R.; Obiedat, A.; Ben-Hur, V.; Lebenthal-Loinger, I.; Stein, I.; Reich, R.; Beenstock, J.; et al. Mnk2 alternative splicing modulates the p38-MAPK pathway and impacts Ras-induced transformation. *Cell Rep.* **2014**, *7*, 501–513. [CrossRef]
155. Moore, C.E.; Pickford, J.; Cagampang, F.R.; Stead, R.L.; Tian, S.; Zhao, X.; Tang, X.; Byrne, C.D.; Proud, C.G. MNK1 and MNK2 mediate adverse effects of high-fat feeding in distinct ways. *Sci. Rep.* **2016**, *6*, 23476. [CrossRef]
156. Deak, M.; Clifton, A.D.; Lucocq, L.M.; Alessi, D.R. Mitogen- and stress-activated protein kinase-1 (MSK1) is directly activated by MAPK and SAPK2/p38, and may mediate activation of CREB. *EMBO J.* **1998**, *17*, 4426–4441. [CrossRef] [PubMed]
157. Galan-Ganga, M.; Garcia-Yague, A.J.; Lastres-Becker, I. Role of MSK1 in the Induction of NF-kappaB by the Chemokine CX3CL1 in Microglial Cells. *Cell. Mol. Neurobiol.* **2019**, *39*, 331–340. [CrossRef]
158. Caivano, M.; Cohen, P. Role of mitogen-activated protein kinase cascades in mediating lipopolysaccharide-stimulated induction of cyclooxygenase-2 and IL-1 beta in RAW264 macrophages. *J. Immunol.* **2000**, *164*, 3018–3025. [CrossRef]
159. Ananieva, O.; Darragh, J.; Johansen, C.; Carr, J.M.; McIlrath, J.; Park, J.M.; Wingate, A.; Monk, C.E.; Toth, R.; Santos, S.G.; et al. The kinases MSK1 and MSK2 act as negative regulators of Toll-like receptor signaling. *Nat. Immunol.* **2008**, *9*, 1028–1036. [CrossRef]
160. Cuadrado, A.; Lafarga, V.; Cheung, P.C.; Dolado, I.; Llanos, S.; Cohen, P.; Nebreda, A.R. A new p38 MAP kinase-regulated transcriptional coactivator that stimulates p53-dependent apoptosis. *EMBO J.* **2007**, *26*, 2115–2126. [CrossRef]
161. Uehara, I.; Tanaka, N. Role of p53 in the Regulation of the Inflammatory Tumor Microenvironment and Tumor Suppression. *Cancers* **2018**, *10*, 219. [CrossRef]
162. Bulavin, D.V.; Saito, S.; Hollander, M.C.; Sakaguchi, K.; Anderson, C.W.; Appella, E.; Fornace, A.J., Jr. Phosphorylation of human p53 by p38 kinase coordinates N-terminal phosphorylation and apoptosis in response to UV radiation. *EMBO J.* **1999**, *18*, 6845–6854. [CrossRef] [PubMed]
163. Puigserver, P.; Rhee, J.; Lin, J.; Wu, Z.; Yoon, J.C.; Zhang, C.Y.; Krauss, S.; Mootha, V.K.; Lowell, B.B.; Spiegelman, B.M. Cytokine stimulation of energy expenditure through p38 MAP kinase activation of PPARgamma coactivator-1. *Mol. Cell* **2001**, *8*, 971–982. [CrossRef]
164. Bost, F.; Kaminski, L. The metabolic modulator PGC-1alpha in cancer. *Am. J. Cancer Res.* **2019**, *9*, 198–211. [PubMed]
165. Rius-Perez, S.; Torres-Cuevas, I.; Millan, I.; Ortega, A.L.; Perez, S. PGC-1alpha, Inflammation, and Oxidative Stress: An Integrative View in Metabolism. *Oxid. Med. Cell. Longev.* **2020**, *2020*, 1452696. [CrossRef]
166. Barger, P.M.; Browning, A.C.; Garner, A.N.; Kelly, D.P. p38 mitogen-activated protein kinase activates peroxisome proliferator-activated receptor alpha: A potential role in the cardiac metabolic stress response. *J. Biol. Chem.* **2001**, *276*, 44495–44501. [CrossRef] [PubMed]

167. Abu Aboud, O.; Donohoe, D.; Bultman, S.; Fitch, M.; Riiff, T.; Hellerstein, M.; Weiss, R.H. PPARalpha inhibition modulates multiple reprogrammed metabolic pathways in kidney cancer and attenuates tumor growth. *Am. J. Physiol. Cell Physiol.* **2015**, *308*, C890–C898. [CrossRef]
168. Stienstra, R.; Saudale, F.; Duval, C.; Keshtkar, S.; Groener, J.E.; van Rooijen, N.; Staels, B.; Kersten, S.; Muller, M. Kupffer cells promote hepatic steatosis via interleukin-1beta-dependent suppression of peroxisome proliferator-activated receptor alpha activity. *Hepatology* **2010**, *51*, 511–522. [CrossRef]
169. Li, J.; Yin, W.H.; Takeuchi, K.; Guan, H.; Huang, Y.H.; Chan, J.K. Inflammatory myofibroblastic tumor with RANBP2 and ALK gene rearrangement: A report of two cases and literature review. *Diagn. Pathol.* **2013**, *8*, 147. [CrossRef]
170. Wang, S.; Nath, N.; Minden, A.; Chellappan, S. Regulation of Rb and E2F by signal transduction cascades: Divergent effects of JNK1 and p38 kinases. *EMBO J.* **1999**, *18*, 1559–1570. [CrossRef]
171. Indovina, P.; Pentimalli, F.; Casini, N.; Vocca, I.; Giordano, A. RB1 dual role in proliferation and apoptosis: Cell fate control and implications for cancer therapy. *Oncotarget* **2015**, *6*, 17873–17890. [CrossRef]
172. Kitajima, S.; Takahashi, C. Intersection of retinoblastoma tumor suppressor function, stem cells, metabolism, and inflammation. *Cancer Sci.* **2017**, *108*, 1726–1731. [CrossRef] [PubMed]
173. Rao, P.S.; Satelli, A.; Zhang, S.; Srivastava, S.K.; Srivenugopal, K.S.; Rao, U.S. RNF2 is the target for phosphorylation by the p38 MAPK and ERK signaling pathways. *Proteomics* **2009**, *9*, 2776–2787. [CrossRef] [PubMed]
174. Rai, K.; Akdemir, K.C.; Kwong, L.N.; Fiziev, P.; Wu, C.J.; Keung, E.Z.; Sharma, S.; Samant, N.S.; Williams, M.; Axelrad, J.B.; et al. Dual Roles of RNF2 in Melanoma Progression. *Cancer Discov.* **2015**, *5*, 1314–1327. [CrossRef]
175. Liu, S.; Jiang, M.; Wang, W.; Liu, W.; Song, X.; Ma, Z.; Zhang, S.; Liu, L.; Liu, Y.; Cao, X. Nuclear RNF2 inhibits interferon function by promoting K33-linked STAT1 disassociation from DNA. *Nat. Immunol.* **2018**, *19*, 41–52. [CrossRef]
176. Huang, L.; Jian, Z.; Gao, Y.; Zhou, P.; Zhang, G.; Jiang, B.; Lv, Y. RPN2 promotes metastasis of hepatocellular carcinoma cell and inhibits autophagy via STAT3 and NF-kappaB pathways. *Aging* **2019**, *11*, 6674–6690. [CrossRef] [PubMed]
177. Ni, L.; Yu, J.; Gui, X.; Lu, Z.; Wang, X.; Guo, H.; Zhou, Y. Overexpression of RPN2 promotes osteogenic differentiation of hBMSCs through the JAK/STAT3 pathway. *FEBS Open Bio* **2020**, *10*, 158–167. [CrossRef]
178. Greenblatt, M.B.; Shim, J.H.; Zou, W.; Sitara, D.; Schweitzer, M.; Hu, D.; Lotinun, S.; Sano, Y.; Baron, R.; Park, J.M.; et al. The p38 MAPK pathway is essential for skeletogenesis and bone homeostasis in mice. *J. Clin. Investig.* **2010**, *120*, 2457–2473. [CrossRef]
179. Ge, C.; Zhao, G.; Li, Y.; Li, H.; Zhao, X.; Pannone, G.; Bufo, P.; Santoro, A.; Sanguedolce, F.; Tortorella, S.; et al. Role of Runx2 phosphorylation in prostate cancer and association with metastatic disease. *Oncogene* **2016**, *35*, 366–376. [CrossRef]
180. Raaz, U.; Schellinger, I.N.; Chernogubova, E.; Warnecke, C.; Kayama, Y.; Penov, K.; Hennigs, J.K.; Salomons, F.; Eken, S.; Emrich, F.C.; et al. Transcription Factor Runx2 Promotes Aortic Fibrosis and Stiffness in Type 2 Diabetes Mellitus. *Circ. Res.* **2015**, *117*, 513–524. [CrossRef]
181. Al-Ayoubi, A.M.; Zheng, H.; Liu, Y.; Bai, T.; Eblen, S.T. Mitogen-activated protein kinase phosphorylation of splicing factor 45 (SPF45) regulates SPF45 alternative splicing site utilization, proliferation, and cell adhesion. *Mol. Cell. Biol.* **2012**, *32*, 2880–2893. [CrossRef]
182. Sampath, J.; Long, P.R.; Shepard, R.L.; Xia, X.; Devanarayan, V.; Sandusky, G.E.; Perry, W.L., 3rd; Dantzig, A.H.; Williamson, M.; Rolfe, M.; et al. Human SPF45, a splicing factor, has limited expression in normal tissues, is overexpressed in many tumors, and can confer a multidrug-resistant phenotype to cells. *Am. J. Pathol.* **2003**, *163*, 1781–1790. [CrossRef]
183. Gianni, M.; Parrella, E.; Raska, I., Jr.; Gaillard, E.; Nigro, E.A.; Gaudon, C.; Garattini, E.; Rochette-Egly, C. P38MAPK-dependent phosphorylation and degradation of SRC-3/AIB1 and RARalpha-mediated transcription. *EMBO J.* **2006**, *25*, 739–751. [CrossRef] [PubMed]
184. Xu, J.; Wu, R.C.; O'Malley, B.W. Normal and cancer-related functions of the p160 steroid receptor co-activator (SRC) family. *Nat. Rev. Cancer* **2009**, *9*, 615–630. [CrossRef]
185. Al-Otaiby, M.; Tassi, E.; Schmidt, M.O.; Chien, C.D.; Baker, T.; Salas, A.G.; Xu, J.; Furlong, M.; Schlegel, R.; Riegel, A.T.; et al. Role of the nuclear receptor coactivator AIB1/SRC-3 in angiogenesis and wound healing. *Am. J. Pathol.* **2012**, *180*, 1474–1484. [CrossRef]

186. Sohn, M.; Shin, S.; Yoo, J.Y.; Goh, Y.; Lee, I.H.; Bae, Y.S. Ahnak promotes tumor metastasis through transforming growth factor-beta-mediated epithelial-mesenchymal transition. *Sci. Rep.* **2018**, *8*, 14379. [CrossRef] [PubMed]
187. Suzuki, S.; Fukuda, T.; Nagayasu, S.; Nakanishi, J.; Yoshida, K.; Hirata-Tsuchiya, S.; Nakao, Y.; Sano, T.; Yamashita, A.; Yamada, S.; et al. Dental pulp cell-derived powerful inducer of TNF-alpha comprises PKR containing stress granule rich microvesicles. *Sci. Rep.* **2019**, *9*, 3825. [CrossRef] [PubMed]
188. Geest, C.R.; Buitenhuis, M.; Laarhoven, A.G.; Bierings, M.B.; Bruin, M.C.; Vellenga, E.; Coffer, P.J. p38 MAP kinase inhibits neutrophil development through phosphorylation of C/EBPalpha on serine 21. *Stem Cells* **2009**, *27*, 2271–2282. [CrossRef] [PubMed]
189. Shi, D.B.; Wang, Y.W.; Xing, A.Y.; Gao, J.W.; Zhang, H.; Guo, X.Y.; Gao, P. C/EBPalpha-induced miR-100 expression suppresses tumor metastasis and growth by targeting ZBTB7A in gastric cancer. *Cancer Lett.* **2015**, *369*, 376–385. [CrossRef]
190. Friedman, A.D. C/EBPalpha induces PU.1 and interacts with AP-1 and NF-kappaB to regulate myeloid development. *Blood Cells Mol. Dis.* **2007**, *39*, 340–343. [CrossRef]
191. Horie, R.; Ishida, T.; Maruyama-Nagai, M.; Ito, K.; Watanabe, M.; Yoneyama, A.; Higashihara, M.; Kimura, S.; Watanabe, T. TRAF activation of C/EBPbeta (NF-IL6) via p38 MAPK induces HIV-1 gene expression in monocytes/macrophages. *Microbes Infect.* **2007**, *9*, 721–728. [CrossRef]
192. Yeh, H.W.; Lee, S.S.; Chang, C.Y.; Lang, Y.D.; Jou, Y.S. A New Switch for TGFbeta in Cancer. *Cancer Res.* **2019**, *79*, 3797–3805. [CrossRef] [PubMed]
193. Zahid, M.D.K.; Rogowski, M.; Ponce, C.; Choudhury, M.; Moustaid-Moussa, N.; Rahman, S.M. CCAAT/enhancer-binding protein beta (C/EBPbeta) knockdown reduces inflammation, ER stress, and apoptosis, and promotes autophagy in oxLDL-treated RAW264.7 macrophage cells. *Mol. Cell. Biochem.* **2020**, *463*, 211–223. [CrossRef] [PubMed]
194. Anbalagan, M.; Rowan, B.G. Estrogen receptor alpha phosphorylation and its functional impact in human breast cancer. *Mol. Cell. Endocrinol.* **2015**, *418*, 264–272. [CrossRef] [PubMed]
195. Jia, M.; Dahlman-Wright, K.; Gustafsson, J.A. Estrogen receptor alpha and beta in health and disease. *Best Pract. Res. Clin. Endocrinol. Metab.* **2015**, *29*, 557–568. [CrossRef] [PubMed]
196. Rouse, J.; Cohen, P.; Trigon, S.; Morange, M.; Alonso-Llamazares, A.; Zamanillo, D.; Hunt, T.; Nebreda, A.R. A novel kinase cascade triggered by stress and heat shock that stimulates MAPKAP kinase-2 and phosphorylation of the small heat shock proteins. *Cell* **1994**, *78*, 1027–1037. [CrossRef]
197. Singh, R.K.; Najmi, A.K. Novel Therapeutic Potential of Mitogen-Activated Protein Kinase Activated Protein Kinase 2 (MK2) in Chronic Airway Inflammatory Disorders. *Curr. Drug Targets* **2019**, *20*, 367–379. [CrossRef]
198. Manke, I.A.; Nguyen, A.; Lim, D.; Stewart, M.Q.; Elia, A.E.; Yaffe, M.B. MAPKAP kinase-2 is a cell cycle checkpoint kinase that regulates the G2/M transition and S phase progression in response to UV irradiation. *Mol. Cell* **2005**, *17*, 37–48. [CrossRef]
199. McLaughlin, M.M.; Kumar, S.; McDonnell, P.C.; Van Horn, S.; Lee, J.C.; Livi, G.P.; Young, P.R. Identification of mitogen-activated protein (MAP) kinase-activated protein kinase-3, a novel substrate of CSBP p38 MAP kinase. *J. Biol. Chem.* **1996**, *271*, 8488–8492. [CrossRef]
200. Reynoird, N.; Mazur, P.K.; Stellfeld, T.; Flores, N.M.; Lofgren, S.M.; Carlson, S.M.; Brambilla, E.; Hainaut, P.; Kaznowska, E.B.; Arrowsmith, C.H.; et al. Coordination of stress signals by the lysine methyltransferase SMYD2 promotes pancreatic cancer. *Genes Dev.* **2016**, *30*, 772–785. [CrossRef]
201. Ronkina, N.; Menon, M.B.; Schwermann, J.; Tiedje, C.; Hitti, E.; Kotlyarov, A.; Gaestel, M. MAPKAP kinases MK2 and MK3 in inflammation: Complex regulation of TNF biosynthesis via expression and phosphorylation of tristetraprolin. *Biochem. Pharmacol.* **2010**, *80*, 1915–1920. [CrossRef]
202. New, L.; Jiang, Y.; Zhao, M.; Liu, K.; Zhu, W.; Flood, L.J.; Kato, Y.; Parry, G.C.; Han, J. PRAK, a novel protein kinase regulated by the p38 MAP kinase. *EMBO J.* **1998**, *17*, 3372–3384. [CrossRef]
203. Al-Mahdi, R.; Babteen, N.; Thillai, K.; Holt, M.; Johansen, B.; Wetting, H.L.; Seternes, O.M.; Wells, C.M. A novel role for atypical MAPK kinase ERK3 in regulating breast cancer cell morphology and migration. *Cell Adh. Migr.* **2015**, *9*, 483–494. [CrossRef] [PubMed]
204. Moens, U.; Kostenko, S.; Sveinbjornsson, B. The Role of Mitogen-Activated Protein Kinase-Activated Protein Kinases (MAPKAPKs) in Inflammation. *Genes* **2013**, *4*, 101–133. [CrossRef] [PubMed]
205. Suelves, M.; Lluis, F.; Ruiz, V.; Nebreda, A.R.; Munoz-Canoves, P. Phosphorylation of MRF4 transactivation domain by p38 mediates repression of specific myogenic genes. *EMBO J.* **2004**, *23*, 365–375. [CrossRef] [PubMed]

206. Arons, E.; Zhou, H.; Sokolsky, M.; Gorelik, D.; Potocka, K.; Davies, S.; Fykes, E.; Still, K.; Edelman, D.C.; Wang, Y.; et al. Expression of the muscle-associated gene MYF6 in hairy cell leukemia. *PLoS ONE* **2020**, *15*, e0227586. [CrossRef] [PubMed]
207. Yang, T.T.; Xiong, Q.; Enslen, H.; Davis, R.J.; Chow, C.W. Phosphorylation of NFATc4 by p38 mitogen-activated protein kinases. *Mol. Cell. Biol.* **2002**, *22*, 3892–3904. [CrossRef] [PubMed]
208. Cole, A.J.; Iyengar, M.; Panesso-Gomez, S.; O'Hayer, P.; Chan, D.; Delgoffe, G.M.; Aird, K.M.; Yoon, E.; Bai, S.; Buckanovich, R.J. NFATC4 promotes quiescence and chemotherapy resistance in ovarian cancer. *JCI Insight* **2020**, *5*. [CrossRef] [PubMed]
209. Kim, H.B.; Kumar, A.; Wang, L.; Liu, G.H.; Keller, S.R.; Lawrence, J.C., Jr.; Finck, B.N.; Harris, T.E. Lipin 1 represses NFATc4 transcriptional activity in adipocytes to inhibit secretion of inflammatory factors. *Mol. Cell. Biol.* **2010**, *30*, 3126–3139. [CrossRef]
210. Sekine, Y.; Takagahara, S.; Hatanaka, R.; Watanabe, T.; Oguchi, H.; Noguchi, T.; Naguro, I.; Kobayashi, K.; Tsunoda, M.; Funatsu, T.; et al. p38 MAPKs regulate the expression of genes in the dopamine synthesis pathway through phosphorylation of NR4A nuclear receptors. *J. Cell Sci.* **2011**, *124*, 3006–3016. [CrossRef]
211. Mohan, H.M.; Aherne, C.M.; Rogers, A.C.; Baird, A.W.; Winter, D.C.; Murphy, E.P. Molecular pathways: The role of NR4A orphan nuclear receptors in cancer. *Clin. Cancer Res.* **2012**, *18*, 3223–3228. [CrossRef]
212. McMorrow, J.P.; Murphy, E.P. Inflammation: A role for NR4A orphan nuclear receptors? *Biochem. Soc. Trans.* **2011**, *39*, 688–693. [CrossRef] [PubMed]
213. Mikkola, I.; Bruun, J.A.; Bjorkoy, G.; Holm, T.; Johansen, T. Phosphorylation of the transactivation domain of Pax6 by extracellular signal-regulated kinase and p38 mitogen-activated protein kinase. *J. Biol. Chem.* **1999**, *274*, 15115–15126. [CrossRef] [PubMed]
214. Zhao, X.; Yue, W.; Zhang, L.; Ma, L.; Jia, W.; Qian, Z.; Zhang, C.; Wang, Y. Downregulation of PAX6 by shRNA inhibits proliferation and cell cycle progression of human non-small cell lung cancer cell lines. *PLoS ONE* **2014**, *9*, e85738. [CrossRef] [PubMed]

International Journal of
Molecular Sciences

Review

p38β and Cancer: The Beginning of the Road

Olga Roche [1,2,†], Diego M. Fernández-Aroca [1,†], Elena Arconada-Luque [1,†], Natalia García-Flores [1], Liliana F. Mellor [1], María José Ruiz-Hidalgo [1,3] and Ricardo Sánchez-Prieto [2,4,*]

1. Laboratorio de Oncología, Unidad de Medicina Molecular, Centro Regional de Investigaciones Biomédicas, Unidad Asociada de Biomedicina UCLM, Unidad Asociada al CSIC, Universidad de Castilla-La Mancha, 02008 Albacete, Spain; Olga.roche@uclm.es (O.R.); DiegoManuel.Fernandez@alu.uclm.es (D.M.F.-A.); elena.arconada@uclm.es (E.A.-L.); natalia.garcia16@alu.uclm.es (N.G.-F.); liliana.mellor@gmail.com (L.F.M.); maria.rhidalgo@uclm.es (M.J.R.-H.)
2. Departamento de Ciencias Médicas, Facultad de Medicina de Albacete, Universidad de Castilla-La Mancha, 02008 Albacete, Spain
3. Departamento de Química Inorgánica, Orgánica y Bioquímica, Área de Bioquímica y Biología Molecular, Facultad de Medicina de Albacete, Universidad de Castilla-La Mancha, 02008 Albacete, Spain
4. Departamento de Biología del Cáncer, Instituto de Investigaciones Biomédicas Alberto Sols (CSIC-UAM), Unidad Asociada de Biomedicina UCLM, Unidad Asociada al CSIC, Consejo Superior de Investigaciones Cientificas, 28029 Madrid, Spain

* Correspondence: rsprieto@iib.uam.es; Tel.: +34-915-854-420

† These authors contributed equal to this work.

Received: 15 September 2020; Accepted: 10 October 2020; Published: 12 October 2020

Abstract: The p38 mitogen-activated protein kinase (MAPK) signaling pathway is implicated in cancer biology and has been widely studied over the past two decades as a potential therapeutic target. Most of the biological and pathological implications of p38MAPK signaling are often associated with p38α (MAPK14). Recently, several members of the p38 family, including p38γ and p38δ, have been shown to play a crucial role in several pathologies including cancer. However, the specific role of p38β (MAPK11) in cancer is still elusive, and further investigation is needed. Here, we summarize what is currently known about the role of p38β in different types of tumors and its putative implication in cancer therapy. All evidence suggests that p38β might be a key player in cancer development, and could be an important therapeutic target in several pathologies, including cancer.

Keywords: p38MAPK; MAPK11; p38β; cancer

---

1. Introduction

Mitogen-activated protein kinases (MAPKs) are an evolutionarily conserved family of enzymes that link extracellular signals to the intracellular machinery in order to control a plethora of cellular processes including proliferation, cell survival, differentiation and apoptosis, among others. In fact, its deregulation is associated with many human diseases including inflammation, neurodegenerative disorders and cancer [1].

In mammals, four conventional MAPK subfamilies have been identified: extracellular signal-regulated protein kinases 1/2 (ERK1/2), c-Jun N-terminal kinases 1-3 (JNK1/2/3), p38MAPKs (α, β, γ and δ), and the most recently discovered and least characterized ERK5 [1]. Each MAPK has its own activators, inhibitory phosphatases, substrates and scaffold proteins that allow the correct function of the different MAPK signaling pathways [2,3]. The diversity and specificity of MAPKs in cellular responses are achieved with a linear architecture, consisting of a module of three protein kinases: a MAPK kinase kinase (MAP3K or MKKKs) at the top, which phosphorylates a MAPK kinase (MAP2K, MKKs or MEKs) on specific serine (S) and threonine (T) residues. Eventually, there is a dual phosphorylation of the T and tyrosine (Y) residues of the conserved T-X-Y motif, located in a loop close to the active site of the terminal MAPK [4].

In addition, there is the group of atypical MAPKs, including ERK3/4, ERK7, ERK8 and Nemo-like kinase (NLK) [5–8], whose regulation and activation is not related to the module of the three kinases described for the conventional ones.

2. The p38MAPK Family

In mammalian cells, the p38MAPK family includes four members: p38α (MAPK14), p38β (MAPK11), p38γ (MAPK12) and p38δ (MAPK13), which have a high degree of sequence homology at the amino acid level (>60%) [9]. p38MAPKs differ in their expression patterns and substrate specificities, suggesting diverse functions. The p38MAPKs are S/T proline-directed kinases with an activation motif, T-G-Y, in which the substrate specificity is not only determined by the targeted amino acids, but also by specific docking domains present on the substrate and by a specific substrate binding motif in the MAPK (for a recent review of the p38MAPK-mediated signaling see [10]). The primary MAP2Ks for the p38MAPKs modules are MKK3 and MKK6 [11], although initially it was also considered the activation through MKK4 [12]. Activation of MAP2Ks occurs by phosphorylation of two conserved S and T residues on their activation loop by a broad range of MAPK3s. The MAP3Ks of this pathway include ASK1 (apoptosis signal-regulating kinase 1), DLK1 (dual-leucine-zipper bearing kinase 1), TAK1 (transforming growth factor β-activated kinase 1), TAO (thousand and one amino acid) 1 and 2, TPL2 (tumor progression loci 2), MLK3 (mixed-lineage kinase 3), MEKK (3 and 4), and ZAK1 (leucine zipper and sterile-α motif kinase 1) [13]. However, it has also been reported activation/inactivation of this signaling pathway by non-canonical mechanisms as in the case of T-cell receptor [14] or GRK2 [15].

The p38MAPK family can be divided into two subsets: on the one hand, p38α and p38β, and on the other hand p38δ and p38γ. This classification is based on their homology and their susceptibility to be inhibited by pyridinyl imidazoles (SB203580 and SB202190 compounds) at low concentrations. p38α and p38β have a higher homology between them (75%) and both can be inhibited by pyridinyl imidazoles, whereas p38δ and p38γ are 61% and 62% identical to p38α, respectively, and are not susceptible to be inhibited by SB203580 and SB202190 [16,17]. The number of specific inhibitors for p38MAPK is rapidly growing, allowing for a better understanding of the biological role of each p38 family member [18]. Important substrates in the p38MAPK signaling pathway include downstream kinases such as MK2/3, PRAK MSK1 or MNK1/2, as well as various transcription factors including ATF1/2/6, c-Myc, c-Fos, GAT4A, MEF2A/C, SRF, STAT1, p53 and CHOP among others [19]. This diversity of factors associated with the p38MAPK signaling pathway gives a glimpse of the plethora of biological processes implicated in this pathway.

3. p38β

p38β (also known as Stress Activating Protein Kinase 2 (SAPK2), Stress Activating Protein Kinase 2b (SAPK2b), MAPK11 or P38β2) was described in 1996 by Jiang and coworkers [20], and is encoded by the *MAPK11* gene that maps to chr22:50,263,713-50,270,380 in the human genome (UCSC genome browser/GRCh38/hg38), and comprises 12 exons (NCBI reference sequence NM_0022751.7). The protein is 364 amino acids long, and has a kinase domain (amino acids 24-308) that includes a T-G-Y (amino acids 180-182) dual phosphorylation motif, which is required for its kinase activity [16,21]. The 3D structure of p38β resembles that of a typical kinase with a smaller β-sheet N-terminal domain and a larger C-terminal domain. The ATP-binding site is located between the two domains, which are linked by a single polypeptide chain (residues T107-G110). This structure is also very similar to that of p38α, although they differ in the relative orientation of the N- and C-terminal domains [22]. This orientation causes a reduction in the size of the ATP-binding pocket of p38β compared to p38α. The difference in size between the two pockets could play a role in their different substrate specificity [20], and could be exploited in order to design selective compounds able to inhibit each p38 protein independently [22].

p38β is ubiquitously expressed, but at lower levels than p38α. p38β is expressed in the human brain, heart, placenta, lung, liver, skeletal muscle, kidney, spleen, testis, ovary, prostate, thymus and pancreas [20,21]. Moreover, p38β is abundant in endothelial cells, but undetectable in other lineages as macrophages or monocytes [23]. Similar to p38α, p38β is activated by pro-inflammatory cytokines and environmental stress, such as IL-1β, TNF, sorbitol, arsenite, anisomycin, high osmolarity, H_2O_2 and UV light [16,24]. The MAP2K that activates p38β is MKK6, whereas p38α is activated by MKK3 and MKK6 [20,25,26]. A unique characteristic of p38β is the ability to modulate its basal activity by autophosphorylation events. p38β is capable of self-activation by cis autophosphorylation of the residue T180 located in the activation loop. This activation occurs spontaneously in vitro, but can be regulated in mammalian cell cultures [27]. Moreover, p38β also autophosphorylates in trans residues T241 and S261 in vivo. Indeed, phosphorylation of S261 reduces the activity of T180-phosphorylated p38β, whereas, T241 phosphorylation reduces its phosphorylation in trans, although these two phosphorylation events do not seem to affect the activity of dually phosphorylated (T180/Y182) p38β [28].

The functions of p38β are mostly redundant with those of p38α. For instance, it has been shown that p38β cannot perform specific functions of p38α during development [29]. In fact, p38α knockout mice are lethal due to placental defects [30], while p38β knockout mice are fertile and viable [31].

The substrates attributed to p38β are mainly based on the use of SB compounds, which inhibit both p38α and p38β, not allowing to determine if they are bona fide substrates of p38β (p38MAPK substrates are reviewed in [10]). However, there are p38β targets that have been confirmed with other approaches. Among the several p38β substrates, there are protein kinases, transcription factors, and transcriptional regulators. Regarding protein kinases, the MAPK-activated protein kinases MAPKAPKs are a group of proteins downstream of MAPKs. A subgroup of MAPKAPKs is composed of MK2, MK3 (also known as 3pK), and MK5 (also designated as p38-regulated/activated protein kinase (PRAK)). These three kinases regulate key cellular processes such as gene expression at the transcriptional and post-transcriptional level, control cytoskeletal architecture and cell cycle progression, and play an important role in pathological processes such as inflammation and cancer (reviewed in [32]). p38α and p38β inhibit mitotic entry through MK2/3 phosphorylation in vivo [33] and MK2/MK3 activation is blocked by the inhibitor SB203580 in vitro [34]. MK5 is activated by p38α and p38β in vitro and in vivo [35], regulating the shuttling of this protein from the nucleus to the cytoplasm [36]. Another substrate of p38β is Protein kinase C ∈(PKC∈), a serine/threonine kinase involved in the regulation of cytokinesis in mitotic cells. This protein is primed to bind 14-3-3 by a series of phosphorylation events initiated by p38MAPK (in S350), GSK3 (in S346) and PKC itself (in S368). In vitro studies have shown that p38α and p38β phosphorylate S350 creating a GSK3 recognition site for the phosphorylation of S346, and that chemical inhibition by SB203580 prevents S346 phosphorylation in cells stimulated by UV-C [37].

Other studies have shown that p38β is also associated with several transcription factors. For example, MEFs (Myocite Enhancer Factors) are a family of transcription factors composed of MEF2A-D that regulates cell differentiation, proliferation, apoptosis, migration, and metabolism [38]. MEF2A and MEF2C are phosphorylated by p38α and p38β in vitro through a MAP kinase docking domain that is specific to these MAPKs, and activates their transcriptional activity in vivo [39]. Moreover, SB202190 inhibits the transcriptional activity of MEF2C induced by LPS or MKK6 in monocytic cells [40]. Another transcription factor targeted by p38β is NFATc4 (Nuclear Factor of activated T cells 4). NFATc4 belongs to the NFAT family of transcription factors, and is involved in cardiac development, mitochondrial function, and in activation of adipocyte specific genes during differentiation [41,42]. NFATC4 is phosphorylated by p38α, β, γ and δ in the presence of an activated MKK6 mutant (MKK6-GLu) in vitro and in vivo, p38α phosphorylates NFATc4 at S168 and S170 in the NFAT homology domain regulating the subcellular distribution of the transcription factor, promoting cytoplasmic localization of the NFATc4, and blocking adipocyte formation under differentiation conditions [43]. Moreover, phosphorylation of S168 and S170 of endogenous NFATc4 by p38MAPK is sensitive to SB203580 [44].

AP-1, a dimeric complex that is composed of members of the JUN, FOS, ATF or MAF protein families, regulates a wide range of cellular processes including cell proliferation, death, survival and differentiation, and has also been shown to be a downstream target of p38β [45]. C-FOS and ATF2 were also shown to be phosphorylated in vitro and in vivo by the four p38MAPKs, increasing its transcriptional activity [16,21,46,47]. Furthermore, it was reported that histone deacetylase 3 interacts specifically with p38β in LPS-stimulated cells, diminishing its phosphorylation, and leading to a repression of ATF-2 transcriptional activity as in the case of TNF gene expression [48]. Another transcription factor targeted by p38β is MafA, a member of the MAF family of basic leucine zipper proteins, that act as an important regulator of development and differentiation in many organs/tissues, and is a key player in Insulin regulation (for a review see [49]). MafA is also phosphorylated by the four p38MAPK isoforms in vitro and in vivo, and this phosphorylation might control MafA function, as it was shown previously in lens differentiation in primary cultures of chicken neuroretinal cells [50].

Other substrates of p38β with different functions that are shared with p38α, have been also reported. This group of miscellaneous substrates includes the BAF 60 protein BAF60c [51,52], E47 [53], P18 (Hamlet) [54,55], Cyclin D3 [56], the variant of the histone H2A, H2AX [57], KH-type splicing regulatory protein (KSRP) [58], and the membrane associated metalloprotease TACE [59]. However, there are two proteins, Glycogen Synthase (GS) and Raptor that seem to be specific substrates of p38β, and are not phosphorylated by any of the other p38MAPK proteins. p38β binds specifically to GS in skeletal muscle, brain and liver, and its efficient phosphorylation allows GSK3 to phosphorylate other residues of GS, causing partial inhibition of its activity [60]. In the case of Raptor, a regulatory-associated protein of mTOR, activated p38β by arsenite interacts with Raptor resulting in the phosphorylation of Raptor on S863 and S771, enhancing mTORC1 activity [61]. Therefore, the search for new specific substrates based on genetic evidence rather than on SB compounds, is a key step in the further understanding of biological functions mediated by p38β.

4. p38β and Cancer

Although p38β has been related to several pathological conditions like Huntington disease [62] and cardiac hypertrophy [63,64], this review will be focused on the role of p38β in cancer.

Since the mid-90s, when the p38MAPK signaling pathway was initially related to the cellular response to DNA damage agents including antitumor treatments [65], up to recent evidence indicating its use as a potential therapeutic target [13], the role of the p38MAPK signaling pathway in cancer has been deeply studied. However, most of the work has been focused on p38α, which has been repeatedly shown to play an important role in cancer biology. Consistent data from experimental models in different pathological conditions [66,67], have allowed us to consider p38α as a biomarker [24,68,69], and also as a putative target for cancer therapy (for a review see [70]). Conversely, much less is known about the role of the other p38 proteins (p38β/γ/δ) in cancer, although recent studies have shown an important role of p38γ/δ in cancer [71,72], however, further studies are needed to elucidate the definitive role of these two p38 proteins in cancer pathology (for a review see [73]).

Little is known about the role of p38β in cancer, although this protein has been associated with key molecules in this disease. For example, p38β has been proposed as a key target of the proto-oncogene Pokemon, a transcription factor known to be implicated in tumorigenesis and metastasis in hepatic cells [74]. Also, it has been reported that p38β could be a critical step in tumor formation through regulation of lipocalin 2 (LCN2) expression, a direct target of Plakophilin 3 (PKP3). In this sense, it has been shown that in different types of tumors, high LCN2 expression correlates with increased invasion, tumor formation and metastasis (for a review see [75]). Interestingly, in the absence of PKP3, p38β is able to control the expression of LCN2, indicating a potential role of p38β in tumor formation [76]. p38β has also been associated with integrin-αv, known to maintain cellular proliferation in keratinocytes by controlling c-Myc translation through FAK, p38β and p90RSK1. Chemical inhibition of p38β or genetic interference of *MAPK11* in keratinocytes promotes a marked decrease in c-Myc levels [77].

It was proposed that p38β could play a key role in biological processes for tumor progression and angiogenesis. For instance, TGF-β1 was shown to induce endothelial cell apoptosis by changing VEGF signaling from p38β, with survival function, to p38α with a pro-apoptotic function [78], in agreement with previous observations in cardiomyocytes [79]. Other studies have shown a direct connection of p38β with VEGF in a murine retinal model [80], further highlighting the importance of p38β in neovascularization and hypoxia-induced cell proliferation. Altogether, these studies suggest that p38β could be a potential target for an anti-angiogenic approach.

Also, p38β has been related to other aspects associated to cancer disease, with important implications in the patient's quality of life. For example, p38β has been related to cachexia through the control exerted onto the autophagic protein ULK1 in both in vitro and in vivo muscle wasting models [81]. Indeed, it is known that p38β functions upstream of FoxO–BNIP3 signaling axis to mediate an energy stress response [82], supporting the role of this MAPK in energy sensing. Another interesting aspect of p38β is its relationship with cancer-associated pain. In an experimental model of rats, pain associated to intra-tibial injection of mammary gland carcinoma cells, showed a marked reduction by intrathecal administration of a p38β antisense oligonucleotide [83]. Furthermore, the reduction of cancer-associated pain by music therapy was also attributed to low expression of p38β and p38α [84].

In addition to the connection with key proteins and biological processes in cancer, there are several examples showing the implication of p38β in different types of tumors. In pancreatic cancer, Singh and coworkers reported that p38β could be a potential biomarker [85]. Furthermore, the authors showed that peptide inhibitors for p38β are able to induce toxicity in pancreatic cell lines such as PANC-1, suggesting a potential therapeutic implication [85]. In hepatocellular carcinoma, recent data showed that p38β is a target of miR-516a-5p, which is controlled by a novel circular RNA, circ-0001955, that increases the expression of p38β, facilitating hepatocellular tumorigenesis [86]. In bladder cancer, p38β has been reported to be a critical player in cell motility through the signaling axis ILK-p38β-Hsp27 [87]. In prostate cancer, p38β has been related to metastases through the control exerted on the Wnt inhibitor Dickkopf-1, indicating the possibility of being considered as a therapeutic target [88]. Another study considered that the p38α/β inhibitor SB202190 could be used as a putative therapy in this type of tumor, in which STK11 could be a critical biomarker for this p38-based therapy, but no genetic evidence supported a critical role for p38β [89]. Therefore, further investigation is necessary to clarify the role of this particular MAPK in the biology and therapy of prostate cancer. In endometrial cancer, p38β has been shown to mediate the proliferation of tumor cells by inhibiting apoptosis. In this case, the anti-apoptotic ability of p38β seems to be controlled by the long non-coding RNA 1220, that controls p38β expression [90]. Interestingly, in lung cancer it has been recently reported that p38α, but not the rest of p38MAPK members, could be a potential biomarker of chemotherapy response [68]. However, overexpression of p38β was shown to be related to a specific subset of lung cancer in non-smokers in China [91]. Other reports indicate that a single-nucleotide polymorphism in p38β (rs2076139) is a potential biomarker associated with progression-free survival in metastatic non-small-cell lung cancer patients receiving platinum-based chemotherapy [92]. In addition, in lung cancer of non or light smokers it was shown that p38β and p38α, could be predictors of the expression levels of the DNA excision repair protein ERCC1, a key protein in DNA damage reparation with implications for the response to platinum compounds [93]. Indeed, chemical inhibition of p38α/β decreased viability of lung cancer cell lines, but genetic interference showed that most of this effect relies on p38β [93]. However, in terms of response to cisplatin, the effect of p38β was not applicable to all the experimental models [93], suggesting a more prominent role for p38α. Nonetheless, further studies are required to fully elucidate the role of p38β in lung cancer and its therapy. In breast cancer, the only connection with p38β has been related to bone metastases, through the up-regulation of the expression and secretion of monocyte chemotactic protein-1, which activates osteoclast differentiation and activity. Interestingly, the authors show how targeting p38β in breast cancer cells could be a novel approach to treat bone destruction associated with bone metastasis [94]. In silico evidence connected triple-negative breast cancer with epirubicin response and p38β overexpression, among other MAPKs, but no experimental data have been provided so far [95]. In Head and Neck Squamous Cell Carcinoma (HNSCC) patients, high expression levels of all p38MAPK isoforms, including p38β, have been detected in sera. Interestingly, these levels are downregulated after therapy, except for p38δ, suggesting that all p38MAPKs could be potential biomarkers in this disease. In addition, the authors indicated a potential role of p38δ as a putative target for HNSCC therapy that cannot be extrapolated to p38β [96].

p38β has also been related to leukemic pathology, for example, in acute myeloid leukemia (AML) and in Sézary syndrome. In AML, an aggressive hematologic malignancy, the overexpression of the SET oncoprotein, able to inhibit the protein phosphatase PP2A, is a key event that correlates with poor prognosis [97]. In this regard, p38β has been associated with the inhibitory effect of SET onto PP2A by two different mechanisms: first, by promoting SET cytoplasmic translocation through CK2 phosphorylation, and second, by direct binding to and stabilization of the SET protein [98]. Therefore, and considering the anticancer activity of several PP2A-activating drugs [99], p38β could be a potential novel target in AML, especially in those cases with SET over-expression. In the Sézary syndrome, a leukemic variant of cutaneous T-cell lymphomas, it has been reported that the overexpression of p38β could be a potential driver gene or a novel biomarker [100]. Indeed, in Sézary syndrome-derived cell lines, inhibition of PKCβ and GSK3 with the small molecules Enzastaurin and AR-A014418 promote a marked decrease in p38β expression without changing p38α levels [100]. Also, SB203580 and SB202190 promote cell death in those cell lines as well as in primary samples from Sézary syndrome patients. However, the genetic interference of p38β does not show any effect in cell viability [100], suggesting that further studies are necessary to fully evaluate the potential therapeutic implications of p38β in Sézary syndrome.

Nevertheless, in other types of tumors, p38β appears not to have any implication or, if so, a marginal role. For example, in melanoma, preliminary evidence in cell lines discard this MAPK, but not p38α, as a key player in this pathology [101]. Another example could be colorectal cancer, in which the 1628A>G (rs2235356) genetic variation in the p38β promoter region may contribute to the susceptibility to colorectal cancer in a Chinese population [102]. However, recent reports discard specifically this result in a Swedish population [103], suggesting that maybe p38β is not a universal biomarker for colorectal cancer. Indeed, other screening study discards p38β and indicates that p38α could be considered as a potential diagnostic marker and a putative therapeutic target for colorectal cancer [104]. In fact, this last observation is in agreement with previous reports using patient-derived xenografts [105]. Altogether, all this evidence suggests a marginal role for p38β in colorectal cancer.

Finally, regarding the implications of p38β in cancer biology, it is important to mention that Stress Activated Protein Kinases signaling pathways, JNK and p38MAPK, have been shown to play a dual role in cancer, both as an oncogene and as a tumor suppressor gene (for a review see [106,107]). This dual role seems to be dependent on several factors, including the experimental model and the stage of cell transformation, among others. Colorectal cancer is a paradigmatic example of this duality, showing that p38α could behave as an oncogene or a tumor suppressor depending on the stage of the carcinogenesis process [66]. It is likely that similar to p38α, p38β may have a dual role in cancer, also playing a tumor suppressor role. For instance, p38 MAPK signaling has been proposed to act as a tumor suppressor gene by controlling oncogenic properties of key molecules such as Ras (reviewed in [108]), Wip1 [109], EGFR [110]), and urokinase plasminogen activator [111] among others. However, most of these studies address the function of p38α specifically, with no reference to p38β, or if so, discarding its implication in the tumor suppressor activity. Future work investigating the potential tumor suppressing activity of p38β is needed to fully understand if p38β can potentially act both as a proto-oncogene and as a tumor suppressor gene in cancer pathology.

From the therapeutic point of view, the implication of p38MAPK signaling pathway in the mechanism of action of several anti-cancer drugs has been widely studied, but most of these studies have focused on the role of p38α [112]. In fact, p38α has been connected to DNA damage-response [113] through its relation with key proteins in DNA damage such as ATM or p53 [114]. p38α has been proposed as a master regulator of the apoptotic effects triggered by genotoxic drugs [115], and also as a central part of the cellular response to ionizing radiation [116]. However, no data involving specifically p38β has been published. There are only few examples suggesting a role for p38β in response to cancer therapy. For instance, in leukemia-derived cell lines, both p38α and p38β have been linked to interferon-α, leading to an inhibition of the cellular growth [117]. Moreover, p38β has been proposed as a key molecule in the stimulation of cell death triggered by the p38α/β inhibitor SB202190, UV, and FasL indicating a role in cytotoxicity [118]. Regarding other commonly used cancer treatments, such as chemo/radiotherapy or immunotherapy, there are no studies addressing the implications of p38β in response to these treatments, except for the one mentioned above in lung cancer [93], and for thymoquinone, a natural compound, in which its antitumor effect has been related to down regulation of p38β [119]. The relevance "per se" of p38β in cancer treatments, as a putative target, has been demonstrated in the previously mentioned experimental model of pancreatic cancer by using specific p38β inhibitory peptides [85] but, unfortunately, no other examples of specific targeted therapy based on p38β have been reported so far.

5. Future Directions

Although p38β is the least studied member of the p38MAPK family, possibly due to its functional redundancy with p38α, recent evidence shows that it may play a differential role with biological and pathological implications, as in the case of cancer. The lack of specific inhibitors for this MAPK has greatly complicated its study, since it involves the use of genetic approaches almost on a mandatory basis. Undoubtedly, the development of specific inhibitors for p38β could accelerate the research of this MAPK. However, there are still aspects to be investigated in the coming years such as the role of p38β in transcriptional regulation, its specific substrates, its involvement in the process of cell transformation and cancer (Figure 1), its implication in the cellular response to chemo and radiotherapy treatments, or even its use as a putative therapeutic target. Our knowledge of p38β is increasing every day, similarly to other members of the p38 family, allowing us to unravel the complexity of p38MAPK signaling, and to further elucidate the specific roles of each p38 family member. However, as most of the current research is focused on p38α, further studies on other p38 proteins, including p38β are needed to fully understand the importance of the p38MAPK signaling in human pathology.

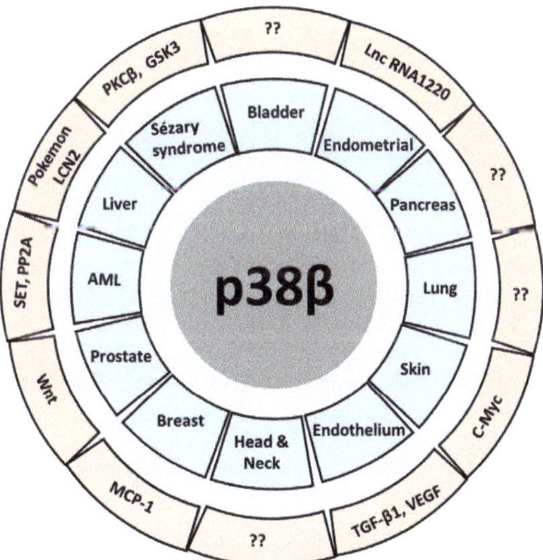

Figure 1. Schematic representation of the involvement of p38β in different types of tumors (blue) and the related molecules (red).

Author Contributions: O.R., E.A.-L. and N.G.-F. collaborated in the composition of Sections 1–3, provided proofreading comments and helped in the edition. D.M.F.-A. and M.J.R.-H. collaborated in the composition of Sections 4 and 5, provided proofreading comments and helped in the edition. L.F.M. provided proofreading comments, helped in the edition and collaborated in Section 4. R.S.-P. designed and coordinated the whole review, collaborated composing all the sections, and thoroughly proofread the manuscript for its submission. All authors have read and agreed to the published version of the manuscript.

Funding: This work was supported by grants from Fundación Leticia Castillejo Castillo, Ministerio de Ciencia, Innovación y Universidades (MCIU), Agencia Estatal de Investigación (AEI) and Fondo Europeo de Desarrollo Regional (FEDER) (RTI2018-094093-B-I00) to RSP and MJRH. OR holds a contract for accessing the Spanish System of Science, Technology and Innovation (SECTI) funded by the University of Castilla-La Mancha (UCLM) and received partial support from the European Social Fund (FSE) through its Operative Program for Castilla-La Mancha (2007–2013). RSP and MJRH's Research Institute, and the work carried out in their laboratory, received partial support from the European Community through the FEDER. EAL holds a research predoctoral contract cofounded by the European Social Fund and the UCLM. We acknowledge support of the publication fee by the CSIC Open Access Publication Support Initiative through its Unit of Information Resources for Research (URICI).

Conflicts of Interest: The authors declare no conflict of interest.

References

1. Kim, E.K.; Choi, E.-J. Pathological roles of MAPK signaling pathways in human diseases. *Biochim. Biophys. Acta* **2010**, *1802*, 396–405. [CrossRef]
2. Morrison, D.K.; Davis, R.J. Regulation of MAP kinase signaling modules by scaffold proteins in mammals. *Annu. Rev. Cell Dev. Biol.* **2003**, *19*, 91–118. [CrossRef] [PubMed]
3. Peti, W.; Page, R. Molecular basis of MAP kinase regulation. *Protein Sci.* **2013**, *22*, 1698–1710. [CrossRef]
4. Cargnello, M.; Roux, P.P. Activation and function of the MAPKs and their substrates, the MAPK-activated protein kinases. *Microbiol. Mol. Biol. Rev.* **2011**, *75*, 50–83. [CrossRef] [PubMed]
5. Coulombe, P.; Meloche, S. Atypical mitogen-activated protein kinases: Structure, regulation and functions. *Biochim. Biophys. Acta* **2007**, *1773*, 1376–1387. [CrossRef] [PubMed]
6. Kuo, W.-L.; Duke, C.J.; Abe, M.K.; Kaplan, E.L.; Gomes, S.; Rosner, M.R. ERK7 expression and kinase activity is regulated by the ubiquitin-proteosome pathway. *J. Biol. Chem.* **2004**, *279*, 23073–23081. [CrossRef] [PubMed]
7. Ishitani, T.; Ishitani, S. Nemo-like kinase, a multifaceted cell signaling regulator. *Cell. Signal.* **2013**, *25*, 190–197. [CrossRef]
8. Abe, M.K.; Saelzler, M.P.; Espinosa, R.; Kahle, K.T.; Hershenson, M.B.; Le Beau, M.M.; Rosner, M.R. ERK8, a new member of the mitogen-activated protein kinase family. *J. Biol. Chem.* **2002**, *277*, 16733–16743. [CrossRef]
9. Yokota, T.; Wang, Y. p38 MAP kinases in the heart. *Gene* **2016**, *575*, 369–376. [CrossRef]
10. Han, J.; Wu, J.; Silke, J. An overview of mammalian p38 mitogen-activated protein kinases, central regulators of cell stress and receptor signaling. *F1000Res* **2020**, *9*, 653. [CrossRef]
11. Remy, G.; Risco, A.M.; Iñesta-Vaquera, F.A.; González-Terán, B.; Sabio, G.; Davis, R.J.; Cuenda, A. Differential activation of p38MAPK isoforms by MKK6 and MKK3. *Cell. Signal.* **2010**, *22*, 660–667. [CrossRef] [PubMed]
12. Dérijard, B.; Raingeaud, J.; Barrett, T.; Wu, I.H.; Han, J.; Ulevitch, R.J.; Davis, R.J. Independent human MAP-kinase signal transduction pathways defined by MEK and MKK isoforms. *Science* **1995**, *267*, 682–685. [CrossRef] [PubMed]
13. Martínez-Limón, A.; Joaquin, M.; Caballero, M.; Posas, F.; de Nadal, E. The p38 Pathway: From Biology to Cancer Therapy. *Int. J. Mol. Sci.* **2020**, *21*, 1913. [CrossRef] [PubMed]
14. Salvador, J.M.; Mittelstadt, P.R.; Guszczynski, T.; Copeland, T.D.; Yamaguchi, H.; Appella, E.; Fornace, A.J.; Ashwell, J.D. Alternative p38 activation pathway mediated by T cell receptor-proximal tyrosine kinases. *Nat. Immunol.* **2005**, *6*, 390–395. [CrossRef] [PubMed]
15. Peregrin, S.; Jurado-Pueyo, M.; Campos, P.M.; Sanz-Moreno, V.; Ruiz-Gomez, A.; Crespo, P.; Mayor, F.; Murga, C. Phosphorylation of p38 by GRK2 at the docking groove unveils a novel mechanism for inactivating p38MAPK. *Curr. Biol.* **2006**, *16*, 2042–2047. [CrossRef] [PubMed]
16. Kumar, S.; McDonnell, P.C.; Gum, R.J.; Hand, A.T.; Lee, J.C.; Young, P.R. Novel homologues of CSBP/p38 MAP kinase: Activation, substrate specificity and sensitivity to inhibition by pyridinyl imidazoles. *Biochem. Biophys. Res. Commun.* **1997**, *235*, 533–538. [CrossRef] [PubMed]

17. Young, P.R.; McLaughlin, M.M.; Kumar, S.; Kassis, S.; Doyle, M.L.; McNulty, D.; Gallagher, T.F.; Fisher, S.; McDonnell, P.C.; Carr, S.A.; et al. Pyridinyl imidazole inhibitors of p38 mitogen-activated protein kinase bind in the ATP site. *J. Biol. Chem.* **1997**, *272*, 12116–12121. [CrossRef] [PubMed]
18. Haller, V.; Nahidino, P.; Forster, M.; Laufer, S.A. An updated patent review of p38 MAP kinase inhibitors (2014–2019). *Expert Opin. Ther. Pat.* **2020**, *30*, 453–466. [CrossRef]
19. Trempolec, N.; Dave-Coll, N.; Nebreda, A.R. SnapShot: p38 MAPK substrates. *Cell* **2013**, *152*, 924–924.e1. [CrossRef]
20. Jiang, Y.; Chen, C.; Li, Z.; Guo, W.; Gegner, J.A.; Lin, S.; Han, J. Characterization of the structure and function of a new mitogen-activated protein kinase (p38beta). *J. Biol. Chem.* **1996**, *271*, 17920–17926. [CrossRef]
21. Stein, B.; Yang, M.X.; Young, D.B.; Janknecht, R.; Hunter, T.; Murray, B.W.; Barbosa, M.S. p38-2, a novel mitogen-activated protein kinase with distinct properties. *J. Biol. Chem.* **1997**, *272*, 19509–19517. [CrossRef] [PubMed]
22. Patel, S.B.; Cameron, P.M.; O'Keefe, S.J.; Frantz-Wattley, B.; Thompson, J.; O'Neill, E.A.; Tennis, T.; Liu, L.; Becker, J.W.; Scapin, G. The three-dimensional structure of MAP kinase p38beta: Different features of the ATP-binding site in p38beta compared with p38alpha. *Acta Crystallogr. D Biol. Crystallogr.* **2009**, *65*, 777–785. [CrossRef] [PubMed]
23. Hale, K.K.; Trollinger, D.; Rihanek, M.; Manthey, C.L. Differential expression and activation of p38 mitogen-activated protein kinase alpha, beta, gamma, and delta in inflammatory cell lineages. *J. Immunol.* **1999**, *162*, 4246–4252. [PubMed]
24. Johnston, S.J.; Ahmad, D.; Aleskandarany, M.A.; Kurozumi, S.; Nolan, C.C.; Diez-Rodriguez, M.; Green, A.R.; Rakha, E.A. Co-expression of nuclear P38 and hormone receptors is prognostic of good long-term clinical outcome in primary breast cancer and is linked to upregulation of DNA repair. *BMC Cancer* **2018**, *18*, 1027. [CrossRef]
25. Enslen, H.; Raingeaud, J.; Davis, R.J. Selective activation of p38 mitogen-activated protein (MAP) kinase isoforms by the MAP kinase kinases MKK3 and MKK6. *J. Biol. Chem.* **1998**, *273*, 1741–1748. [CrossRef] [PubMed]
26. Keesler, G.A.; Bray, J.; Hunt, J.; Johnson, D.A.; Gleason, T.; Yao, Z.; Wang, S.W.; Parker, C.; Yamane, H.; Cole, C.; et al. Purification and activation of recombinant p38 isoforms alpha, beta, gamma, and delta. *Protein Expr. Purif.* **1998**, *14*, 221–228. [CrossRef]
27. Beenstock, J.; Ben-Yehuda, S.; Melamed, D.; Admon, A.; Livnah, O.; Ahn, N.G.; Engelberg, D. The p38β Mitogen-activated Protein Kinase Possesses an Intrinsic Autophosphorylation Activity, Generated by a Short Region Composed of the α-G Helix and MAPK Insert. *J. Biol. Chem.* **2014**, *289*, 23546–23556. [CrossRef]
28. Beenstock, J.; Melamed, D.; Mooshayef, N.; Mordechay, D.; Garfinkel, B.P.; Ahn, N.G.; Admon, A.; Engelberg, D. p38β Mitogen-Activated Protein Kinase Modulates Its Own Basal Activity by Autophosphorylation of the Activating Residue Thr180 and the Inhibitory Residues Thr241 and Ser261. *Mol. Cell. Biol.* **2016**, *36*, 1540–1554. [CrossRef]
29. del Barco Barrantes, I.; Coya, J.M.; Maina, F.; Arthur, J.S.C.; Nebreda, A.R. Genetic analysis of specific and redundant roles for p38alpha and p38beta MAPKs during mouse development. *Proc. Natl. Acad. Sci. USA* **2011**, *108*, 12764–12769. [CrossRef]
30. Adams, R.H.; Porras, A.; Alonso, G.; Jones, M.; Vintersten, K.; Panelli, S.; Valladares, A.; Perez, L.; Klein, R.; Nebreda, A.R. Essential role of p38alpha MAP kinase in placental but not embryonic cardiovascular development. *Mol. Cell* **2000**, *6*, 109–116. [CrossRef]
31. Beardmore, V.A.; Hinton, H.J.; Eftychi, C.; Apostolaki, M.; Armaka, M.; Darragh, J.; McIlrath, J.; Carr, J.M.; Armit, L.J.; Clacher, C.; et al. Generation and characterization of p38beta (MAPK11) gene-targeted mice. *Mol. Cell. Biol.* **2005**, *25*, 10454–10464. [CrossRef] [PubMed]
32. Gaestel, M. MAPKAP kinases-MKs-two's company, three's a crowd. *Nat. Rev. Mol. Cell Biol.* **2006**, *7*, 120–130. [CrossRef] [PubMed]
33. Llopis, A.; Salvador, N.; Ercilla, A.; Guaita-Esteruelas, S.; Barrantes, I.D.B.; Gupta, J.; Gaestel, M.; Davis, R.J.; Nebreda, A.R.; Agell, N. The stress-activated protein kinases p38α/β and JNK1/2 cooperate with Chk1 to inhibit mitotic entry upon DNA replication arrest. *Cell Cycle* **2012**, *11*, 3627–3637. [CrossRef] [PubMed]
34. Clifton, A.D.; Young, P.R.; Cohen, P. A comparison of the substrate specificity of MAPKAP kinase-2 and MAPKAP kinase-3 and their activation by cytokines and cellular stress. *FEBS Lett.* **1996**, *392*, 209–214. [CrossRef]

35. New, L.; Jiang, Y.; Zhao, M.; Liu, K.; Zhu, W.; Flood, L.J.; Kato, Y.; Parry, G.C.; Han, J. PRAK, a novel protein kinase regulated by the p38 MAP kinase. *EMBO J.* **1998**, *17*, 3372–3384. [CrossRef]
36. New, L.; Jiang, Y.; Han, J. Regulation of PRAK subcellular location by p38 MAP kinases. *Mol. Biol. Cell* **2003**, *14*, 2603–2616. [CrossRef]
37. Saurin, A.T.; Durgan, J.; Cameron, A.J.; Faisal, A.; Marber, M.S.; Parker, P.J. The regulated assembly of a PKCepsilon complex controls the completion of cytokinesis. *Nat. Cell Biol.* **2008**, *10*, 891–901. [CrossRef]
38. Pon, J.R.; Marra, M.A. MEF2 transcription factors: Developmental regulators and emerging cancer genes. *Oncotarget* **2015**, *7*, 2297–2312. [CrossRef]
39. Yang, S.-H.; Galanis, A.; Sharrocks, A.D. Targeting of p38 Mitogen-Activated Protein Kinases to MEF2 Transcription Factors. *Mol. Cell. Biol.* **1999**, *19*, 4028–4038. [CrossRef]
40. Han, J.; Jiang, Y.; Li, Z.; Kravchenko, V.V.; Ulevitch, R.J. Activation of the transcription factor MEF2C by the MAP kinase p38 in inflammation. *Nature* **1997**, *386*, 296–299. [CrossRef]
41. Ho, I.C.; Kim, J.H.; Rooney, J.W.; Spiegelman, B.M.; Glimcher, L.H. A potential role for the nuclear factor of activated T cells family of transcriptional regulatory proteins in adipogenesis. *Proc. Natl. Acad. Sci. USA* **1998**, *95*, 15537–15541. [CrossRef] [PubMed]
42. Bushdid, P.B.; Osinska, H.; Waclaw, R.R.; Molkentin, J.D.; Yutzey, K.E. NFATc3 and NFATc4 are required for cardiac development and mitochondrial function. *Circ. Res.* **2003**, *92*, 1305–1313. [CrossRef] [PubMed]
43. Yang, T.T.C.; Xiong, Q.; Enslen, H.; Davis, R.J.; Chow, C.-W. Phosphorylation of NFATc4 by p38 mitogen-activated protein kinases. *Mol. Cell. Biol.* **2002**, *22*, 3892–3904. [CrossRef] [PubMed]
44. Yang, T.T.C.; Yu, R.Y.L.; Agadir, A.; Gao, G.-J.; Campos-Gonzalez, R.; Tournier, C.; Chow, C.-W. Integration of Protein Kinases mTOR and Extracellular Signal-Regulated Kinase 5 in Regulating Nucleocytoplasmic Localization of NFATc4. *Mol. Cell. Biol.* **2008**, *28*, 3489–3501. [CrossRef] [PubMed]
45. Shaulian, E.; Karin, M. AP-1 as a regulator of cell life and death. *Nat. Cell Biol.* **2002**, *4*, E131–E136. [CrossRef]
46. Tanos, T.; Marinissen, M.J.; Leskow, F.C.; Hochbaum, D.; Martinetto, H.; Gutkind, J.S.; Coso, O.A. Phosphorylation of c-Fos by members of the p38 MAPK family. Role in the AP-1 response to UV light. *J. Biol. Chem.* **2005**, *280*, 18842–18852. [CrossRef] [PubMed]
47. Cuenda, A.; Cohen, P.; Buée-Scherrer, V.; Goedert, M. Activation of stress-activated protein kinase-3 (SAPK3) by cytokines and cellular stresses is mediated via SAPKK3 (MKK6); comparison of the specificities of SAPK3 and SAPK2 (RK/p38). *EMBO J.* **1997**, *16*, 295–305. [CrossRef]
48. Mahlknecht, U.; Will, J.; Varin, A.; Hoelzer, D.; Herbein, G. Histone deacetylase 3, a class I histone deacetylase, suppresses MAPK11-mediated activating transcription factor-2 activation and represses TNF gene expression. *J. Immunol.* **2004**, *173*, 3979–3990. [CrossRef]
49. Kaneto, H.; Matsuoka, T. Role of pancreatic transcription factors in maintenance of mature β-cell function. *Int. J. Mol. Sci.* **2015**, *16*, 6281–6297. [CrossRef]
50. Sii-Felice, K.; Pouponnot, C.; Gillet, S.; Lecoin, L.; Girault, J.-A.; Eychène, A.; Felder-Schmittbuhl, M.-P. MafA transcription factor is phosphorylated by p38 MAP kinase. *FEBS Lett.* **2005**, *579*, 3547–3554. [CrossRef]
51. Simone, C.; Forcales, S.V.; Hill, D.A.; Imbalzano, A.N.; Latella, L.; Puri, P.L. p38 pathway targets SWI-SNF chromatin-remodeling complex to muscle-specific loci. *Nat. Genet.* **2004**, *36*, 738–743. [CrossRef] [PubMed]
52. Forcales, S.V.; Albini, S.; Giordani, L.; Malecova, B.; Cignolo, L.; Chernov, A.; Coutinho, P.; Saccone, V.; Consalvi, S.; Williams, R.; et al. Signal-dependent incorporation of MyoD–BAF60c into Brg1-based SWI/SNF chromatin-remodelling complex. *EMBO J.* **2012**, *31*, 301–316. [CrossRef] [PubMed]
53. Page, J.L.; Wang, X.; Sordillo, L.M.; Johnson, S.E. MEKK1 signaling through p38 leads to transcriptional inactivation of E47 and repression of skeletal myogenesis. *J. Biol. Chem.* **2004**, *279*, 30966–30972. [CrossRef]
54. Cuadrado, A.; Lafarga, V.; Cheung, P.C.F.; Dolado, I.; Llanos, S.; Cohen, P.; Nebreda, A.R. A new p38 MAP kinase-regulated transcriptional coactivator that stimulates p53-dependent apoptosis. *EMBO J.* **2007**, *26*, 2115–2126. [CrossRef]
55. Cuadrado, A.; Corrado, N.; Perdiguero, E.; Lafarga, V.; Muñoz-Canoves, P.; Nebreda, A.R. Essential role of p18Hamlet/SRCAP-mediated histone H2A.Z chromatin incorporation in muscle differentiation. *EMBO J.* **2010**, *29*, 2014–2025. [CrossRef] [PubMed]
56. Casanovas, O.; Jaumot, M.; Paules, A.-B.; Agell, N.; Bachs, O. P38SAPK2 phosphorylates cyclin D3 at Thr-283 and targets it for proteasomal degradation. *Oncogene* **2004**, *23*, 7537–7544. [CrossRef]
57. Lu, C.; Shi, Y.; Wang, Z.; Song, Z.; Zhu, M.; Cai, Q.; Chen, T. Serum starvation induces H2AX phosphorylation to regulate apoptosis via p38 MAPK pathway. *FEBS Lett.* **2008**, *582*, 2703–2708. [CrossRef] [PubMed]

58. Briata, P.; Forcales, S.V.; Ponassi, M.; Corte, G.; Chen, C.-Y.; Karin, M.; Puri, P.L.; Gherzi, R. p38-dependent phosphorylation of the mRNA decay-promoting factor KSRP controls the stability of select myogenic transcripts. *Mol. Cell* **2005**, *20*, 891–903. [CrossRef] [PubMed]
59. Xu, P.; Derynck, R. Direct activation of TACE-mediated ectodomain shedding by p38 MAP kinase regulates EGF receptor-dependent cell proliferation. *Mol. Cell* **2010**, *37*, 551–566. [CrossRef] [PubMed]
60. Kuma, Y.; Campbell, D.G.; Cuenda, A. Identification of glycogen synthase as a new substrate for stress-activated protein kinase 2b/p38beta. *Biochem. J.* **2004**, *379*, 133–139. [CrossRef]
61. Wu, X.-N.; Wang, X.-K.; Wu, S.-Q.; Lu, J.; Zheng, M.; Wang, Y.-H.; Zhou, H.; Zhang, H.; Han, J. Phosphorylation of Raptor by p38beta participates in arsenite-induced mammalian target of rapamycin complex 1 (mTORC1) activation. *J. Biol. Chem.* **2011**, *286*, 31501–31511. [CrossRef] [PubMed]
62. Yu, M.; Fu, Y.; Liang, Y.; Song, H.; Yao, Y.; Wu, P.; Yao, Y.; Pan, Y.; Wen, X.; Ma, L.; et al. Suppression of MAPK11 or HIPK3 reduces mutant Huntingtin levels in Huntington's disease models. *Cell Res.* **2017**, *27*, 1441–1465. [CrossRef]
63. Zhang, S.; Weinheimer, C.; Courtois, M.; Kovacs, A.; Zhang, C.E.; Cheng, A.M.; Wang, Y.; Muslin, A.J. The role of the Grb2-p38 MAPK signaling pathway in cardiac hypertrophy and fibrosis. *J. Clin. Investig.* **2003**, *111*, 833–841. [CrossRef] [PubMed]
64. Sharma, B.; Chaube, U.; Patel, B.M. Beneficial Effect of Silymarin in Pressure Overload Induced Experimental Cardiac Hypertrophy. *Cardiovasc. Toxicol.* **2019**, *19*, 23–35. [CrossRef] [PubMed]
65. Pandey, P.; Raingeaud, J.; Kaneki, M.; Weichselbaum, R.; Davis, R.J.; Kufe, D.; Kharbanda, S. Activation of p38 mitogen-activated protein kinase by c-Abl-dependent and -independent mechanisms. *J. Biol. Chem.* **1996**, *271*, 23775–23779. [CrossRef] [PubMed]
66. Gupta, J.; del Barco Barrantes, I.; Igea, A.; Sakellariou, S.; Pateras, I.S.; Gorgoulis, V.G.; Nebreda, A.R. Dual function of p38α MAPK in colon cancer: Suppression of colitis-associated tumor initiation but requirement for cancer cell survival. *Cancer Cell* **2014**, *25*, 484–500. [CrossRef] [PubMed]
67. Vitos-Faleato, J.; Real, S.M.; Gutierrez-Prat, N.; Villanueva, A.; Llonch, E.; Drosten, M.; Barbacid, M.; Nebreda, A.R. Requirement for epithelial p38α in KRAS-driven lung tumor progression. *Proc. Natl. Acad. Sci. USA* **2020**, *117*, 2588–2596. [CrossRef]
68. Sahu, V.; Mohan, A.; Dey, S. p38 MAP kinases: Plausible diagnostic and prognostic serum protein marker of non small cell lung cancer. *Exp. Mol. Pathol.* **2019**, *107*, 118–123. [CrossRef]
69. Roseweir, A.K.; Halcrow, E.S.; Chichilo, S.; Powell, A.G.; McMillan, D.C.; Horgan, P.G.; Edwards, J. ERK and p38MAPK combine to improve survival in patients with BRAF mutant colorectal cancer. *Br. J. Cancer* **2018**, *119*, 323–329. [CrossRef]
70. Igea, A.; Nebreda, A.R. The Stress Kinase p38α as a Target for Cancer Therapy. *Cancer Res.* **2015**, *75*, 3997–4002. [CrossRef]
71. Del Reino, P.; Alsina-Beauchamp, D.; Escós, A.; Cerezo-Guisado, M.I.; Risco, A.; Aparicio, N.; Zur, R.; Fernandez-Estévez, M.; Collantes, E.; Montans, J.; et al. Pro-oncogenic role of alternative p38 mitogen-activated protein kinases p38γ and p38δ, linking inflammation and cancer in colitis-associated colon cancer. *Cancer Res.* **2014**, *74*, 6150–6160. [CrossRef] [PubMed]
72. Zur, R.; Garcia-Ibanez, L.; Nunez-Buiza, A.; Aparicio, N.; Liappas, G.; Escós, A.; Risco, A.; Page, A.; Saiz-Ladera, C.; Alsina-Beauchamp, D.; et al. Combined deletion of p38γ and p38δ reduces skin inflammation and protects from carcinogenesis. *Oncotarget* **2015**, *6*, 12920–12935. [CrossRef] [PubMed]
73. Cuenda, A.; Sanz-Ezquerro, J.J. p38γ and p38δ: From Spectators to Key Physiological Players. *Trends Biochem. Sci.* **2017**, *42*, 431–442. [CrossRef] [PubMed]
74. Chen, Z.; Liu, F.; Zhang, N.; Cao, D.; Liu, M.; Tan, Y.; Jiang, Y. p38β, a novel regulatory target of Pokemon in hepatic cells. *Int. J. Mol. Sci.* **2013**, *14*, 13511–13524. [CrossRef] [PubMed]
75. Santiago-Sánchez, G.S.; Pita-Grisanti, V.; Quiñones-Díaz, B.; Gumpper, K.; Cruz-Monserrate, Z.; Vivas-Mejía, P.E. Biological Functions and Therapeutic Potential of Lipocalin 2 in Cancer. *Int. J. Mol. Sci.* **2020**, *21*, 4365. [CrossRef] [PubMed]
76. Basu, S.; Chaudhary, N.; Shah, S.; Braggs, C.; Sawant, A.; Vaz, S.; Thorat, R.; Gupta, S.; Dalal, S.N. Plakophilin3 loss leads to an increase in lipocalin2 expression, which is required for tumour formation. *Exp. Cell Res.* **2018**, *369*, 251–265. [CrossRef]
77. Duperret, E.K.; Dahal, A.; Ridky, T.W. Focal-adhesion-independent integrin-αv regulation of FAK and c-Myc is necessary for 3D skin formation and tumor invasion. *J. Cell Sci.* **2015**, *128*, 3997–4013. [CrossRef]

78. Ferrari, G.; Terushkin, V.; Wolff, M.J.; Zhang, X.; Valacca, C.; Poggio, P.; Pintucci, G.; Mignatti, P. TGF-β1 induces endothelial cell apoptosis by shifting VEGF activation of p38(MAPK) from the prosurvival p38β to proapoptotic p38α. *Mol. Cancer Res.* **2012**, *10*, 605–614. [CrossRef]
79. Kim, J.K.; Pedram, A.; Razandi, M.; Levin, E.R. Estrogen prevents cardiomyocyte apoptosis through inhibition of reactive oxygen species and differential regulation of p38 kinase isoforms. *J. Biol. Chem.* **2006**, *281*, 6760–6767. [CrossRef]
80. Kumar, R.; Janjanam, J.; Singh, N.K.; Rao, G.N. A new role for cofilin in retinal neovascularization. *J. Cell. Sci.* **2016**, *129*, 1234–1249. [CrossRef]
81. Liu, Z.; Sin, K.W.T.; Ding, H.; Doan, H.A.; Gao, S.; Miao, H.; Wei, Y.; Wang, Y.; Zhang, G.; Li, Y.-P. p38β MAPK mediates ULK1-dependent induction of autophagy in skeletal muscle of tumor-bearing mice. *Cell Stress* **2018**, *2*, 311–324. [CrossRef] [PubMed]
82. Lin, A.; Yao, J.; Zhuang, L.; Wang, D.; Han, J.; Lam, E.W.-F.; TCGA Research Network; Gan, B. The FoxO-BNIP3 axis exerts a unique regulation of mTORC1 and cell survival under energy stress. *Oncogene* **2014**, *33*, 3183–3194. [CrossRef] [PubMed]
83. Dong, H.; Xiang, H.-B.; Ye, D.-W.; Tian, X.-B. Inhibitory effects of intrathecal p38β antisense oligonucleotide on bone cancer pain in rats. *Int. J. Clin. Exp. Pathol.* **2014**, *7*, 7690–7698. [PubMed]
84. Gao, J.; Chen, S.; Lin, S.; Han, H. Effect of music therapy on pain behaviors in rats with bone cancer pain. *J. BUON* **2016**, *21*, 466–472. [PubMed]
85. Singh, A.K.; Pandey, R.; Gill, K.; Singh, R.; Saraya, A.; Chauhan, S.S.; Yadav, S.; Pal, S.; Singh, N.; Dey, S. p38β MAP kinase as a therapeutic target for pancreatic cancer. *Chem. Biol. Drug Des.* **2012**, *80*, 266–273. [CrossRef]
86. Yao, Z.; Xu, R.; Yuan, L.; Xu, M.; Zhuang, H.; Li, Y.; Zhang, Y.; Lin, N. Circ_0001955 facilitates hepatocellular carcinoma (HCC) tumorigenesis by sponging miR-516a-5p to release TRAF6 and MAPK11. *Cell Death Dis.* **2019**, *10*, 945. [CrossRef] [PubMed]
87. Yu, L.; Yuan, X.; Wang, D.; Barakat, B.; Williams, E.D.; Hannigan, G.E. Selective regulation of p38β protein and signaling by integrin-linked kinase mediates bladder cancer cell migration. *Oncogene* **2014**, *33*, 690–701. [CrossRef]
88. Browne, A.J.; Göbel, A.; Thiele, S.; Hofbauer, L.C.; Rauner, M.; Rachner, T.D. p38 MAPK regulates the Wnt inhibitor Dickkopf-1 in osteotropic prostate cancer cells. *Cell Death Dis.* **2016**, *7*, e2119. [CrossRef]
89. Grossi, V.; Lucarelli, G.; Forte, G.; Peserico, A.; Matrone, A.; Germani, A.; Rutigliano, M.; Stella, A.; Bagnulo, R.; Loconte, D.; et al. Loss of STK11 expression is an early event in prostate carcinogenesis and predicts therapeutic response to targeted therapy against MAPK/p38. *Autophagy* **2015**, *11*, 2102–2113. [CrossRef]
90. Li, Y.; Kong, C.; Wu, C.; Wang, Y.; Xu, B.; Liang, S.; Ying, X. Knocking down of LINC01220 inhibits proliferation and induces apoptosis of endometrial carcinoma through silencing MAPK11. *Biosci. Rep.* **2019**, *39*. [CrossRef]
91. Wu, H.; Meng, S.; Xu, Q.; Wang, X.; Wang, J.; Gong, R.; Song, Y.; Duan, Y.; Zhang, Y. Gene expression profiling of lung adenocarcinoma in Xuanwei, China. *Eur. J. Cancer Prev.* **2016**, *25*, 508–517. [CrossRef] [PubMed]
92. Sullivan, I.; Riera, P.; Andrés, M.; Altés, A.; Majem, M.; Blanco, R.; Capdevila, L.; Barba, A.; Barnadas, A.; Salazar, J. Prognostic effect of VEGF gene variants in metastatic non-small-cell lung cancer patients. *Angiogenesis* **2019**, *22*, 433–440. [CrossRef] [PubMed]
93. Planchard, D.; Camara-Clayette, V.; Dorvault, N.; Soria, J.-C.; Fouret, P. p38 Mitogen-activated protein kinase signaling, ERCC1 expression, and viability of lung cancer cells from never or light smoker patients. *Cancer* **2012**, *118*, 5015–5025. [CrossRef] [PubMed]
94. He, Z.; He, J.; Liu, Z.; Xu, J.; Yi, S.F.; Liu, H.; Yang, J. MAPK11 in breast cancer cells enhances osteoclastogenesis and bone resorption. *Biochimie* **2014**, *106*, 24–32. [CrossRef]
95. Huang, J.; Luo, Q.; Xiao, Y.; Li, H.; Kong, L.; Ren, G. The implication from RAS/RAF/ERK signaling pathway increased activation in epirubicin treated triple negative breast cancer. *Oncotarget* **2017**, *8*, 108249–108260. [CrossRef]
96. Sahu, V.; Nigam, L.; Agnihotri, V.; Gupta, A.; Shekhar, S.; Subbarao, N.; Bhaskar, S.; Dey, S. Diagnostic Significance of p38 Isoforms (p38α, p38β, p38γ, p38δ) in Head and Neck Squamous Cell Carcinoma: Comparative Serum Level Evaluation and Design of Novel Peptide Inhibitor Targeting the Same. *Cancer Res. Treat.* **2019**, *51*, 313–325. [CrossRef]

97. Cristóbal, I.; Garcia-Orti, L.; Cirauqui, C.; Cortes-Lavaud, X.; García-Sánchez, M.A.; Calasanz, M.J.; Odero, M.D. Overexpression of SET is a recurrent event associated with poor outcome and contributes to protein phosphatase 2A inhibition in acute myeloid leukemia. *Haematologica* **2012**, *97*, 543–550. [CrossRef]
98. Arriazu, E.; Vicente, C.; Pippa, R.; Peris, I.; Martínez-Balsalobre, E.; García-Ramírez, P.; Marcotegui, N.; Igea, A.; Alignani, D.; Rifón, J.; et al. A new regulatory mechanism of protein phosphatase 2A activity via SET in acute myeloid leukemia. *Blood Cancer J.* **2020**, *10*, 3. [CrossRef]
99. Arriazu, E.; Pippa, R.; Odero, M.D. Protein Phosphatase 2A as a Therapeutic Target in Acute Myeloid Leukemia. *Front. Oncol.* **2016**, *6*, 78. [CrossRef]
100. Bliss-Moreau, M.; Coarfa, C.; Gunaratne, P.H.; Guitart, J.; Krett, N.L.; Rosen, S.T. Identification of p38β as a therapeutic target for the treatment of Sézary syndrome. *J. Investig. Dermatol.* **2015**, *135*, 599–608. [CrossRef] [PubMed]
101. Wen, S.-Y.; Cheng, S.-Y.; Ng, S.-C.; Aneja, R.; Chen, C.-J.; Huang, C.-Y.; Kuo, W.-W. Roles of p38α and p38β mitogen-activated protein kinase isoforms in human malignant melanoma A375 cells. *Int. J. Mol. Med.* **2019**, *44*, 2123–2132. [CrossRef] [PubMed]
102. Huang, Q.; Chen, D.; Song, S.; Fu, X.; Wei, Y.; Lu, J.; Wang, L.; Wang, J. A genetic variation of the p38β promoter region is correlated with an increased risk of sporadic colorectal cancer. *Oncol. Lett.* **2013**, *6*, 3–8. [CrossRef] [PubMed]
103. Dimberg, J.; Olsen, R.S.; Skarstedt, M.; Löfgren, S.; Zar, N.; Matussek, A. Polymorphism of the p38β gene in patients with colorectal cancer. *Oncol. Lett.* **2014**, *8*, 1093–1095. [CrossRef] [PubMed]
104. Tian, X.; Sun, D.; Zhao, S.; Xiong, H.; Fang, J. Screening of potential diagnostic markers and therapeutic targets against colorectal cancer. *Onco Targets Ther.* **2015**, *8*, 1691–1699. [CrossRef] [PubMed]
105. Gupta, J.; Igea, A.; Papaioannou, M.; Lopez-Casas, P.P.; Llonch, E.; Hidalgo, M.; Gorgoulis, V.G.; Nebreda, A.R. Pharmacological inhibition of p38 MAPK reduces tumor growth in patient-derived xenografts from colon tumors. *Oncotarget* **2015**, *6*, 8539–8551. [CrossRef] [PubMed]
106. Engelberg, D. Stress-activated protein kinases-tumor suppressors or tumor initiators? *Semin. Cancer Biol.* **2004**, *14*, 271–282. [CrossRef]
107. Wagner, E.F.; Nebreda, A.R. Signal integration by JNK and p38 MAPK pathways in cancer development. *Nat. Rev. Cancer* **2009**, *9*, 537–549. [CrossRef]
108. Loesch, M.; Chen, G. The p38 MAPK stress pathway as a tumor suppressor or more? *Front. Biosci. A J. Virtual Libr.* **2008**, *13*, 3581–3593. [CrossRef]
109. Bulavin, D.V.; Phillips, C.; Nannenga, B.; Timofeev, O.; Donehower, L.A.; Anderson, C.W.; Appella, E.; Fornace, A.J. Inactivation of the Wip1 phosphatase inhibits mammary tumorigenesis through p38 MAPK-mediated activation of the p16(Ink4a)-p19(Arf) pathway. *Nat. Genet.* **2004**, *36*, 343–350. [CrossRef]
110. Frey, M.R.; Dise, R.S.; Edelblum, K.L.; Polk, D.B. p38 kinase regulates epidermal growth factor receptor downregulation and cellular migration. *EMBO J.* **2006**, *25*, 5683–5692. [CrossRef]
111. Han, Q.; Leng, J.; Bian, D.; Mahanivong, C.; Carpenter, K.A.; Pan, Z.K.; Han, J.; Huang, S. Rac1-MKK3-p38-MAPKAPK2 pathway promotes urokinase plasminogen activator mRNA stability in invasive breast cancer cells. *J. Biol. Chem.* **2002**, *277*, 48379–48385. [CrossRef] [PubMed]
112. García-Cano, J.; Roche, O.; Cimas, F.J., Pascual-Serra, R.; Ortega Mucleo, M.; Fernández-Aroca, D.M.; Sánchez-Prieto, R. p38MAPK and Chemotherapy: We Always Need to Hear Both Sides of the Story. *Front. Cell Dev. Biol.* **2016**, *4*, 69. [CrossRef] [PubMed]
113. Cánovas, B.; Igea, A.; Sartori, A.A.; Gomis, R.R.; Paull, T.T.; Isoda, M.; Pérez-Montoyo, H.; Serra, V.; González-Suárez, E.; Stracker, T.H.; et al. Targeting p38α Increases DNA Damage, Chromosome Instability, and the Anti-tumoral Response to Taxanes in Breast Cancer Cells. *Cancer Cell* **2018**, *33*, 1094–1110.e8. [CrossRef]
114. Roy, S.; Roy, S.; Rana, A.; Akhter, Y.; Hande, M.P.; Banerjee, B. The role of p38 MAPK pathway in p53 compromised state and telomere mediated DNA damage response. *Mutat. Res. Genet. Toxicol. Environ. Mutagen.* **2018**, *836*, 89–97. [CrossRef]
115. Sui, X.; Kong, N.; Ye, L.; Han, W.; Zhou, J.; Zhang, Q.; He, C.; Pan, H. p38 and JNK MAPK pathways control the balance of apoptosis and autophagy in response to chemotherapeutic agents. *Cancer Lett.* **2014**, *344*, 174–179. [CrossRef] [PubMed]
116. Munshi, A.; Ramesh, R. Mitogen-activated protein kinases and their role in radiation response. *Genes Cancer* **2013**, *4*, 401–408. [CrossRef] [PubMed]

117. Lee, W.-H.; Liu, F.-H.; Lee, Y.-L.; Huang, H.-M. Interferon-alpha induces the growth inhibition of human T-cell leukaemia line Jurkat through p38alpha and p38beta. *J. Biochem.* **2010**, *147*, 645–650. [CrossRef] [PubMed]
118. Nemoto, S.; Xiang, J.; Huang, S.; Lin, A. Induction of apoptosis by SB202190 through inhibition of p38beta mitogen-activated protein kinase. *J. Biol. Chem.* **1998**, *273*, 16415–16420. [CrossRef] [PubMed]
119. Abdelfadil, E.; Cheng, Y.-H.; Bau, D.-T.; Ting, W.-J.; Chen, L.-M.; Hsu, H.-H.; Lin, Y.-M.; Chen, R.-J.; Tsai, F.-J.; Tsai, C.-H.; et al. Thymoquinone induces apoptosis in oral cancer cells through p38β inhibition. *Am. J. Chin. Med.* **2013**, *41*, 683–696. [CrossRef]

Article

Transcriptomic Evaluation of Pulmonary Fibrosis-Related Genes: Utilization of Transgenic Mice with Modifying p38 Signal in the Lungs

Shuichi Matsuda [1,2], Jun-Dal Kim [3], Fumihiro Sugiyama [4], Yuji Matsuo [2], Junji Ishida [3], Kazuya Murata [3,4], Kanako Nakamura [5], Kana Namiki [6], Tatsuhiko Sudo [7], Tomoyuki Kuwaki [8], Masahiko Hatano [1], Koichiro Tatsumi [2], Akiyoshi Fukamizu [3] and Yoshitoshi Kasuya [1,6,*]

1 Department of Biomedical Science, Graduate School of Medicine, Chiba University, Chiba City, Chiba 260-8670, Japan; elliptical-full.shuichi@nifty.com (S.M.); hatanom@faculty.chiba-u.jp (M.H.)
2 Department of Respirology, Graduate School of Medicine, Chiba University, Chiba City, Chiba 260-8670, Japan; myfortune@nifty.com (Y.M.); tatsumi@faculty.chiba-u.jp (K.T.)
3 Life Science Center for Survival Dynamics, Tsukuba Advanced Research Alliance (TARA), University of Tsukuba, Tsukuba, Ibaraki 305-8577, Japan; kimjd75@tara.tsukuba.ac.jp (J.-D.K.); jishida@tara.tsukuba.ac.jp (J.I.); kmurata@md.tsukuba.ac.jp (K.M.); akif@tara.tsukuba.ac.jp (A.F.)
4 Laboratory Animal Resource Center and Trans-Border Medical Research Center, University of Tsukuba, Tsukuba, Ibaraki 305-8575, Japan; bunbun@md.tsukuba.ac.jp
5 Graduate School of Sciences and Technology, University of Tsukuba, Tsukuba, Ibaraki 305-8572, Japan; s2021036@u.tsukuba.ac.jp
6 Department of Biochemistry and Molecular Pharmacology, Graduate School of Medicine, Chiba University, Chiba City, Chiba 260-8670, Japan; cada0191@chiba-u.jp
7 Chemical Biology Core Facility and Antibiotics Laboratory, RIKEN Advanced Science Institute, Wako, Saitama 351-0198, Japan; sudo@riken.jp
8 Department of Physiology, Graduate School of Medical and Dental Sciences, Kagoshima University, Kagoshima City, Kagoshima 890-8544, Japan; kuwaki@m3.kufm.kagoshima-u.ac.jp
* Correspondence: kasuya@faculty.chiba-u.jp; Tel.: +81-43-226-2193; Fax: +81-43-226-2196

Received: 31 July 2020; Accepted: 8 September 2020; Published: 14 September 2020

Abstract: Idiopathic pulmonary fibrosis (IPF) is a progressive fibrosing lung disease that is caused by the dysregulation of alveolar epithelial type II cells (AEC II). The mechanisms involved in the progression of IPF remain incompletely understood, although the immune response accompanied by p38 mitogen-activated protein kinase (MAPK) activation may contribute to some of them. This study aimed to examine the association of p38 activity in the lungs with bleomycin (BLM)-induced pulmonary fibrosis and its transcriptomic profiling. Accordingly, we evaluated BLM-induced pulmonary fibrosis during an active fibrosis phase in three genotypes of mice carrying stepwise variations in intrinsic p38 activity in the AEC II and performed RNA sequencing of their lungs. Stepwise elevation of p38 signaling in the lungs of the three genotypes was correlated with increased severity of BLM-induced pulmonary fibrosis exhibiting reduced static compliance and higher collagen content. Transcriptome analysis of these lung samples also showed that the enhanced p38 signaling in the lungs was associated with increased transcription of the genes driving the p38 MAPK pathway and differentially expressed genes elicited by BLM, including those related to fibrosis as well as the immune system. Our findings underscore the significance of p38 MAPK in the progression of pulmonary fibrosis.

Keywords: p38 mitogen-activated protein kinase; bleomycin-induced pulmonary fibrosis; idiopathic pulmonary fibrosis; RNA sequencing; alveolar epithelial type II cells

1. Introduction

Pulmonary fibrosis is the result of the end-stage pathological development of existing lung diseases caused by infection, autoimmunity, chronic inflammation, and idiopathy. Idiopathic pulmonary fibrosis (IPF), one of the most common causes of interstitial pneumonia, is characterized by progressive and irreversible fibrotic scar formation in the gas exchange regions of the lung, resulting in organ malfunction. IPF is a devastating lung disease as patients show poor prognosis, with a median survival of 2–5 years as well as increased risks of pulmonary hypertension and lung cancer [1]. A chronic inflammatory process of the lung has long been considered a main potential mechanism underlying IPF [2]. Moreover, innate and adaptive inflammation may contribute to determining the rate of disease progression in patients with IPF [3]. However, the mortality of patients with IPF is correlated with the extent of fibrotic focus formation, which results from the abnormal and excessive accumulation of extracellular matrix (ECM) components, including collagen, fibronectin, and elastin [4]. Hence, recent studies focusing on the behaviors of ECM-producing myofibroblasts in pulmonary fibrosis may also inform the identification of therapeutic options for IPF [5–8]. In terms of current pharmacological therapies for IPF, while nintedanib and pirfenidone have been approved by the Food and Drug Administration, neither can improve the survival of patients with IPF [9]. Indeed, new beneficial strategies that enable patients with IPF to survive longer and with improved quality of life have been long-awaited.

Among mitogen-activated protein kinases (MAPKs), members of the p38 MAPK family are activated in response to environmental stresses such as inflammatory stimuli by cytokines and Toll-like receptor ligands, osmolality shock, ultraviolet irradiation, oxidative stress, chemotherapeutic drugs, etc. Of the four isoforms (α, β2, γ, and δ) of p38, p38α is ubiquitously expressed in adult tissues and its physiological and pathological roles have been well investigated [10]. p38 MAPKs are activated by dual phosphorylation of the TGY motif within their activation loop by two upstream MAPK kinases (MAP2Ks)—mitogen-activated protein kinase kinase (MKK)-3 and MKK6—that are activated by various types of MAPKK kinases (MAP3Ks) [11]. In addition to this canonical activation pathway, specific binding of transforming growth factor (TGF)-β-activated kinase 1-binding protein 1 to p38α leads to p38α autophosphorylation and activation [12]. TGF-β signaling is one of the most crucial factors in the murine pulmonary fibrosis model and may be potentiated in the pathogenesis of IPF [13,14]. These findings strongly suggest the involvement of p38 signaling in the development of pulmonary fibrosis. In fact, several studies have reported that p38 inhibitors, SB239063 and FR-167653, can ameliorate bleomycin (BLM)-induced pulmonary fibrosis [15,16]. Lipopolysaccharide-induced epithelial-mesenchymal transition (EMT), in the early pulmonary fibrosis process, may be associated with p38 and TGF-β/smad3 signaling pathways [17]. Additionally, macrophage-specific loss of function of forkhead box M1, which inhibits the p38 signaling pathway, exacerbates BLM-induced pulmonary fibrosis [18]. Furthermore, pirfenidone was originally recognized as a small molecule p38γ inhibitor that blocks the synthesis of TGF-β [19]. Hence, the involvement of p38 signaling in the pathogenesis of pulmonary fibrosis is indubitable.

Here, we designed the study to elucidate new therapeutic target genes for IPF based on the notion that p38 positively regulates the development of pulmonary fibrosis. Mice with stepwise changes in the intrinsic activity of p38, specifically in alveolar epithelial type II cells (AEC II), were subjected to the pulmonary fibrosis model by BLM because AEC II could play a critical role in the progression of IPF [20]. RNA sequencing of total RNA derived from the lungs followed by transcriptome analysis was performed.

2. Results

2.1. Aggravation of BLM-Induced Murine Pulmonary Fibrosis Correlated with Increased Intrinsic p38 Activity in the Lungs

Histopathological assessment revealed that worsening severity of pulmonary fibrosis was associated with increased intrinsic p38 activity in the lungs (Figure 1A,B). At 8 days post-instillation (dpi) of BLM, tissue infiltration of inflammatory cells and thickening of the alveolar interstitium involved in aberrant collagen accumulation were observed. Moreover, these changes proceeded to diffuse and multifocal distributions at 15 dpi. These histopathological findings of BLM-induced pulmonary fibrosis in the MKK6-constitutive active (MKK6-CA) group were more severe and extensive than those in the wild-type (WT) group, whereas those in the p38-dominant negative (p38-DN) group were less severe and extensive than those in the WT group. Moreover, the distinct severity of pulmonary fibrosis was evident in semi-quantitative evaluation assessed by a modified Ashcroft score and stratified by three mouse groups. In contrast, no apparent inflammatory and fibrotic changes were observed in phosphate-buffered saline (PBS)-treated groups.

In addition, total cell counts in bronchoalveolar lavage fluid (BALF) at 8 dpi of BLM tended to increase with increased intrinsic p38 activity in the lungs (Figure 1C). Regardless of mouse genotype, macrophages and lymphocytes accounted for approximately 50% and 30% of the total cells in the BALF of BLM-treated groups, respectively (Supplementary Figure S1). Similarly, comprehensive protein analysis in BALF by western blot array revealed 77 upregulated molecules that were evoked by BLM and were associated with increased p38 activity in the lungs (Supplementary Table S1). These upregulated molecules included many pro-inflammatory and pro-fibrotic mediators such as interleukin (IL)-13, IL-17, stromal cell-derived factor 1(SDF-1)/C-X-C motif chemokine ligand (CXCL)-12, interferon (IFN)-γ, keratinocyte chemoattractant (KC)/CXCL1, monokine induced by gamma interferon (MIG)/CXCL9, macrophage inflammatory protein-1α (MIP-1α)/CC chemokine ligand (CCL)-3, and regulated upon activation normal T cell expressed and secreted (RANTES)/CCL5.

Measurements of collagen and static compliance in the lungs also supported the morphological alterations in the three mouse groups treated with BLM (Figure 1D,E). The amount of collagen in the left lungs at 8 dpi in the MKK6-CA group was significantly higher than that in the WT and p38-DN groups, although the difference between the WT and p38-DN groups was not significant. The decrease in static compliance in the MKK6-CA and WT groups was significantly larger than that in the p38-DN group, although the difference was not significant between the MKK6-CA and WT groups. Additionally, a reduction in body weight tended to increase with increased p38 signaling in the lungs, exhibiting the systemic effect implicated in the severity of BLM-treated mice (Figure 1F).

We verified the differences in intrinsic p38 activity in the lungs that underlie the different severity of BLM induced fibrosis among three mouse genotypes. Immunofluorescence staining revealed the presence of AEC II, macrophages, and other parenchymal cells expressing p38 in the PBS-treated lungs (Supplementary Figure S2A). The differences in p38 expression in these lung cells were not observed among three mouse groups. In contrast, the proportion of the lung cells showing the nuclear localization of phospho-p38 (P-p38) was increased by BLM treatment (Supplementary Figure S2B,C). Additionally, the increased proportion corresponded with the theoretical stepwise upregulation of intrinsic p38 activity in the lungs, and this finding was most prominent in AEC II. Although p38 is ubiquitously expressed in the cytoplasm of resting cells, activated p38 is represented by the phosphorylation-dependent nuclear localization of p38 in response to various types of stimulation such as DNA damage [21,22]. Therefore, these results demonstrate the three graded intensities of p38 activation induced by BLM among three different mouse genotypes.

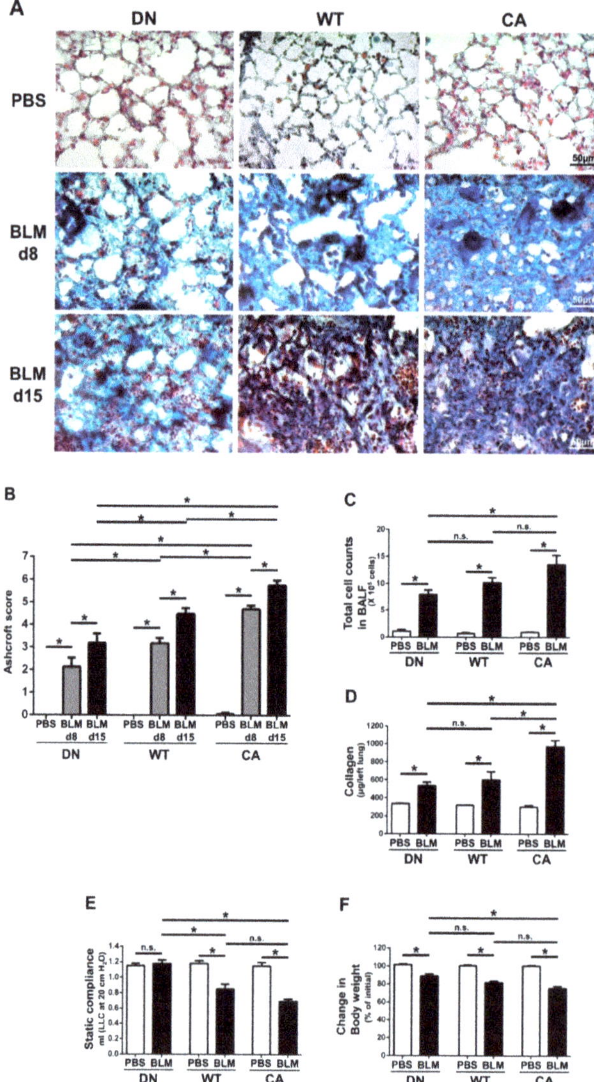

Figure 1. Bleomycin (BLM)-induced pulmonary fibrosis in mice bearing three different abilities of p38 in the lungs. The three mouse genotype groups; namely, MKK6 constitutive active group (CA), wild type group (WT), and p38 dominant negative group (DN), were intratracheally administrated BLM and phosphate-buffered saline (PBS). (**A**) Representative histopathological images of the lung sections stained by Masson's trichrome (scale bar = 50 μm). Lungs were collected at 8 days post-instillation (dpi) of BLM and PBS, and 15 dpi of BLM. (**B**) Quantification of the fibrotic severity using modified Ashcroft scoring was evaluated in six different lesions at 8 dpi of BLM and PBS, and 15 dpi of BLM (n = 9). (**C**) The numbers of total cells in bronchoalveolar lavage fluid were measured at 8 dpi of BLM and PBS (n = 7). (**D**) The collagen contents of the left lung lobes were measured at 8 dpi of BLM and PBS and normalized to the weight of each left lung (n = 4). (**E**) The static lung compliances were measured at 8 dpi of BLM and PBS (n = 4). (**F**) Proportions of body weight at 8 dpi of BLM and PBS to that before administration (n = 14). All data are represented as means ± standard error of the mean (SEM). * $p < 0.05$, n.s., no significant difference (measured by one-way an analysis of variance (ANOVA) followed by Tukey's test or unpaired Student's *t*-test).

2.2. Comparative Transcriptome Analysis of a BLM-Induced Pulmonary Fibrosis Model Exhibiting Different Severity Due to p38 Activity in the Lungs

RNA sequencing (RNA-seq) was performed using lung samples at 8 dpi when the severity of BLM-induced pulmonary fibrosis was apparently different among the three groups and transcriptomic changes in the BLM-induced fibrosis model are more likely to be correlated with the progression of IPF [23,24]. Principal component analysis (PCA) showed a relationship in the expression of genes among the three mouse groups treated with BLM and PBS, while hierarchical clustering analysis visualized using a heatmap highlighted the trend of differentially expressed genes (DEGs) between the BLM- and PBS-treated groups (Figure 2A). In the PCA plot, the BLM-treated groups were all well separated from the PBS-treated groups and the variance in BLM-treated groups was less than that in PBS-treated groups, indicating the assembly of distinct clusters following BLM exposure. Consistent with this observation, hierarchical clustering analysis identified DEGs between the BLM- and PBS-treated groups. Gene set enrichment analysis (GSEA) in the p38 MAPK pathway revealed that genes involved in regulating this pathway were significantly upregulated in the BLM-treated WT and MKK6-CA groups compared to those in the PBS-treated group (false discovery rate [FDR] q value < 0.25) but not in the p38-DN group (Figure 2B). Moreover, volcano plots in the three mouse groups showed that the increased number of DEGs between the BLM- and PBS-treated groups was associated with an increase in p38 signaling in the lungs (Figure 2C and Supplementary Tables S2–S4). BLM treatment upregulated approximately two-folds more DEGs and downregulated 2.5-folds more DEGs in the MKK6-CA group than those in the p38-DN group.

Next, we performed transcriptome analysis to detect the enriched functions of DEGs driven by BLM treatment. K-means clustering followed by Kyoto Encyclopedia of Genes and Genomes (KEGG) pathway enrichment analysis revealed the enriched pathways of four clusters in DEGs between the BLM- and PBS-treated groups of three mouse genotypes (Figure 3A). These pathways included fibrosis-related pathways such as protein processing in endoplasmic reticulum (ER) (yellow cluster), cytokine–cytokine receptor interaction (purple cluster), and ECM–receptor interaction (purple and green clusters) in addition to pathways related to immune systems such as hematopoietic cell lineage and leukocyte transendothelial migration. In contrast, we identified 493 common DEGs upregulated by BLM among the three mouse groups (Figure 3B and Supplementary Table S5). Regarding 493 common upregulated DEGs, enrichment analysis by gene ontology (GO) revealed three ECM-related annotations among the top five enriched molecular functions and all five terms associated with immune systems among the top five enriched biological processes. Additionally, the cytokine–cytokine receptor interaction was the most significantly enriched pathway among the top five KEGG pathways.

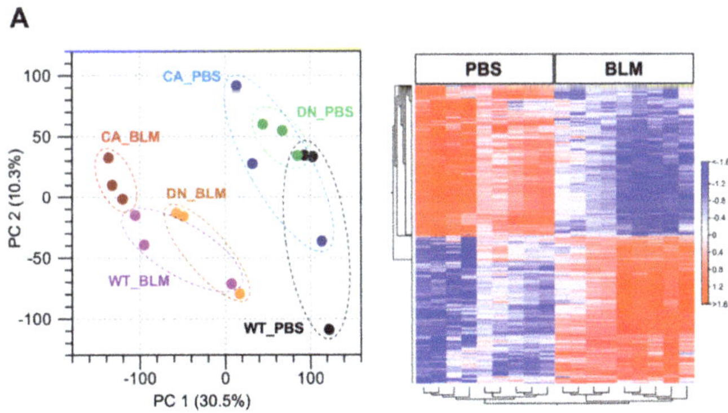

Figure 2. *Cont.*

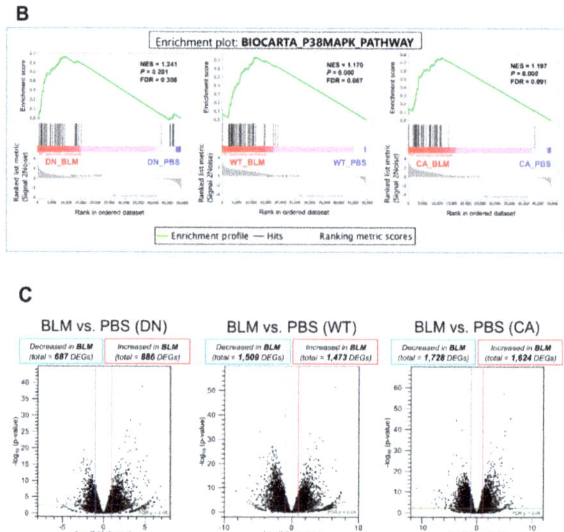

Figure 2. Expression profiling of BLM- and PBS-treated lungs in three different mouse genotypes. Three samples from each group were sequenced. (**A**) Principal component analysis of RNA sequencing datasets among the BLM- and PBS-treated three mouse groups (left). Hierarchical clustering shown in a heatmap of gene expression profiles between the BLM- and PBS-treated groups (right). The red and blue strips represent upregulated and downregulated genes in each group, respectively. (**B**) Gene set enrichment analysis of differential expression in the p38 MAPK pathway between the BLM- and PBS-treated groups of three mouse genotypes. The normalized enrichment scores (NES), normal *p*-values, and false discovery rate (FDR) q values are indicated. (**C**) Volcano plot of differentially expressed genes altered by BLM treatment among three mouse groups. Upregulated and downregulated genes are discriminated based on log2 fold-change and adjusted FDR *p*-value (<0.05).

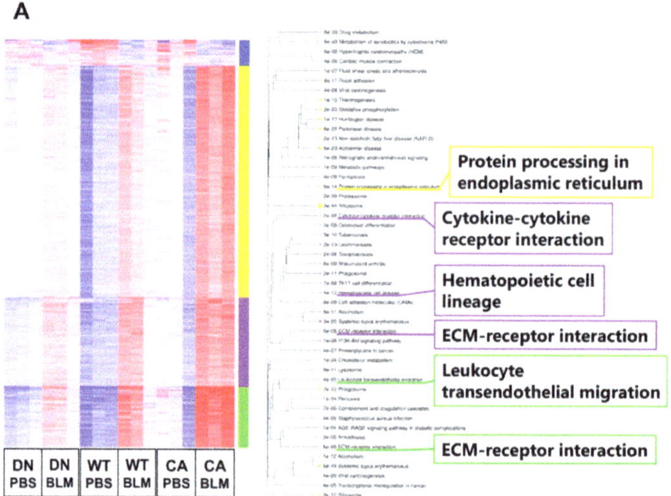

Figure 3. *Cont.*

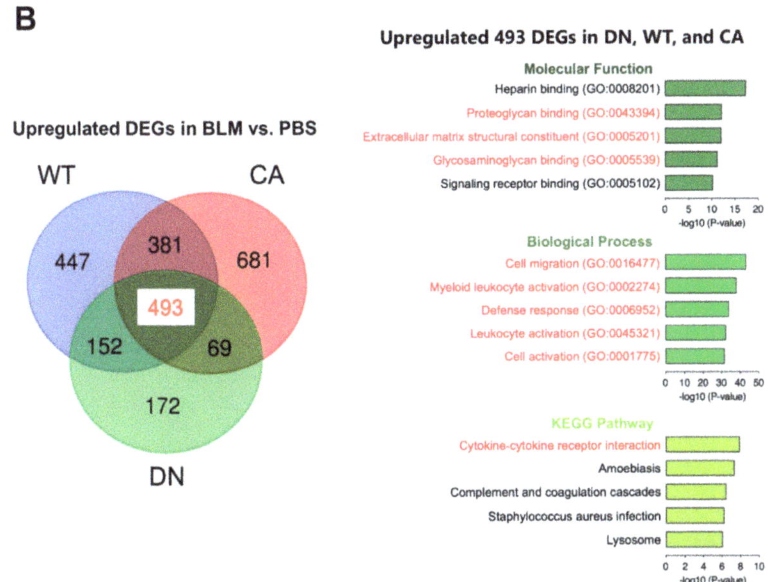

Figure 3. Gene ontology (GO) and Kyoto Encyclopedia of Genes and Genomes (KEGG) pathway enrichment analysis of differentially expressed genes (DEGs) altered by BLM treatment among three mouse groups. (**A**) K-means clustering shown in a heatmap based on the gene expression profiles between the BLM- and PBS-treated groups of three mouse genotypes (left). The colored bars on the right of the diagram indicate clusters. The trees represent enriched KEGG pathways corresponding to each cluster (right). (**B**) Venn diagram showing the overlap of DEGs upregulated by BLM among three mouse groups, with the numbers of DEGs indicated in each area (left). GO and KEGG pathway enrichment analysis of 493 common upregulated DEGs among the three genotypes (right). Top five enriched GO terms associated with molecular function (upper) and biological process (middle), and KEGG pathway analysis (bottom).

2.3. Exploration of Novel Potential Genes Contributing to the Progression of Pulmonary Fibrosis

To identify the pathogenetically relevant genes in the progression of IPF, we investigated the correlation of upregulated genes between BLM-induced fibrotic lungs showing the three different severity levels and human IPF lungs (Figure 4). In the BLM-treated groups, K-means clustering analysis identified a cluster of 2722 genes that their mean reads per kilobase of exon per million mapped sequence reads (RPKM) values increased along with stepwise elevation of p38 signaling in the lungs (Supplementary Table S6). We verified 137 DEGs that were included in this cluster and upregulated in common with the three BLM-treated mouse groups. Additionally, human RNA-seq data that provided 475 upregulated DEGs in IPF lung tissue compared to healthy lung tissue was obtained from the Gene Expression Omnibus website (accession ID: GSE52463). Finally, comparison of our data with human RNA-seq data identified four overlapping DEGs; namely, EPH receptor A3 (*EPHA3*), POU class 2 homeobox associating factor 1 (*POU2AF1*), SAM domain, SH3 domain and nuclear localization signals 1 (*SAMSN1*), and ectodysplasin A2 receptor (*EDA2R*).

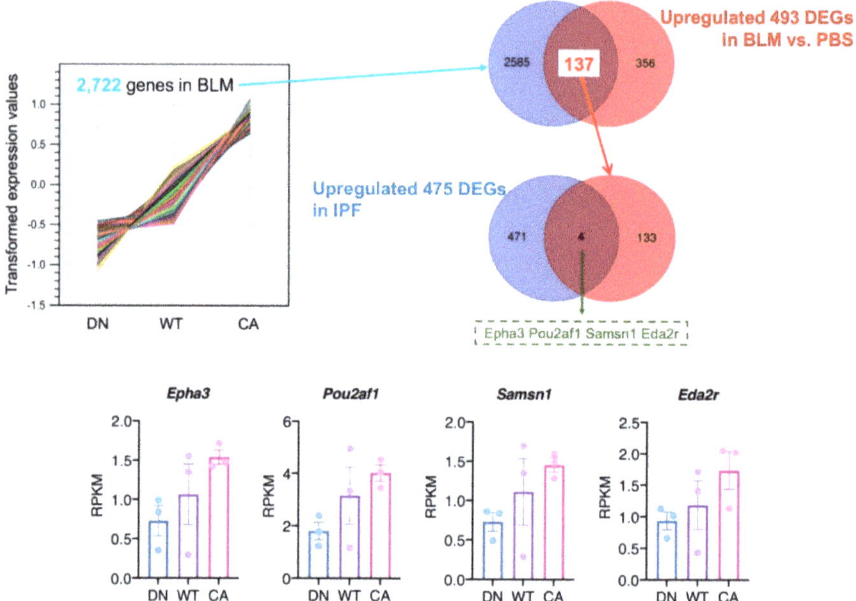

Figure 4. Identification of potential target genes by comparison to a publicly available idiopathic pulmonary fibrosis (IPF) dataset. K-means cluster analysis among three mouse groups treated with BLM revealed a cluster of 2722 genes showing correlation between variations of their mean expression values and stepwise changes in intrinsic p38 activity in the lungs (**left upper**). The Venn diagram in the right upper tier shows the overlap of 2722 genes in this cluster and 493 common upregulated DEGs among the three mouse groups. Likewise, the Venn diagram in the right middle tier shows the overlap of 137 genes identified in our study and 475 upregulated DEGs in human IPF lungs from dataset GSE52463. The four overlapping genes identified in these analyses were EPH receptor A3 (*EPHA3*), POU class 2 homeobox associating factor 1 (*POU2AF1*), SAM domain, SH3 domain and nuclear localization signals 1 (*SAMSN1*), and ectodysplasin A2 receptor (*EDA2R*) (**bottom**). Each bar and plot represent mean reads per kilobase of exon per million mapped sequence reads (RPKM) ± SEM and RPKM value of each sample, respectively.

3. Discussion

This study investigated the molecular mechanisms involved in the progressive worsening of BLM-induced pulmonary fibrosis in three genotypes of mice carrying stepwise variations of p38 activity in AEC II. We demonstrated that BLM-induced severe inflammation and fibrosis that was correlated with increased p38 activity in the lungs. Transcriptome analysis of this model provided a connection between the progression of pulmonary fibrosis and genes driving ER functions, ECM-cell interaction, and the immune system. Moreover, we identified candidate genes associated with IPF progression in comparison to a publicly available IPF dataset. The results of these comprehensive analyses suggest that the progression of pulmonary fibrosis occurs concurrently with increased p38 activity in AEC II, which provokes the enhancement of inflammation and immune systems. Therefore, this novel model of pulmonary fibrosis serves as a tool for understanding IPF progression.

Addressing the mechanisms contributing to IPF progression can lead to improved prognosis as this complex multi-pathway disease shows heterogeneity in its clinical course [1,4]. The present study applied a severe model of BLM-induced pulmonary fibrosis to study disease progression. Although the BLM-induced pulmonary fibrosis model is insufficient to mimic the pathogenesis of IPF, it has shown high reproducibility and the important mechanisms of pulmonary fibrosis, such as epithelial-mesenchymal crosstalk and TGF-β signaling pathway [25,26]. BLM-induced pulmonary fibrosis shows a transition from the inflammatory to the fibrotic phase at around 7 dpi, establishment of fibrosis at 14 dpi, and subsequent formation of reversible lesions [27,28]. In this context, the analyses were mainly conducted at 8 dpi as the optimal timing to evaluate the progression of pulmonary fibrosis. Additionally, the transcriptome profiling approach enables us to reveal the molecular mechanisms regulating fibrosis in this model and compare them to the profiles in the lung samples of IPF patients. Recent studies have shown that variations in gene expression of BLM-induced pulmonary fibrosis were correlated with changes in IPF severity [23,24]. Notably, the genes differentially expressed in BLM-induced pulmonary fibrosis are most abundant in the active fibrotic phase (7–14 dpi), which shows the highest correlation with IPF lung samples [23,24]. This finding explains the rationale that the gene expression profiles of BLM-treated lungs are altered before remarkable changes in morphology and function occur. Taken together, these findings are compatible with our study, which has implications for the development of pulmonary fibrosis in transcriptome analysis.

Although apoptosis and reprogramming of lung epithelial cells play a prominent role in IPF, the molecular details remain uncertain [20,29]. p38 is required for maintaining AEC II homeostasis as a physiological function, whereas extracellular stimuli-mediated enhancement of p38 is attributed to lung inflammation and immune responses and is associated with apoptosis in AEC II [30,31]. We focused on p38 activity in AEC II to examine pulmonary fibrosis progression and performed transcriptome analysis. The results showed distinct expression of p38 MAPK pathway genes that was positively correlated with stepwise changes in intrinsic p38 activity in the lungs and the contribution of the immune system and ER functions to the development of pulmonary fibrosis mediated by activation of the p38 MAPK pathway. Regarding lung inflammation, the exacerbation correlated with increased p38 activity in the AEC II manifested as increases in inflammatory cells and pro-inflammatory cytokines in BALF and the enrichment of genes facilitating immune cell infiltration and cytokine interaction pathways. Augmentation of pro-fibrotic cytokines and immune response arising from inflammation leads to progression of tissue remodeling and fibrosis in the lungs [32]. In particular, the TGF-β signaling pathway driven by p38 induces EMT and fibroblast proliferation and activation through epithelial–mesenchymal crosstalk [33–35]. In this study, TGF-β1 was included in the 137 overlapping genes that showed correlations with intrinsic p38 activity in the lungs and upregulation among the three mouse groups treated with BLM (Figure 4 and Supplementary Tables S5 and S6). Additionally, IL-13 and IL-17, which in BALF were upregulated with a concomitant increase in intrinsic p38 activity in the lungs, can promote TGF-β signaling pathway-dependent EMT and fibroblast proliferation and resistance to apoptosis [36,37]. These cytokines originate from immune cells such as T cells, suggesting an association between the immune system and fibrosis [38,39]. A previous study showed that increased immune cells and aberrant regenerating epithelial cells express inflammatory mediators, including IL-17, in active fibrotic lesions of IPF lungs [40]. Furthermore, single-cell RNA-seq analysis of epithelial cells displaying atypical phenotypes in IPF lungs showed that these epithelial cells modulated the expression of inflammatory response- and TGF-β signaling pathway-related genes, leading to fibrotic remodeling [41]. Collectively, these findings strongly suggest that inflammation and immune response enhanced by increased p38 activity in AEC II may contribute to the fibrotic process in the lungs.

Another possible explanation for the mechanism affecting fibrosis is that maladaptive ER stress response and its mediated apoptosis occurred concomitantly with increased p38 activity in AEC II. ER functions to retain cellular homeostasis by conducting posttranslational modification of proteins, with an adaptive process called unfold protein response under various stress conditions, although an excess of ER stress disrupting this adaptation elicits apoptosis [42]. The enrichment analysis in our study revealed that the ER protein processing pathway activated by BLM-induced reactive oxygen species was correlated with increased p38 activity in the AEC II. This result is consistent with the fundamental principle that ER stress can function in concert with the p38 MAPK pathway [43]. Simultaneously, AEC II homeostasis is sustained by its interaction with ECM, and p38 may participate in it [44].

A recent study using human fibroblasts revealed that the p38 MAPK pathway mediated the acquired resistance of ER stress modified by ECM metabolism through cell-to-ECM interaction [45]. In our study, ECM-receptor interaction was also a pathway enriched in accord with increased p38 activity in AEC II. Moreover, upregulation of ECM-related genes in enrichment analysis and matrix metalloproteases (MMPs; MMP-2, -3, and -9) in BALF (Supplementary Table S1), in addition to higher amounts of collagen, were connected to increased p38 activity in AEC II. These results are consistent with the study that MK2, a downstream substrate of p38, engaged in fibroblast activation and ECM production potentiated by fibroblast activation [46]. In addition, MMPs, which degrade all components of the ECM, are regulated by p38, while their upregulation leads to apoptosis and abnormal regeneration of lung epithelial cells [47]. Hence, ER stress in AEC II can be augmented by not only BLM-induced cytotoxicity, but also by the accumulation of ECM and AEC II to ECM interaction, controlled by the p38 MAPK pathway. These results emphasize the importance of p38 activity in AEC II and its related molecules in the progression of pulmonary fibrosis. In contrast, this murine model created by the intratracheal administration of BLM was not followed up after the establishment of pulmonary fibrosis. Therefore, further studies are required to determine whether p38 activity in AEC II influences the restoration of BLM-induced pulmonary fibrosis.

We validated four therapeutic target genes by comparing our data with publicly available data from IPF patients. First, *EPHA3* is expressed predominantly in lymphocytes and encodes a receptor tyrosine kinase implicated in regulating cell adhesion and cellular motility [48]. A recent study demonstrated that a novel epithelial cell population derived from IPF lungs co-expressed EPHA3 and CC chemokine receptor (CCR)-10 and facilitated the development of lung remodeling [49]. CCL28, a chemokine ligand for CCR10, was upregulated in BALF in accordance with increased p38 activity (Supplementary Table S1) [50]. These findings suggest that coordination between CCL28-CCR10 chemokine signaling and the p38 MAPK pathway has important implications for reprogramming of epithelial cells, a speculation that warrants further investigation. Second, *POU2AF1* encodes a transcriptional coactivator that regulates B cell maturation and humoral immunity and is expressed in both airway epithelial and B cells [51,52]. A prior study using IPF lungs documented that transcriptome analysis identified *POU2AF1* as a promoter of pulmonary fibrosis and it is highly expressed in aggregates of B cells [53]. Third, *EDA2R* regulates ectodermal tissue development; its expression in lung epithelial cells such as AEC II was ascertained by single-cell RNA-seq data set in IPF lungs [41,54]. Genome-wide association study using human lung tissue identified *EDA2R* as a candidate gene involved in lung aging [55]. In addition, this gene accelerates the apoptotic process in two different types of epithelial cells by activation of p53 signaling and caspase cascade [56,57]. These findings suggest the involvement of *EDA2R* in AEC II senescence and apoptosis. Lastly, *SAMSN1* is expressed in healthy lung epithelial cells but not in lung cancer cells [58]. Although its functions in the lungs remain unknown, this gene is pivotal in regulating B cell activation and differentiation [59]. Thus, changes in the expression levels of these candidate genes by p38 activity may be involved in promoting fibrosis through molecular interactions between epithelial and immune cells in the IPF lung. This hypothesis is supported by two previous reports showing an association of lymphocytes and epithelial cells with progressive fibrosis in transcriptome analysis of IPF lungs [60,61]. Therefore, the interplay between these genes and the p38 MAPK pathway may be key to understanding the immunological mechanisms underlying IPF progression. However, further studies are needed to confirm the clinical significance of these genes in the patients with rapidly progressive IPF.

4. Materials and Methods

4.1. Mice

All animal procedures conformed to the Japanese regulations for animal care and use, following the Guidelines for Animal Experimentation of the Japanese Association for Laboratory Animal Science. Male and female C57BL/6J mice were purchased from Clea Japan (Tokyo, Japan). Using the 3.7SP-C/SV40 vector kindly provided by Dr. Jeffrey A. Whitsett (Children's Hospital Medical Center, Division of Pulmonary Biology, Cincinnati, Ohio), we generated the following transgenic mice: C57BL/6J-hSP-C-M2 flag-p38α dominant-negative (d.n., dual mutations in wild mouse p38α: Thr180 to Ala; Tyr182 to Phe) TG (p38-DN) mice [62] and C57BL/6J-hSP-C-3HA-tag-MKK6 constitutive-active (c.a., dual mutations in wild human MKK6: Ser207 to Asp; Tyr211 to Asp) TG (MKK6-CA) mice [63]. We confirmed that each transgene-derived product was expressed at least in surfactant protein C (SP-C)-positive AEC II in the lung by using anti-M2-Flag or anti-HA tag antibody. Male heterozygous TG mice and WT littermates aged 10–12 weeks were used for the experiments. The animals were housed in standard laboratory cages and allowed food and water throughout the experiments. The studies were performed according to a protocol approved by the Committee of Animal Welfare of Chiba University.

4.2. BLM-Induced Pulmonary Fibrosis Model

Mice were anesthetized and the neck skin of each was cut longitudinally to expose the trachea. After a single intratracheal instillation of BLM hydrochloride (3 mg/kg; Nippon Kayaku, Tokyo, Japan) dissolved in PBS using a repeating syringe dispenser (Hamilton, Reno, NV, USA), the skin was sutured. Control mice were administered a sham treatment with PBS. Then, changes in body weight were measured daily. To evaluate the histopathological changes in the lung samples at 8 and 15 dpi of BLM, freshly cut lung sections (5 μm thick) were placed on adhesive glass slides (Matsunami Glass Ltd., Osaka, Japan) and stained with Masson's trichrome. The changes in the fibrotic lung samples were evaluated semi-quantitatively according to the modified Ashcroft method with a scoring grade of 0 to 8 [64]. In addition, the collagen content of the left lung was measured using the Sicol Soluble Collagen Assay Kit (Biocolor Life Science Assays, Carrickfergus, United Kingdom) according to the manufacturer's protocol.

4.3. Evaluation of Inflammatory Cells in BALF

At 8 dpi, the trachea was exposed and lavaged three times with 1 mL ice-cold PBS using a 20-gauge catheter. The BALF was centrifuged at 400× *g* for 10 min and the resulting supernatants were stored at −80 °C for protein array analysis. The resulting cell pellets were resuspended in PBS and subjected to cell counting using a hemocytometer in combination with Diff-Quick (Sysmex Corporation, Kobe, Japan) staining.

4.4. Measurement of Left Lung Compliance

As described previously [65], the lung compliance of the mice was measured by drawing static air pressure–volume relationships in a mixture of medetomidine, midazolam, and butorphanol (M/M/B: 0.3/4/5 mg/kg)-anesthetized mice tracheotomized with polyethylene tubing (O.D. = 0.8 mm). Total lung capacity was defined as the lung volume of full inflation judged by visual inspection of the lung that fully occupied the chest cavity. Functional residual volume was defined as deflation at 0 cm H_2O. Lung volumes at an airway pressure of 20 cm H_2O were estimated between mice at 8 dpi with BLM and PBS in the three genotypes (WT, p38-DN, and MKK6-CA mice).

4.5. Immunofluorescence Staining

The lung sections were pretreated with 1:10 FcR blocking agent (Miltenyi Biotech, Gladbach, Germany) for 10 min. They were then treated with primary antibodies (1:100 dilution) as follows: goat anti-proSP-C polyclonal antibody (sc-7706; Santa Cruz Biotech, Dallas, TX, USA), rabbit anti-p38 polyclonal antibody (original production [66]), rabbit anti-proSP-C antibody (customized production [60]; Sigma-Aldrich Japan Genosys, Ishikari, Japan), and mouse anti-phospho-p38 MAPK (pT180/pY182) (clone30, 612281; BD Biosciences, NJ, USA), followed by staining with appropriate fluorescein-conjugated secondary antibodies (1:200 dilution), and 4′,6-diamidino-2-phenylindole (DAPI) was used for nuclear staining. The stained sections were observed under a fluorescence microscope (Axio Imager A2; Zeiss, Oberkochen, Germany).

4.6. RNA Sequencing

At 8 dpi, mice under anesthesia were intracardially perfused with ice-cold PBS to wash out blood cells in the lungs and sacrificed. The left lung lobes were homogenized in ISOGEN plus (TaKaRa Bio, Kusatsu, Japan), and total RNA was extracted. Thereafter, 500 ng of total RNA was ribosomal RNA-depleted using a NEBNext rRNA Depletion Kit (New England Biolabs) and was converted to Illumina sequencing library using NEBNext Ultra Directional RNA Library Prep Kit (New England Biolabs). The library was validated to determine the size distribution and concentration using a Bioanalyzer (Agilent Technologies). Sequencing was performed on a NextSeq 500 (Illumina) instrument with paired-end 36-base read options. Reads were mapped on the mm10 mouse reference genome and quantified using CLC Genomics Workbench version 12.0 (QIAGEN). All RNA-seq data sets were deposited in the Gene Expression Omnibus database at the National Center for Biotechnology Information with accession number GSE154074.

4.7. Identification of Differentially Expressed Genes (DEGs)

To estimate the expression patterns of transcripts among the three genotypes (WT, p38α d.n.-TG and MKK6 c.a.-TG mice) with or without BLM instillation, the read counts were normalized by calculating the number of reads per kilobase per million for each transcript in individual samples using CLC Genomics Workbench version 12.0 (QIAGEN) [67]. Filtering characteristics of fold change −2 to 2 (FDR at $p < 0.05$) were used to identify the DEGs. Subsequently, the distinct gene expression patterns were analyzed comparatively through PCA and clustering heatmaps using CLC Genomics Workbench. GSEA for p38 MAPK pathways in the BLM-treated group among the three genotypes was also performed using GSEA_4.0.3. [68]. K-means functional enrichment analysis of DEGs was analyzed using integrated differential expression and pathway analysis (iDEP) online tools [69]. A volcano plot was used to compare the gene expression levels in terms of the $\log_2$ fold change. The GO (molecular function and biological process) and KEGG pathway analyses of DEGs between BLM- and PBS-treated groups were performed using ToppGene Suite (https://toppgene.cchmc.org) [70]. Finally, K-means cluster analysis was performed to identify BLM-upregulated genes that depended on the theoretical intrinsic activity of the p38 signal (p38-DN < WT < MKK6-CA) using CLC Genomics Workbench.

4.8. Statistical Analysis

Data are expressed as means ± standard error of the mean (SEM). Statistical analysis was conducted using GraphPad Prism Version 6 (GraphPad Software, San Diego, CA, USA). Statistical significance was determined by one-way analysis of variance (ANOVA) followed by Tukey's or Student's *t*-tests, and *p*-values < 0.05 were considered significant.

Supplementary Materials: Supplementary Materials can be found at http://www.mdpi.com/1422-0067/21/18/6746/s1.

Author Contributions: S.M., J.-D.K., and Y.K. developed the concept and designed the experiments. S.M., J.-D.K., F.S., Y.M., J.I., K.M., K.N. (Kanako Nakamura), K.N. (Kana Namiki), T.S., T.K., and Y.K. performed the experiments. M.H., K.T., and A.F. provided conceptual advice throughout the project. S.M., J.-D.K., and Y.K. wrote the paper. All authors discussed the results and implications and commented on the manuscript at all stages. All authors have read and agreed to the published version of the manuscript.

Funding: This work was supported in part by Grants-in-Aid for Scientific Research ((B), 24390137 to Yoshitoshi Kasuya) from the Ministry of Education, Science, Sports and Culture of Japan, a Therapeutics Research Initiative Grant from Chiba University, School of Medicine (2018-G7 to Yoshitoshi Kasuya and 2019-Y10 to Shuichi Matsuda) and the Cooperative Research Project Program of Life Science Center for Survival Dynamics, Tsukuba Advanced Research Alliance (TARA Center), University of Tsukuba, Japan.

Acknowledgments: We would like to thank Ben-Shiang Deng and Hiromi Takahashi for guidance in lung compliance measuring, Tomoko Misawa for technical support of pathologically staining, Kento Yoshioka for assistance with mouse genotyping.

Conflicts of Interest: The authors declare no conflict of interest.

Abbreviations

AEC II	alveolar epithelial type II cells
ANOVA	analysis of variance
BALF	bronchoalveolar lavage fluids
BLM	bleomycin
CCL	CC chemokine ligand
CCR	CC chemokine receptor
CXCL	C-X-C motif chemokine ligand
DAPI	4′,6-diamidinoA-phenylindole
DEG	differentially expressed gene
dpi	days post-instillation
ECM	extracellular matrix
EDA2R	ectodysplasin A2 receptor
EMT	epithelial-mesenchymal transition
EPHA3	EPH receptor A3
ER	endoplasmic reticulum
FDR	false discovery rate
GO	gene ontology
GSEA	gene set enrichment analysis
HA	hemagglutinin
iDEP	integrated Differential Expression and Pathway analysis
IFN	interferon
IL	interleukin
IPF	idiopathic pulmonary fibrosis
KC	keratinocyte chemoattractant
KEGG	Kyoto Encyclopedia of Genes and Genomes
MAPK	mitogen-activated protein kinase
MAP2K	MAPK kinase
MAP3K	MAPKK kinase
MIP-1α	macrophage inflammatory protein-1α
MIG	monokine induced by gamma interferon
MKK3/6	mitogen-activated protein kinase kinase 3/6
MKK6-CA	MKK6-constitutive active
MMP	matrix metalloprotease
NES	normalized enrichment scores
p38-DN	p38-dominant negative
P-p38	phospho-p38
PBS	phosphate-buffered saline
PCA	principal component analysis
POU2AF1	POU class 2 homeobox associating factor 1
RANTES	regulated upon activation normal T cell expressed and secreted
RNA-seq	RNA sequencing
RPKM	reads per kilobase of exon per million mapped sequence reads
SAMSN1	SAM domain, SH3 domain and nuclear localization signals 1
SDF-1	stromal cell-derived factor 1
SEM	standard error of the mean
SP-C	surfactant protein C
SV40	sarcoma virus 40
TGF	transforming growth factor
WT	wild-type

References

1. Lederer, D.J.; Martinez, F.J. Idiopathic pulmonary fibrosis. *N. Engl. J. Med.* **2018**, *378*, 1811–1823. [CrossRef] [PubMed]
2. Crystal, R.G.; Bitterman, P.B.; Rennard, S.I.; Hance, A.J.; Keogh, B.A. Interstitial lung diseases of unknown cause. Disorders characterized by chronic inflammation of the lower respiratory tract. *N. Engl. J. Med.* **1984**, *310*, 154–166. [CrossRef] [PubMed]
3. Balestro, E.; Calabrese, F.; Turato, G.; Lunardi, F.; Bazzan, E.; Marulli, G.; Biondini, D.; Rossi, E.; Sanduzzi, A.; Rea, F.; et al. Immune inflammation and disease progression in idiopathic pulmonary fibrosis. *PLoS ONE* **2016**, *11*, e0154516. [CrossRef] [PubMed]
4. Wynn, T.A.; Ramalingam, T.R. Mechanisms of fibrosis: Therapeutic translation for fibrotic disease. *Nat. Med.* **2012**, *18*, 1028–1040. [CrossRef] [PubMed]
5. Yamazaki, R.; Kasuya, Y.; Fujita, T.; Umezawa, H.; Yanagihara, M.; Nakamura, H.; Yoshino, I.; Tatsumi, K.; Murayama, T. Antifibrotic effects of cyclosporine A on TGF-β1–treated lung fibroblasts and lungs from bleomycin-treated mice: Role of hypoxia-inducible factor-1α. *FASEB J.* **2018**, *31*, 3359–3371. [CrossRef] [PubMed]
6. Rangarajan, S.; Bone, N.B.; Zmijewska, A.A.; Jiang, S.; Park, D.W.; Bernard, K.; Locy, M.L.; Ravi, S.; Deshane, J.; Mannon, R.B.; et al. Metformin reverses established lung fibrosis in a bleomycin model. *Nat. Med.* **2018**, *24*, 1121–1127. [CrossRef] [PubMed]
7. Penke, L.R.; Speth, J.M.; Dommeti, V.L.; White, E.S.; Bergin, I.L.; Peters-Golden, M. FOXM1 is a critical driver of lung fibroblast activation and fibrogenesis. *J. Clin. Investig.* **2018**, *128*, 2389–2405. [CrossRef]
8. Suzuki, K.; Kim, J.D.; Ugai, K.; Matsuda, S.; Mikami, H.; Yoshioka, K.; Ikari, J.; Hatano, M.; Fukamizu, A.; Tatsumi, K.; et al. Transcriptomic changes involved in the dedifferentiation of myofibroblasts derived from the lung of a patient with idiopathic pulmonary fibrosis. *Mol. Med. Rep.* **2020**, *22*, 1518–1526. [CrossRef]
9. Raghu, G.; Selman, M. Nintedanib and pirfenidone. New antifibrotic treatments indicated for idiopathic pulmonary fibrosis offer hopes and raises questions. *Am. J. Respir. Crit. Care Med.* **2018**, *191*, 252–254. [CrossRef]
10. Rincón, M.; Davis, R.J. Regulation of the immune response by stress-activated protein kinases. *Immunol. Rev.* **2009**, *228*, 212–224. [CrossRef]
11. Kasuya, Y.; Umezawa, H.; Hatano, M. Stress-activated protein kinases in spinal cord injury: Focus on roles of p38. *Int. J. Mol. Sci.* **2018**, *19*, 867. [CrossRef] [PubMed]
12. Ge, B.; Gram, H.; Di Padova, F.; Huang, B.; New, L.; Ulevitch, R.J.; Luo, Y.; Han, J. MAPKK-independent activation of p38α mediated by TAB1-dependent autophosphorylation of p38α. *Science* **2002**, *295*, 1291–1294. [CrossRef] [PubMed]
13. Li, M.; Krishnaveni, M.S.; Li, C.; Zhou, B.; Xing, Y.; Banfalvi, A.; Li, A.; Lombardi, V.; Akbari, O.; Borok, Z.; et al. Epithelium-specific deletion of TGF-β receptor type II protects mice from bleomycin-induced pulmonary fibrosis. *J. Clin. Investig.* **2011**, *121*, 277–287. [CrossRef]
14. Coker, R.K.; Laurent, G.J.; Jeffery, P.K.; du Bois, R.M.; Black, C.M.; McAnulty, R.J. Localization of transforming growth factor β1 and β3 mRNA transcripts in normal and fibrotic human lung. *Thorax* **2001**, *56*, 549–556. [CrossRef]
15. Underwood, D.C.; Osborn, R.R.; Bochnowicz, S.; Webb, E.F.; Rieman, D.J.; Lee, J.C.; Romanic, A.M.; Adams, J.L.; Hay, D.W.; Griswold, D.E. SB 239063, a p38 MAPK inhibitor, reduces neutrophilia, inflammatory cytokines, MMP-9, and fibrosis in lung. *Am. J. Physiol. Lung Cell. Mol. Physiol.* **2000**, *279*, L895–L902. [CrossRef]
16. Matsuoka, H.; Arai, T.; Mori, M.; Goya, S.; Kida, H.; Morishita, H.; Fujiwara, H.; Tachibana, I.; Osaki, T.; Hayashi, S. A p38 MAPK inhibitor, FR-167653, ameliorates murine bleomycin-induced pulmonary fibrosis. *Am. J. Physiol. Lung Cell. Mol. Physiol.* **2002**, *283*, L103–L112. [CrossRef]
17. Cao, Y.; Liu, Y.; Ping, F.; Yi, L.; Zeng, Z.; Li, Y. miR-200b/c attenuates lipopolysaccharide-induced early pulmonary fibrosis by targeting ZEB1/2 via p38 MAPK and TGF-β/smad3 signaling pathways. *Lab. Investig.* **2018**, *98*, 339–359. [CrossRef]
18. Goda, C.; Balli, D.; Black, M.; Milewski, D.; Le, T.; Ustiyan, V.; Ren, X.; Kalinichenko, V.V.; Kalin, T.V. Loss of FOXM1 in macrophages promotes pulmonary fibrosis by activating p38 MAPK signaling pathway. *PLoS Genet.* **2020**, *16*, e1008692. [CrossRef] [PubMed]

19. Moran, N. p38 kinase inhibitor approved for idiopathic pulmonary fibrosis. *Nat. Biotechnol.* **2011**, *29*, 301. [CrossRef] [PubMed]
20. Parimon, T.; Yao, C.; Stripp, B.R.; Noble, P.W.; Chen, P. Alveolar epithelial type II cells as drivers of lung fibrosis in idiopathic pulmonary fibrosis. *Int. J. Mol. Sci.* **2020**, *21*, 2269. [CrossRef] [PubMed]
21. Gong, X.; Ming, X.; Deng, P.; Jiang, Y. Mechanisms regulating the nuclear translocation of p38 MAP kinase. *J. Cell Biochem.* **2010**, *110*, 1420–1429. [CrossRef] [PubMed]
22. Wood, C.D.; Thornton, T.M.; Sabio, G.; Davis, R.A.; Rincon, M. Nuclear localization of p38 MAPK in response to DNA damage. *Int. J. Biol. Sci.* **2009**, *5*, 428–437. [CrossRef] [PubMed]
23. Peng, R.; Sridhar, S.; Tyagi, G.; Phillips, E.J.; Garrido, R.; Harris, P.; Burns, L.; Renteria, L.; Woods, J. Bleomycin induces molecular changes directly relevant to idiopathic pulmonary fibrosis: A model for "active" disease. *PLoS ONE* **2013**, *8*, e59348. [CrossRef]
24. Bauer, Y.; Tedrow, J.; de Bernard, S.; Birker-Robaczewska, M.; Gibson, K.F.; Guardela, B.J.; Hess, P.; Klenk, A.; Lindell, K.O.; Sylvie Poirey, S.; et al. A novel genomic signature with translational significance for human idiopathic pulmonary fibrosis. *Am. J. Respir. Cell Mol. Biol.* **2015**, *52*, 217–231. [CrossRef] [PubMed]
25. Moeller, A.; Ask, K.; Warburton, D.; Gauldie, J.; Kolb, M. The bleomycin animal model: A useful tool to investigate treatment options for idiopathic pulmonary fibrosis? *Int. J. Biochem. Cell Biol.* **2008**, *40*, 362–382. [CrossRef] [PubMed]
26. Latta, V.D.; Cecchettini, A.; Ry, S.D.; Morales, M.A. Bleomycin in the setting of lung fibrosis induction: From biological mechanisms to counteractions. *Pharmacol. Res.* **2015**, *97*, 122–130. [CrossRef]
27. Izbicki, G.; Segel, M.J.; Christensen, T.G.; Conner, M.W.; Breuer, R. Time course of bleomycin-induced lung fibrosis. *Int. J. Exp. Pathol.* **2002**, *83*, 111–119. [CrossRef]
28. Fernandez, I.E.; Amarie, O.V.; Mutze, K.; Königshoff, M.; Yildirim, A.Ö.; Eickelberg, O. Systematic phenotyping and correlation of biomarkers with lung function and histology in lung fibrosis. *Am. J. Physiol. Lung Cell Mol. Physiol.* **2016**, *310*, L919–L927. [CrossRef]
29. Selman, M.; Pardo, A. Role of epithelial cells in idiopathic pulmonary fibrosis: From innocent targets to serial killers. *Proc. Am. Thorac. Soc.* **2006**, *3*, 364–372. [CrossRef]
30. Yoshida, K.; Kuwano, K.; Hagimoto, N.; Watanabe, K.; Matsuba, T.; Fujita, M.; Inoshima, I.; Hara, N. MAP kinase activation and apoptosis in lung tissues from patients with idiopathic pulmonary fibrosis. *J. Pathol.* **2002**, *198*, 388–396. [CrossRef]
31. Ventura, J.J.; Tenbaum, S.; Perdiguero, E.; Huth, M.; Guerra, C.; Barbacid, M.; Pasparakis, M.; Nebreda, A.R. p38alpha MAP kinase is essential in lung stem and progenitor cell proliferation and differentiation. *Nat. Genet.* **2007**, *39*, 750–758. [CrossRef] [PubMed]
32. Kolahian, S.; Fernandez, I.E.; Eickelberg, O.; Hartl, D. Immune Mechanisms in Pulmonary Fibrosis. *Am. J. Respir. Cell Mol. Biol.* **2016**, *55*, 309–322. [CrossRef] [PubMed]
33. Bakin, A.V.; Rinehart, C.; Tomlinson, A.K.; Arteaga, C.L. p38 mitogen-activated protein kinase is required for TGFbeta-mediated fibroblastic transdifferentiation and cell migration. *J. Cell Sci.* **2002**, *115*, 3193–3206.
34. Khalil, N.; Xu, Y.D.; O'Connor, R.; Duronio, V. Proliferation of pulmonary interstitial fibroblasts is mediated by transforming growth factor-beta1-induced release of extracellular fibroblast growth factor-2 and phosphorylation of p38 MAPK and JNK. *J. Biol. Chem.* **2005**, *280*, 43000–43009. [CrossRef]
35. Kolosova, I.; Nethery, D.; Kern, J.A. Role of Smad2/3 and p38 MAP kinase in TGF-β1-induced epithelial-mesenchymal transition of pulmonary epithelial cells. *J. Cell. Physiol.* **2011**, *226*, 1248–1254. [CrossRef]
36. Lee, C.G.; Homer, R.J.; Zhu, Z.; Lanone, S.; Wang, X.; Koteliansky, V.; Shipley, J.M.; Gotwals, P.; Noble, P.; Chen, Q.; et al. Interleukin-13 induces tissue fibrosis by selectively stimulating and activating transforming growth factor β1. *J. Exp. Med.* **2001**, *194*, 809–822. [CrossRef]
37. Mi, S.; Li, Z.; Yang, H.Z.; Liu, H.; Wang, J.P.; Ma, Y.G.; Wang, X.X.; Liu, H.Z.; Sun, W.; Hu, Z.W. Blocking IL-17A promotes the resolution of pulmonary inflammation and fibrosis via TGF-beta1-dependent and -independent mechanisms. *J. Immunol.* **2011**, *187*, 3003–3014. [CrossRef]
38. Passalacqua, G.; Mincarini, M.; Colombo, D.; Troisi, G.; Marta Ferrari, M.; Bagnasco, D.; Balbi, F.; Riccio, A.; Canonica, G.W. IL-13 and idiopathic pulmonary fibrosis: Possible links and new therapeutic strategies. *Pulm. Pharmacol. Ther.* **2017**, *45*, 95–100. [CrossRef] [PubMed]
39. Gurczynski, S.J.; Moore, B.B. IL-17 in the lung: The good, the bad, and the ugly. *Am. J. Physiol. Lung Cell Mol. Physiol.* **2018**, *314*, L6–L16. [CrossRef]

40. Nuovo, G.J.; Hagood, J.S.; Magro, C.M.; Chin, N.; Kapil, R.; Davis, L.; Marsh, C.B.; Folcik, V.A. The distribution of immunomodulatory cells in the lungs of patients with idiopathic pulmonary fibrosis. *Mod. Pathol.* **2012**, *25*, 416–433. [CrossRef]
41. Xu, Y.; Mizuno, T.; Sridharan, A.; Du, Y.; Guo, M.; Tang, J.; Wikenheiser-Brokamp, K.A.; Perl, A.T.; Funari, V.A.; Gokey, J.J.; et al. Single-cell RNA sequencing identifies diverse roles of epithelial cells in idiopathic pulmonary fibrosis. *JCI Insight.* **2016**, *1*, e90558. [CrossRef] [PubMed]
42. Korfei, M.; Ruppert, C.; Mahavadi, P.; Henneke, I.; Markart, P.; Koch, M.; Lang, G.; Fink, L.; Bohle, R.M.; Seeger, W.; et al. Epithelial endoplasmic reticulum stress and apoptosis in sporadic idiopathic pulmonary fibrosis. *Am. J. Respir. Crit. Care Med.* **2008**, *178*, 838–846. [CrossRef] [PubMed]
43. Hotamisligil, G.S.; Davis, R.J. Cell signaling and stress responses. *Cold Spring Harb. Perspect. Biol.* **2016**, *8*, a006072. [CrossRef]
44. Liang, J.; Zhang, Y.; Xie, T.; Liu, N.; Chen, H.; Geng, Y.; Kurkciyan, A.; Mena, J.M.; Stripp, B.R.; Jiang, D.; et al. Hyaluronan and TLR4 promote surfactant-protein-C-positive alveolar progenitor cell renewal and prevent severe pulmonary fibrosis in mice. *Nat. Med.* **2016**, *22*, 1285–1293. [CrossRef]
45. Schinzel, R.T.; Higuchi-Sanabria, R.; Shalem, O.; Moehle, E.A.; Webster, B.M.; Joe, L.; Bar-Ziv, R.; Frankino, P.A.; Durieux, J.; Pender, C.; et al. The hyaluronidase, TMEM2, promotes ER homeostasis and longevity independent of the UPR ER. *Cell* **2019**, *179*, 1306–1318.e18. [CrossRef]
46. Liang, J.; Liu, N.; Liu, X.; Mena, J.M.; Xie, T.; Geng, Y.; Huan, C.; Zhang, Y.; Taghavifar, F.; Huang, G.; et al. Mitogen-activated protein kinase-activated protein kinase 2 inhibition attenuates fibroblast invasion and severe lung fibrosis. *Am. J. Respir. Cell Mol. Biol.* **2019**, *60*, 41–48. [CrossRef] [PubMed]
47. Craig, V.J.; Zhang, L.; Hagood, J.S.; Owen, C.A. Matrix metalloproteinases as therapeutic targets for idiopathic pulmonary fibrosis. *Am. J. Respir. Cell Mol. Biol.* **2015**, *53*, 585–600. [CrossRef]
48. Darling, T.K.; Lamb, T.J. Emerging roles for Eph receptors and ephrin ligands in immunity. *Front. Immunol.* **2019**, *10*, 1473. [CrossRef]
49. Habiel, D.M.; Espindola, M.S.; Jones, I.C.; Coelho, A.L.; Stripp, B.; Hogaboam, C.M. CCR10+ epithelial cells from idiopathic pulmonary fibrosis lungs drive remodeling. *JCI Insight.* **2018**, *3*, e122211. [CrossRef]
50. Mohan, T.; Deng, L.; Wang, B.Z. CCL28 chemokine: An anchoring point bridging innate and adaptive immunity. *Int. Immunopharmacol.* **2017**, *51*, 165–170. [CrossRef]
51. Zhou, H.; Brekman, A.; Zuo, W.L.; Ou, X.; Shaykhiev, R.; Agosto-Perez, F.J.; Wang, R.; Walters, M.S.; Salit, J.; Strulovici-Barel, Y.; et al. POU2AF1 Functions in the human airway epithelium to regulate expression of host defense genes. *J. Immunol.* **2016**, *196*, 3159–3167. [CrossRef] [PubMed]
52. Sun, X.; Hou, T.; Cheung, E.; Iu, T.N.; Tam, V.W.; Chu, I.M.; Tsang, M.S.; Chan, P.K.; Lam, C.W.; Wong, C. Anti-inflammatory mechanisms of the novel cytokine interleukin-38 in allergic asthma. *Cell Mol. Immunol.* **2020**, *17*, 631–646. [CrossRef] [PubMed]
53. McDonough, J.E.; Ahangari, F.; Li, Q.; Jain, S.; Verleden, S.E.; Herazo-Maya, J.; Vukmirovic, M.; DeIuliis, G.; Tzouvelekis, A.; Tanabe, N.; et al. Transcriptional regulatory model of fibrosis progression in the human lung. *JCI Insight.* **2019**, *4*, e131597. [CrossRef] [PubMed]
54. Du, Y.; Guo, M.; Whitsett, J.A.; Xu, Y. 'LungGENS': A web-based tool for mapping single-cell gene expression in the developing lung. *Thorax* **2015**, *70*, 1092–1094. [CrossRef] [PubMed]
55. De Vries, M.; Faiz, A.; Woldhuis, R.R.; Postma, D.S.; de Jong, T.V.; Sin, D.D.; Bossé, Y.; Nickle, D.C.; Guryev, V.; Timens, W.; et al. Lung tissue gene-expression signature for the ageing lung in COPD. *Thorax* **2017**. [CrossRef]
56. Brosh, R.; Sarig, R.; Natan, E.B.; Molchadsky, A.; Madar, S.; Bornstein, C.; Buganim, Y.; Shapira, T.; Goldfinger, N.; Paus, R.; et al. p53-dependent transcriptional regulation of EDA2R and its involvement in chemotherapy-induced hair loss. *FEBS Lett.* **2010**, *584*, 2473–2477. [CrossRef]
57. Sisto, M.; Lorusso, L.; Lisi, S. X-linked ectodermal dysplasia receptor (XEDAR) gene silencing prevents caspase-3-mediated apoptosis in Sjögren's syndrome. *Clin. Exp. Med.* **2017**, *17*, 111–119. [CrossRef]
58. Yamada, H.; Yanagisawa, K.; Tokumaru, S.; Taguchi, A.; Nimura, Y.; Osada, H.; Nagino, M.; Takahashi, T. Detailed characterization of a homozygously deleted region corresponding to a candidate tumor suppressor locus at 21q11-21 in human lung cancer. *Genes Chromosomes Cancer* **2008**, *47*, 810–818. [CrossRef]
59. Zhu, Y.X.; Benn, S.; Li, Z.H.; Wei, E.; Masih-Khan, E.; Trieu, Y.; Bali, M.; McGlade, C.J.; Claudio, J.O.; Stewart, A.K. The SH3–SAM adaptor HACS1 is up-regulated in B cell activation signaling cascades. *J. Exp. Med.* **2004**, *200*, 737–747. [CrossRef]

60. DePianto, D.J.; Chandriani, S.; Abbas, A.R.; Jia, G.; N'Diaye, E.N.; Caplazi, P.; Kauder, S.E.; Biswas, S.; Karnik, S.K.; Ha, C.; et al. Heterogeneous gene expression signatures correspond to distinct lung pathologies and biomarkers of disease severity in idiopathic pulmonary fibrosis. *Thorax* **2015**, *70*, 48–56. [CrossRef]
61. Sivakumar, P.; Thompson, J.R.; Ammar, R.; Porteous, M.; McCoubrey, C.; Cantu, E., III; Ravi, K.; Zhang, Y.; Luo, Y.; Streltsov, D.; et al. RNA sequencing of transplant-stage idiopathic pulmonary fibrosis lung reveals unique pathway regulation. *ERJ Open Res.* **2019**, *5*, 00117-2019. [CrossRef] [PubMed]
62. Tanaka, K.; Fujita, T.; Umezawa, H.; Namiki, K.; Yoshioka, K.; Hagihara, M.; Sudo, T.; Kimura, S.; Tatsumi, T.; Kasuya, Y. Therapeutic effect of lung mixed culture-derived epithelial cells on lung fibrosis. *Lab. Investig.* **2014**, *94*, 1247–1259. [CrossRef] [PubMed]
63. Amano, H.; Murata, K.; Matsunaga, H.; Tanaka, K.; Yoshioka, K.; Kobayashi, T.; Ishida, I.; Fukamizu, A.; Sugiyama, F.; Sudo, T.; et al. p38 Mitogen-activated protein kinase accelerates emphysema in mouse model of chronic obstructive pulmonary disease. *J. Recept. Signal Transduct.* **2014**, *34*, 299–306. [CrossRef]
64. Hubner, R.H.; Gitter, W.; El Mokhtari, N.E.; Mathiak, M.; Both, M.; Bolte, H.; Freitag-Wolf, S.; Bewig, B. Standardized quantification of pulmonary fibrosis in histological samples. *Biotechniques* **2008**, *44*, 507–511, 514–517. [CrossRef] [PubMed]
65. Wang, X.; Inoue, S.; Gu, J.; Miyoshi, E.; Noda, K.; Li, W.; Mizuno-Horikawa, Y.; Nakano, M.; Asahi, M.; Takahashi, M.; et al. Dysregulation of TGF-beta1 receptor activation leads to abnormal lung development and emphysema-like phenotype in core fucose-deficient mice. *Proc. Natl. Acad. Sci. USA* **2005**, *102*, 15791–15796. [CrossRef]
66. Maruyama, M.; Sudo, T.; Kasuya, Y.; Shiga, T.; Hu, B.; Osada, H. Immunolocalization of p38 MAP kinase in mouse brain. *Brain Res.* **2000**, *887*, 350–358. [CrossRef]
67. Ohkuro, M.; Kim, J.D.; Kuboi, Y.; Hayashi, Y.; Mizukami, H.; Kobayashi-Kuramochi, H.; Muramoto, K.; Shirato, M.; Michikawa-Tanaka, F.; Moriya, J.; et al. Calreticulin and integrin alpha dissociation induces anti-inflammatory programming in animal models of inflammatory bowel disease. *Nat. Commun.* **2018**, *9*, 1982. [CrossRef]
68. Subramanian, A.; Tamayo, P.; Mootha, V.K.; Mukherjee, S.; Ebert, B.L.; Gillette, M.A.; Paulovich, A.; Pomeroy, S.L.; Golub, T.R.; Lander, E.S.; et al. Gene set enrichment analysis: A knowledge-based approach for interpreting genome-wide expression profiles. *Proc. Natl. Acad. Sci. USA* **2005**, *102*, 15545–15550. [CrossRef]
69. Ge, S.X.; Son, E.W.; Yao, R. iDEP: An integrated web application for differential expression and pathway analysis of RNA-Seq data. *BMC Bioinform.* **2018**, *19*, 534. [CrossRef]
70. Chen, J.; Bardes, E.E.; Aronow, B.J.; Jegga, A.G. ToppGene Suite for gene list enrichment analysis and candidate gene prioritization. *Nucleic Acids Res.* **2009**, *37*, W305–W311. [CrossRef]

International Journal of
Molecular Sciences

Review

p38 MAPK Pathway in the Heart: New Insights in Health and Disease

Rafael Romero-Becerra, Ayelén M. Santamans, Cintia Folgueira and Guadalupe Sabio *

Centro Nacional de Investigaciones Cardiovasculares (CNIC), 28029 Madrid, Spain; rafael.romero@cnic.es (R.R.-B.); ayelenmelina.santamans@cnic.es (A.M.S.); cintia.folgueira@cnic.es (C.F.)
* Correspondence: guadalupe.sabio@cnic.es; Tel.: +34-91453-12-00 (ext. 2004)

Received: 3 August 2020; Accepted: 5 October 2020; Published: 8 October 2020

Abstract: The p38 mitogen-activated kinase (MAPK) family controls cell adaptation to stress stimuli. p38 function has been studied in depth in relation to cardiac development and function. The first isoform demonstrated to play an important role in cardiac development was p38α; however, all p38 family members are now known to collaborate in different aspects of cardiomyocyte differentiation and growth. p38 family members have been proposed to have protective and deleterious actions in the stressed myocardium, with the outcome of their action in part dependent on the model system under study and the identity of the activated p38 family member. Most studies to date have been performed with inhibitors that are not isoform-specific, and, consequently, knowledge remains very limited about how the different p38s control cardiac physiology and respond to cardiac stress. In this review, we summarize the current understanding of the role of the p38 pathway in cardiac physiology and discuss recent advances in the field.

Keywords: MAPK; p38; physiology; metabolism; signaling; hypoxia; arrhythmia

1. Introduction

p38α was identified by three groups in 1994 as a 38 kDa polypeptide that was phosphorylated after exposure to lipopolysaccharide (LPS), hyperosmolarity, or interleukin 1 (IL-1), that directly phosphorylates and activates the upstream kinase MAPKAP-K2 [1–3]. Later, three additional isoforms were described: p38β [4], p38γ (also called SAPK3 and ERK6) [5], and p38δ (also called SAPK4) [6]. The p38 family members are encoded by different genes, located tandemly in two chromosomes. The p38β (*Mapk11*) and p38γ (*Mapk12*) genes are located together on one chromosome (15 in mice and 22 in humans), whereas p38δ (*Mapk13*) and p38α (*Mapk14*) genes are located on another (17 in mice and six in humans). *Mapk12* is proposed to have arisen from tandem duplication of *Mapk11*, and the *Mapk13-Mapk14* gene unit is thought to have originated in a segmental duplication of the *Mapk11-Mapk12* unit [7].

The p38 family can be subdivided into two subsets, with p38α and p38β in one group and p38γ and p38δ in the other. This classification is based partly on amino-acid sequence identity; p38α and p38β are 75% identical, whereas p38γ and p38δ are 62% and 61% identical to p38α, respectively, while sharing 70% sequence identity with each other. The two p38 subsets also differ in their susceptibility to inhibitors, with in vitro and in vivo assays demonstrating that only p38α and p38β are inhibited by pyridinyl imidazoles (SB202190 and SB203580). A third difference between the p38 subgroups is substrate selectivity, with p38γ sharing common substrates with p38δ, and p38α with p38β [8–10].

p38 activity is regulated by phosphorylation at the end of a cascade composed of a MAPK kinase (MKK) and an MKK kinase (MEKK) [11–13]. The cascade is initiated by one of several MKK-phosphorylating MAP3Ks in cell-type- and stimulus-dependent manner. These MAP3Ks include mixed-lineage kinases (MLK), TGF β-activated kinase 1 (TAK1), MAPK/ERK kinase kinases (MEKK),

TAO1 and TAO2, and apoptosis signal-regulating kinase-1 (ASK1) [14]. The p38s are activated by MKK-mediated dual phosphorylation of tyrosine and threonine residues in the conserved Thr–Xaa–Tyr motif (in p38, kinases Xaa is glycine, whereas in JNKs, it is proline in ERKs glutamic acid) [14]. Phosphorylation by MKKs is highly selective due to the specificity of the phosphorylation motif and the interaction of the MKK N-terminal region with different docking sites on the p38s. In addition, in T cells, p38 is activated by autophosphorylation [15] and also through AMPK-TAB1 [16], an alternative pathway that has been shown also in adipose tissue [17].

p38 activation is further tightly regulated by a group of inactivating phosphatases [11,14]. All p38 family members are widely expressed and considered ubiquitous, although p38β is most abundantly expressed in brain and adipose tissue, p38γ in skeletal muscle, and p38δ in secretory glands [5,6,18,19]. While all four p38s are expressed in the heart, the predominant family members in cardiomyocytes are p38α and p38γ. Extensive research into cardiac p38 function has suggested both protective and deleterious roles in the stressed myocardium. Which outcome predominates seems to depend in part on the model system under study and on the identity of the activated p38 family member. However, understanding remains limited of how the different p38 family members control cardiac physiology and respond to cardiac stress. In this review, we summarize current knowledge of p38 function in the heart and discuss recent advances.

2. Cardiovascular Development

In 2000, three groups independently showed that p38α is essential for normal cardiovascular development. Allen M. et al. demonstrated that genetic disruption of the p38α gene *Mapk14* was embryonically lethal [20]. Four months later, Adams R. et al. confirmed the essential requirement for p38α during early mouse development, showing that p38α deletion correlated with a massive reduction in myocardium formation and the appearance of blood-vessel malformations in the head region [21]. These authors suggested that p38α is necessary for placental organogenesis but is not necessary for other aspects of mammalian embryonic development [21]. Mudgett, J. et al. showed that p38α is required for the vascular remodeling associated with placental angiogenesis and trophoblast development [22].

Although p38 has been shown to play a key role in skeletal muscle development [23], less attention has been paid to its role in cardiac development. Several in vitro studies point to a possible role of p38 in cardiac development. For example, p38α activity is required for cardiomyocyte differentiation of P19CL6 cells, which is mediated via the activation of the transcription factor AP-1 [24]. p38α has also been shown to promote cardiogenesis over neurogenesis in ES cells [25]. Unfortunately, despite the strong suggestion of a cardiogenic role of p38α from cell-culture studies, in vivo data supporting this hypothesis are scarce. While embryos lacking p38α die due to defects in placental angiogenesis, cardiac-specific deletion of p38α results in normal development of the heart [26].

Several studies have shown that p38 kinases play an important role in different aspects of cardiogenesis, such as the regulation of cardiomyocyte differentiation and apoptosis. The role of p38 in cardiomyocyte differentiation was first suggested by studies using a specific inhibitor of p38α and p38β (SB203580), which demonstrated that p38 activity regulates important mitotic genes in cardiomyocytes. Neonatal mice lacking p38α have increased cardiomyocyte mitosis, suggesting that p38α acts as a negative regulator of cardiomyocyte proliferation [2]. In adult cardiomyocytes, SB203580 and fibroblast growth factor 1 (FGF1) act synergistically to induce the expression of genes involved in proliferation and regeneration [27,28], indicating that the combination of FGF1 stimulation and p38α inhibition might rescue cardiac structure and function after injury [28]. The importance of p38 in cardiac differentiation was evident by the finding that p38α inhibition or gene deletion were sufficient to block cardiomyogenesis, suggesting that p38α activation constitutes an early switch in embryonic stem cell commitment to cardiomyogenesis [25]. The deletion or inhibition of p38α reduces expression of myocyte enhancer factor 2C (MEF2C), an important transcription factor acting on many genes encoding cardiac structural proteins [29]. The inhibition of p38 correlates with decreases in other cardiac transcription factors and MEF2C targets, such as atrial natriuretic factor (ANF) and myocardin, all of which contribute to the proper activation of the cardiac differentiation program during the early stages of development. p38α also regulates sarcomere assembly through the phosphorylation of ventricular myosin light chain 2 (MLC-2v), as well as the accumulation of α-actinin and its incorporation into sarcomeric units [29]. The lack of MEF2C activation upon p38 inhibition suppresses the expression of bone morphogenetic protein 2 (BMP-2), a key regulator of early cardiac cell development [30]. Most studies of cardiovascular development have focused on p38α; however, it is important to also define the role of other p38 family members in cardiac development. Mice with combined deletion of p38α and p38β display diverse developmental defects at mid-gestation, together with major cardiovascular abnormalities [31]. Embryos that express p38β only under the control of the p38α promoter display a similar heart phenotype as the double-knockout embryos, suggesting that heart development requires endogenous p38β expression [31]. Moreover, p38α and p38β have synergistic roles and specific functions in the regulation of cardiac gene expression during development, suggesting that some specific functions could be explained by differences in expression patterns [31]. It has been demonstrated the selective activation of p38 in the right ventricle during neonatal development and simultaneous inactivation in the left ventricle in neonatal mouse heart [32]. Cardiac-specific deletion of p38α and p38β in mice showed an abnormal gross morphology of the heart, developed right ventricle-specific enlargement dilation and, in consequence, a significant increase in cardiomyocyte proliferation, hypertrophy and a reduction in apoptosis without changes in the left ventricle. Furthermore, p38 inactivation induces XBP1 activity via IREα in the regulation of neonatal cardiomyocyte proliferation [32]. Finally, the role of p38γ and p38δ in cardiomyocyte development have not been assessed; however, animals lacking these kinases have smaller hearts at birth while conserving a normal number of cardiomyocytes, suggesting that these kinases contribute to the control of cardiomyocyte hypertrophy [33].

3. Cardiac Hypertrophy

The role of p38 kinases in cardiac hypertrophy was first suggested by the hypertrophic responses induced upon overexpression in cardiomyocytes of active forms of the upstream activators MKK3 and MMK6 [34–36]. The p38 pathway is also activated in cardiomyocytes exposed to hypertrophic stimuli, and hypertrophic growth is blocked by the SB203580-mediated inhibition of p38α/β [35]. However, MKK3 overexpression in cardiomyocytes also increased apoptosis [34]. The differences between the effects of MKK3 and MKK6 might point to distinct roles of p38 family members, a possibility supported by the finding that p38β activation in cultured cardiomyocytes induces characteristic features of hypertrophy [36], whereas p38α activation promotes cardiomyocyte apoptosis [37]. Consistent with this result, cardiac-specific knockout revealed a critical role for p38α in the cardiomyocyte survival pathway triggered by pressure overload, whereas hypertrophic growth was unaffected [37]. Although some groups have reported a role of p38α in the regulation of cardiac hypertrophy, most of those studies were based on the non-physiological overexpression of dominant-negative p38α or indirect strategies that might alter the function of other p38 family members [38,39].

The cardiomyocyte expression and subcellular localization of p38 isoforms was characterized by Dharmendra Dingar et al. in physiological conditions and in response to chronic pressure overload induced by transverse aorta constriction (TAC) [40]. TAC did not induce changes in the amount of p38α mRNA, whereas p38β and p38δ mRNA increased within 1 day of pressure overload and remained, while p38γ mRNA increased initially before returning to baseline levels by day 7. Despite these increases in mRNA abundance, the overall protein levels of p38β, p38γ, and p38δ were unaltered [40]. Confocal immunofluorescence analysis detected p38α and p38γ in the cytoplasm and nucleus at baseline; however, after chronic pressure overload, p38γ accumulated in the nucleus, whereas p38α distribution was unaffected. These localization differences would result in access to different substrates, and hence distinct functional effects [40].

The lack of pharmacological inhibitors of p38γ and p38δ has limited the study of these family members, though they were recently shown to control physiological and pathological cardiomyocyte growth. p38γ and p38δ are activated by angiotensin II and phosphorylate the mTOR inhibitor DEPTOR, inducing its degradation by the proteasome. Once DEPTOR is degraded, mTOR is released and activated, triggering protein synthesis and cardiomyocyte growth [33]. Mice lacking p38γ and p38δ have smaller hearts than controls and a reduced cardiomyocyte area [33]. Reduced hypertrophy capacity in these mice resulted in partial protection against angiotensin treatment [33]. Further experiments are needed to assess the therapeutic effects of p38γ and p38δ modulation during pathological hypertrophy.

4. Cardiac Regeneration

The best model for studying the molecular mechanisms of cardiac regeneration is the zebrafish, in which injury induces a cardiomyocyte proliferation that can overcome scar formation, thus allowing cardiac muscle regeneration [41]. Little attention has been paid to the role of p38 in zebrafish cardiac regeneration. p38α activation negatively regulates the proliferation of adult zebrafish cardiomyocytes [42], as also occurs in mammals [27]. During heart regeneration in adult zebrafish, the induction of p38α activity blocks cardiomyocyte proliferation, suggesting that p38α activity must be switched off in order to trigger cardiomyocyte proliferation and myocardial regeneration [42].

Mammalian hearts have a very low or non-existent regenerative capacity after cardiac injury. Nevertheless, in principle, signals that acutely trigger cardiomyocyte survival or modulate myoblast activity could be manipulated to promote cardiac regeneration and avoid heart failure. There is evidence implicating FGF1-upregulated genes in cardiac regeneration and cell-cycle control. The inhibition of p38 and stimulation of FGF1 act together to induce the expression of specific genes involved in proliferation and regeneration, such as cytokinesis regulator Ect2, cell-cycle-regulated protein 1 (CRP1), Ki67, cdc2, cyclin A, and the cell-cycle inhibitor p27 [27] (Figure 1A). Moreover, p38 inhibition and FGF1 induction lead to the phosphorylation of the key cell-cycle regulator Rb. These findings suggest that the promotion of cardiomyocyte proliferation by combined treatment with FGF1 and a p38 inhibitor could provide an alternative approach to the rescue of cardiac function after injury [28]. Studies of p38 isoforms in muscle regeneration have shown that p38β, p38γ, and p38δ are not required for efficient adult muscle regeneration and growth after injury [43], suggesting that p38α, in the absence of the other isoforms, especially p38γ, is sufficient to maintain satellite-cell-mediated myogenesis in vivo and in vitro [43].

The actions of p38 kinases in skeletal muscle regeneration have received much more attention than their roles in the heart. However, given the similarities between cardiac and skeletal muscle, the results of skeletal muscle studies can shed valuable light on the role of p38 in cardiac regeneration. Much research has focused on the potential role in adult myogenesis of p38γ, which is highly expressed in skeletal muscle. p38γ-deficient mice have a low muscle regeneration capacity after injury, with a reduced number of satellite cells that express myogenin prematurely and proliferate poorly. Mechanistically, p38γ phosphorylates MyoD, enhancing its occupancy of the myogenin promoter and thereby suppressing its expression. p38γ thus acts in opposition to p38α, blocking premature differentiation by inducing a repressive MyoD transcriptional complex during satellite-cell–mediated muscle growth and muscle regeneration [44] (Figure 1B). MyoD activates proliferation-associated genes but not differentiation genes, whose regulatory regions are repressed by ZEB1. Macrophages present in the injured muscles of Zeb1-deficient mice have low phosphorylation levels of p38, and forced p38 activation alleviates muscle damage and improves muscle regeneration [45]. In a later study, Brien P et al. demonstrated that a lack of p38α results in increased p38γ activation [46], suggesting that p38γ hyperactivation is involved in muscle regeneration.

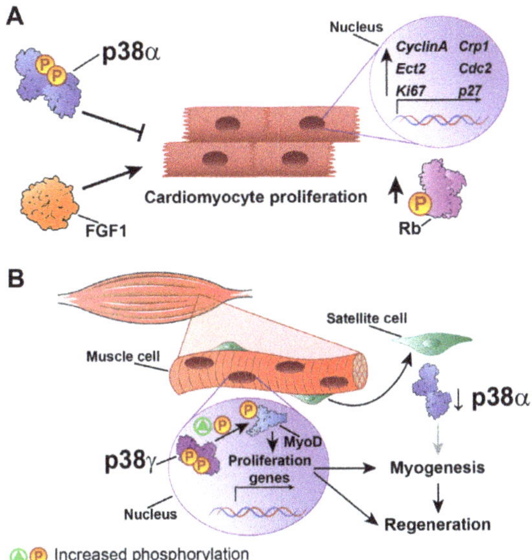

Figure 1. p38 in cardiovascular regeneration. **(A)** p38α blocks cardiovascular regeneration by inhibiting the expression of genes involved in cardiomyocyte proliferation and regeneration, such as *Ect2, Crp1, ki67, cdc2, cyclin A*, and p27, and reducing Rb phosphorylation to block cell-cycle progression. FGF1 stimulation has the opposite effect. **(B)** p38γ activates differentiation and myogenesis in satellite cells by phosphorylating MyoD and activating proliferation. p38α prevents p38γ activation in satellite cells, blocking regeneration. ↑ increase, ↓ decrease.

Aging is characterized by a general decline in metabolic activity and function, with a loss of both skeletal and cardiac muscle accompanied by marked functional and structural impairment. Skeletal muscle has an outstanding regenerative capacity provided by its resident satellite cells. These normally quiescent cells are activated after injury to promote skeletal muscle regeneration [47]. Cardiomyocytes and cardiac muscle have long been thought to have lost these satellite cells and appear to lack the capacity for self-renewal and repair, preventing full recovery. Given the fundamental role of p38 in skeletal muscle regeneration, it is reasonable to postulate a causal link between cardiac muscle loss-of-function and p38.

To identify potential regeneration strategies based on satellite cells, a number of studies have investigated the effect of manipulating p38 signaling on aged satellite cells. For example, the manipulation of satellite cells enhances fibroblast growth factor receptor 1 (FGFR1) signaling and reduces p38α/β activation in satellite cells, increasing self-renewal and stimulating skeletal muscle regeneration [48]. Another study improved skeletal muscle repair and regeneration in rats using gold and gold-silver nanoparticles (AuNPs and Au-AgNPs, respectively) [49]. These nanoparticles regulate MyoD gene expression and activate p38α signaling, enhancing myoblast myogenic differentiation and promoting skeletal muscle regeneration. The effects of AuNPs and AuAgNPs were blocked with SB203580, suggesting that p38α is essential for myogenic differentiation [49]. These studies indicate an important role for p38 signaling in muscle regeneration; however, the underlying molecular mechanisms need further investigation to define the antagonic roles between p38α and p38γ. It will also be important to determine whether this antagonistic relationship operates in cardiac muscle.

5. p38 in Ischemia–Reperfusion Injury

Ischemic heart disease, the leading cause of death worldwide, is normally produced by a coronary artery occlusion that impairs cardiac blood flow. The ensuing decrease in oxygen delivery can lead to myocardial infarction. Reperfusion consists of the restoration of blood flow after the ischemia, and although this step is essential to avoid cardiomyocyte death, it also causes further damage associated with increased oxidative stress and inflammation [50,51]. Short ischemic episodes have been shown to protect the heart from a later ischemic insult, a process known as preconditioning or postconditioning, depending on whether it happens before or after the ischemia [52].

The p38 pathway is activated in response to ischemia–reperfusion and during preconditioning, producing a variety of results and cardiovascular scenarios [53]. In perfused rat hearts, p38 activation during ischemia and reperfusion is associated with a poor cardiac outcome. In contrast, during repetitive preconditioning treatments, p38 is maximally activated in the first episode, and activation gradually is reduced during sustained ischemia–reperfusion, improving cardiac functional recovery [53]. Interestingly, p38 activation is compartmentalized, with ischemia activating p38 in mitochondria, whereas during reperfusion p38 is activated in all cell compartments [54]. It would be interesting to determine whether different p38 family members are activated and localized in different cell compartments during these processes. Experiments performed in PC12 cells showed that moderate hypoxia (5% O_2) increases p38γ and p38α phosphorylation, suggesting that ischemia might activate more than one p38 family member [55]. Most studies have examined p38α and β, both of which are inhibited by the widely used inhibitor SB203580. There is evidence to suggest that these two kinases have opposite roles, with p38α activation during ischemia triggering apoptosis, whereas p38β is responsible for pro-survival signaling during preconditioning [56]. Therefore, SB203580-mediated blockade of both isoforms during preconditioning results in loss of cardioprotective effects, whereas inhibition during I/R is beneficial [53].

The activation of p38 during preconditioning seems to be a consequence of adenosine release, which triggers the opening of ATP-sensitive potassium channels (KATP) during hypoxia [57,58]. The cardioprotective effects of p38 are thought to be due to phosphorylation of the downstream target Hsp27 and the subsequent enhancement of cytoskeletal stabilization during hypoxic stress [59]. In line with this idea, reactive oxygen species (ROS)-induced activation of p38α during hypoxia stabilizes hypoxia-inducible factor 1 (HIF-1) [60]. Moreover, the lack of dual specificity protein phosphatase 4 (DUSP4) leads to p38 hyper-phosphorylation and apoptosis [61]. Preclinical and clinical studies [62–64] with antioxidants highlight the importance of ROS as mediators of cardiac stress injury during I/R. ROS are potent p38 activators, but in vitro results with HL-1 cardiomyocytes also showed that p38 activation drives elevated ROS levels during ischemia–reperfusion [65]. In this analysis, the damaging effects of ROS were abolished by the p38 pan inhibitor BIRB796 [65]. Moreover, p38 inhibition with an antioxidant during ischemia–reperfusion is associated with improved cardiac recovery, decreased infarct size, and reduced apoptosis [66]. This was related to increased endogenous anti-oxidative enzyme activity and inhibition of oxidative stress [67]. The antioxidant Peroxiredoxin 1 and the ROS scavenger N-acetyl-l-cysteine have been also shown to decrease oxidative stress and block the activation of p38 and JNK, thus reducing apoptosis during ischemia–reperfusion [68]. Moreover, oxidative stress can directly regulate p38α activity and protein interactions by affecting the oxidation state of cysteines. Treatment with H_2O_2 induced p38-MKK3 disulfide dimer formation in isolated rat hearts and in an ischemia–reperfusion model, and dimer formation was abolished when the redox-sensitive cysteines were mutated or sterically inaccessible [69]. Further research is needed to determine whether p38 activation triggers ROS production, as proposed by Ashraf et al. [53,65], whether ROS induce p38 activation, or, more likely, there is reciprocal ROS–p38 regulation.

p38 has also been linked to cardiac inflammation through its promotion of the expression of regenerating islet-derived 3γ (Reg3γ), a protein associated with cardiac inflammatory signaling [70]. Moreover, treatment with the anti-inflammatory compound gamboge protects against infarction-induced inflammation by targeting the NF-κB–p38 pathway [71]. p38α activation is also associated with increased fibrosis during ischemia-induced cardiac remodeling [72]. Supporting this idea, inhibition of the p38 substrate MK2 impairs fibrotic scar formation after myocardial infarction [73]. The profibrotic effect of p38α was confirmed by conditional p38α deletion in myofibroblasts, which demonstrated that a lack of p38α blocks cardiac fibroblast differentiation into myofibroblasts, reducing fibrosis in response to ischemic injury [74]. p38 pathway activation by MKK6 overexpression results in interstitial and perivascular cardiac fibrosis [74], and p38 inhibition may underlie the beneficial effects of some statins on cardiac remodeling after myocardial infarction [75,76]. Supporting this idea, protease inhibitors induce cardioprotection in models of ischemia–reperfusion, in part by attenuating p38 phosphorylation, leading to reductions in injury, ROS levels, and infarct size [77].

The ischemia–reperfusion injury response is crucially determined by mitochondrial function and activity because mitochondria control cell metabolic status, intracellular calcium influx, oxidative stress, and apoptotic pathways, among other processes. p38 inhibition during ischemia–reperfusion decreases mitochondrial swelling, protects against ultra-structure alterations, and mitigates mitochondrial membrane depolarization [78]. There is also evidence that p38 activation during ischemia–reperfusion contributes to cardiac damage by triggering intracellular Ca^{2+} overload [79]. Pharmacological inhibition of p38 during ischemia–reperfusion induces cardioprotection by promoting phospholamban phosphorylation, increasing the activity of sarcoplasmic reticulum Ca^{2+}-ATPase (SERCA2), and decreasing Ca^{2+} overload [80]. The role of p38 in the control of intracellular Ca^{2+} was corroborated in H9c2 cells: phosphodiesterase-inhibitor–mediated reduction in ERK1/2, JNK, and p38 activation reduced ischemia-induced apoptosis and restored normal calcium influx, oxidative stress levels, and eNOS expression [81]. p38 inhibition also potentiates the metformin-induced reduction in myocardial ischemia–reperfusion injury in non-obese type 2 diabetic rats [82].

Ischemia–reperfusion injury also increases the risk of arrhythmia. p38 inhibition with SB203580 decreases ventricular tachycardia and ventricular fibrillation when administered to adult Wistar rats before or during ischemia, but not at the onset of reperfusion [83]. Mechanistically, this could be due to the ROS-dependent activation of p38 by ASK1 [84]. ASK1 would respond to the moderate increase in ROS during ischemia, but not to the higher levels of ROS observed in ischemia/reperfusion, acting as a redox sensor to mediate ROS-dependent signaling to p38 [84]. These findings highlight the importance of determining the optimal timing of p38 inhibition in order to achieve an efficient therapeutic response (Figure 2).

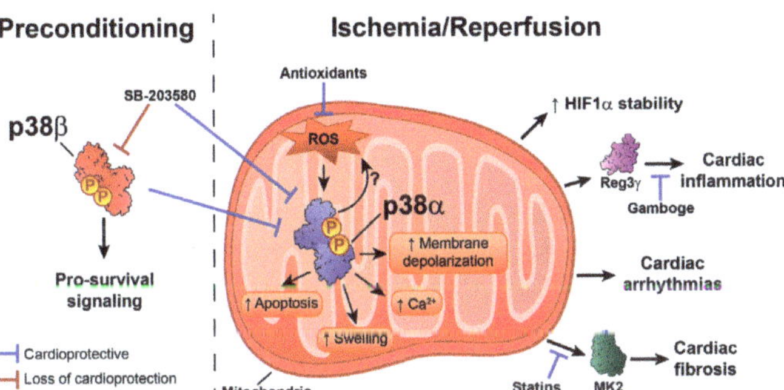

Figure 2. Dual role of p38 activation during preconditioning and ischemia–reperfusion injury. Activation of p38β during preconditioning triggers pro-survival signaling pathways, whereas decreased p38α activation during the ischemic episode leads to cardioprotection. On the other hand, ROS-induced p38α activation during the ischemic insult triggers HIF1-α stabilization; increases (↑) fibrosis, arrhythmias, and inflammation; and disrupts mitochondrial homeostasis. SB203580 administration during preconditioning increases myocardial injury, whereas administration during ischemia–reperfusion improves cardiac outcome. Indirect p38 downregulators, such as gamboge, statins, and antioxidants, seem to have beneficial effects when administered during or after the ischemia. Further research is needed to determine the precise reciprocity of ROS–p38 regulation. ↑ increase.

6. p38 in Heart Failure and Cardiac Arrhythmia

Heart failure (HF) is a major cardiac pathology and a global pandemic that affects more than 37 million people worldwide [85,86]. The p38 pathway is activated in HF, and specifically in the pathological cardiac remodeling that can lead to cardiac arrhythmia in the failing heart [87–90]. p38 plays an important role in the regulation of cardiac remodeling and cardiac contractility. Most studies suggest a negative role of p38 activation in extracellular matrix remodeling and the development of cardiac fibrosis, processes related to the development of HF [91–93]. Studies using transgenic animals with cardiac-specific expression of the activated p38 upstream kinases MKK3bE and MKK6bE showed that p38 pathway activation promotes cardiac interstitial fibrosis and increased expression of embryonic gene markers, similar to the expression profile observed in HF [89,94]. The profibrotic effect of p38 activation may be due to the induction in cardiomyocytes of TNF-α and IL-6 [89], which are closely associated with the development of fibrosis, adverse cardiac remodeling, and HF [95,96]. The effect of p38 activation on cardiac fibrosis is not limited to cardiomyocytes and also affects cardiac fibroblasts. The specific activation of the p38 pathway in cardiac fibroblasts leads to maladaptive cardiac remodeling with a profibrotic and hypertrophic phenotype and the activation of TGF-β signaling [97], a key cytokine involved in cardiac fibrosis and HF [95,96,98,99]. p38 is also necessary for the differentiation of fibroblasts into myofibroblasts, and specific deletion of p38α in cardiac fibroblasts or myofibroblasts reduces cardiac fibrosis in response to cardiac injury [74]. This is consistent with results showing that p38 inhibition decreases cardiac fibrosis and pro-inflammatory cytokine production [74,90], suggesting that p38 blockade is a possible treatment in HF. However, mice with cardiac-specific p38α deletion have a worse outcome to TAC-induced pressure overload, characterized by extensive cardiac fibrosis, dysfunction, and dilatation [37]. These opposite results might indicate that another family member is responsible for the protective effects of inhibitors or that p38 has opposite roles in cardiomyocytes versus other cardiac cells. Moreover, by promoting non-specific phosphorylations, overexpression of activated kinases may produce artificial cardiac structural and functional phenotypes. Further research is needed to determine the specific roles played by the different p38 family members in cardiac fibrosis and HF, since most studies have focused on p38α or the p38 pathway in general.

p38 can also control cardiomyocyte contractility, a predominant target of therapeutic strategies to treat HF [100]. The existing evidence indicates that p38 activation has an anti-inotropic effect in cardiac muscle [101–106]. Different mechanisms have been proposed for the negative effect of p38 on cardiac contractility. For example, p38 has been proposed to mediate the anti-inotropic effects of angiotensin II and ROS, which are also increased during HF, desensitizing the response of myofilaments to Ca^{2+} [102,105,107]. The mechanism underlying p38-mediated dampening of Ca^{2+} responsiveness is unknown, but two main possibilities have been proposed: modification of intracellular pH or phosphorylation of contractile proteins. However, studies have disproved the involvement of pH modification in myofilament sensitivity to Ca^{2+} [105,107], leaving phosphorylation of contractile proteins as the more likely mechanism. Analysis by Liao et al. did not detect increased p38-mediated troponin I phosphorylation, which is known to reduce myofilament responsiveness to Ca^{2+} [102]. Later work by Vahebi et al. showed that, rather than phosphorylation, p38 activation promotes the dephosphorylation of α-tropomyosin and troponin I. This was accompanied by a depression of cardiac and myofilament function and a decrease in maximum ATPase activity [106]. In agreement with this finding, p38 inhibition was found to promote troponin I phosphorylation [108]. p38-mediated dephosphorylation of α-tropomyosin and troponin I appears to be mediated by the protein phosphatases PP2C-α and PP2C-β, since p38, PP2C-α, and PP2C-β were found in the same protein complex in the sarcomere [106].

Another mechanism through which p38 might affect cardiac contractile function is the regulation of proteins involved in cardiomyocyte Ca^{2+} handling. During cardiac contraction, a depolarizing action potential promotes Ca^{2+} release from the sarcoplasmic reticulum, a process known as excitation-contraction coupling. Ca^{2+} enters the cell via L-type Ca^{2+} channels and, in much lower amounts, via the Na^+/Ca^{2+} exchanger (NCX) [109]. This Ca^{2+} activates further sarcoplasmic reticulum Ca^{2+} release via the

Ca^{2+}-triggered Ca^{2+} release channel ryanodine receptor (RyR). The incoming cytosolic Ca^{2+} binds to the thin-filament protein troponin C, initiating myocyte contraction. For relaxation, calcium is removed from the cytosol by the action of Ca^{2+} transporters, mainly the sarcoplasmic/endoplasmic reticulum Ca^{2+}-ATPase 2 (SERCA2) and sarcolemmal NCX [109]. Human HF is characterized by the reduced expression or activity of SERCA2 [110], resulting in increased diastolic Ca^{2+} concentrations due to defects in cytosolic Ca^{2+} removal, leading to reduced contractile force and impaired relaxation [110]. SERCA2 expression and activity are affected by p38 activation. In rat cardiomyocytes, p38 activation reduces SERCA2 mRNA expression and protein levels and reduces the activity of the SERCA2 gene promoter [111,112]. Scharf et al. showed that MK2/3 deletion results in increased protein levels and activity of SERCA2 in the heart and that SERCA2 gene expression is regulated by MK2-dependent Egr-1 transcription factor expression and promoter binding [113]. The p38 pathway regulates not only SERCA2 expression, but also its activity. For example, p38 inhibition increases the inotropic effect of endothelin-1 (ET-1) by modifying the SERCA2 inhibitory protein phospholamban (PLN). ET-1 treatment induces p38 phosphorylation [110,114] and promotes PLN phosphorylation at Ser-16 in the presence of p38 inhibition [80]. Similar results were obtained in a model of ischemia–reperfusion, which activated p38 and reduced SERCA2 expression and activity, as well as PLN phosphorylation, whereas these effects were partially reversed by p38 inhibition [113]. In MK2/3 double knockout mice, PLN phosphorylation is increased in the heart at the Ca^{2+}/calmodulin-dependent protein kinase II (CaMKII) site (Thr-17) [113]. In the dephosphorylated state, PLN binds to and inhibits SERCA2. Phosphorylation at Ser-16 or Thr-17 relieves this inhibition, promoting Ca^{2+} reuptake to the sarcoplasmic reticulum and thus increasing cardiac contractility [115]. By promoting PLN dephosphorylation, p38 activation can therefore limit cardiac contractility. Although the main phosphatase responsible for PLN dephosphorylation is protein phosphatase 1 (PP1), PLN dephosphorylation can also be mediated by protein phosphatase 2A (PP2A) [116,117]. p38 activation has been reported to promote PP2A translocation and activation in myocytes [118], and p38 inhibition reduces PP2A-mediated dephosphorylation of PLN [108,118]. Furthermore, PP2A activation could be the mechanism by which p38 activation inhibits the β-adrenergic receptor-mediated contractile response in cardiomyoctes [104,118]. Kaikkonen et al. also proposed that decreased PP2A activity upon p38 inhibition could also promote PP1 inhibition, and that the p38 isoform responsible for these effects on SERCA2 regulation would likely be p38α [108]. Given that p38 also affects the activity of PP2C-α and PP2C-β, is feasible that p38 also affects the activity of PP2Ce, a novel member of the PP2C family that has been reported to be a specific and potent PLN phosphatase [116].

p38α also appears to participate in α-adrenergic-mediated *Ncx1* gene upregulation [119]. The overexpression of active MKK3 and MKK6 was sufficient to induce NCX1 upregulation in isolated cardiomyocytes, and this effect was mediated primarily by p38α [119,120]. Additionally, chemical inhibition of NCX1 promotes the formation of an NCX1-p38 complex and p38 activation. p38 activation induced by NCX1 inhibition has been suggested to be a physiological mechanism to compensate for loss of NCX1 activity by promoting *Ncx1* gene expression [121]. NCX1 activity and expression is also increased in heart failure [110] and the increased NCX1 activity increases Ca^{2+} extrusion to preserve the reduced diastolic Ca^{2+}. This may compensate in part for the reduced SERCA2 function [110]; however, it can also promote other negative effects on cardiac contractility. For example, high NCX1 activity increases Ca^{2+} release from the cell, reducing sarcoplasmic reticulum Ca^{2+} stores and inducing contractile dysfunction. The elevated translocation of Ca^{2+} across the plasma membrane also results in a higher risk of delayed afterdepolarizations, which can cause arrhythmia and sudden death [110,122]. By reducing SERCA2 and increasing NCX1 function, p38 would seem to play an important role in the development of cardiac arrhythmia. Indeed, p38 activation has been linked to arrhythmogenic cardiac and ionic channel remodeling [83,87,88,123–125]. The pharmacological inhibition of p38 reduces the incidence of arrhythmia after ischemia–reperfusion by increasing the levels of phosphorylated connexin 43 (Cx43) [83]. Connexin clusters in the plasma membrane form gap junctions, which regulate cell-to-cell electrical and metabolic coupling and are essential for normal cardiac impulse transmission [126]. Cx43 is the most abundantly expressed connexin in cardiac myocytes, and its alteration has been linked to increased susceptibility to cardiac arrhythmia by altering action potential propagation in the heart [126,127]. There is strong evidence indicating a role for p38 in the regulation of Cx43. Several processes are involved in Cx43 regulation, including synthesis/degradation, phosphorylation/dephosphorylation at different residues, and cell membrane localization [128]. Cx43 expression is increased by p38 activation induced by several stimuli [129–133], but p38 is also implicated in Cx43 degradation [134]. The effect of p38 activation on Cx43 expression might depend on the activating stimulus and the cell type being studied. In cardiomyocytes, p38 activation appears to promote an increase in Cx43 mRNA and protein expression [129–133]. p38 can form a complex with Cx43 [135,136], and its activation has been reported to promote both Cx43 phosphorylation [135,137,138] and dephosphorylation (via PP2A activity) [136]. Adding further complexity, the increased phosphorylation of Cx43 upon p38 activation has been suggested to promote Cx43 degradation [137]. More research is needed to clarify the role of p38 activation in the regulation of Cx43 and gap junctions. However, despite the varying effects of p38 activation on Cx43 phosphorylation, the outcome of p38 activation is consistent across studies. p38-induced phosphorylation and dephosphorylation of Cx43 both lead to reduced cell-to-cell communication, impaired propagation of the action potential, and the development of cardiac arrhythmia [136–138]. Furthermore p38, inhibition improves cell-to-cell communication and reduces the incidence of arrhythmia [83,138].

Many aspects of the role of p38 in the development of HF and cardiac arrhythmia remain to be clarified; however, most of the evidence points to a negative effect of p38 activation on the onset of HF and arrhythmias. The mechanisms involved include the development of cardiac fibrosis, alterations to Ca^{2+} handling proteins, and the modulation of gap junctions in the cardiomyocyte (Figure 3). Future research will need to address the role of the different p38 family members in these processes, since most studies have focused on p38α or p38α/β.

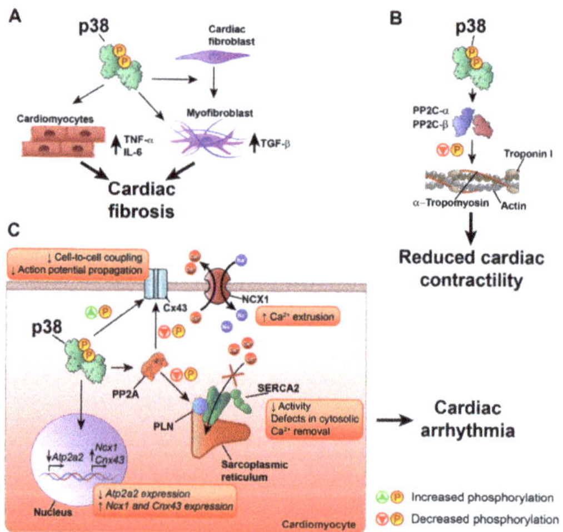

Figure 3. p38 in heart failure and cardiac arrhythmia. p38 activation participates in the development of heart failure and cardiac arrhythmia through three main mechanisms: **(A)** Increased cardiac fibrosis by induction of TNF-α and IL-6 in cardiomyocytes, differentiation of fibroblasts, and induction of TGF-β in cardiac myofibroblasts; **(B)** Reduced cardiac contractility due to dephosphorylation of a-tropomyosin and troponin I via PP2C-α/PP2C-β; **(C)** Promotion of cardiac arrhythmias due to reduced expression and activity of SERCA2 (*Atp2a2*), increased expression of NCX1 (*Ncx1*) and Cx43 (*Cnx43*), and altered Cx43 phosphorylation, inducing cardiac contractile dysfunction and altered action potential propagation. ↑ increase, ↓ decrease.

7. p38 Inhibitors in Clinical Trials

Despite the abundance of experimental evidence for the potential benefits of p38 inhibitors, clinical trials have failed to show improved cardiac outcomes after ischemia–reperfusion. The new anti-inflammatory medication losmapimod inhibits p38, and its administration to patients with non-ST-segment elevation myocardial infarction was well tolerated and improved the cardiac outcome [139]. However, in another study in acute myocardial infarction patients, losmapimod did not reduce the risk of major ischemic cardiovascular events, resulting in the withdrawal of the clinical trial [140].

Alternative approaches to p38 inhibition have been suggested in order to avoid undesirable side effects. The inhibition of MK2 in activated rheumatoid arthritis fibroblast-like synoviocytes avoided the modification of the secretion of chemokines like TNF-alpha, normally associated with the activation of other pro-inflammatory pathways like ERK and JNK during direct p38 inhibition [141]. Moreover, $MK2^{-/-}$ mice showed improved ventricular recovery after ischemia–reperfusion, as well as reductions in infarct size and apoptosis [142]. MK2 inhibition with MMI-0100 after acute myocardial infarction inhibited cardiac fibrosis by enhancing primary cardiac fibroblast cell death while inhibiting cardiomyocyte apoptosis [73].

The lack of efficacy of p38 inhibitors in clinical trials might be due to the high similarity among p38 family members. The lack of isoform-specific inhibitors promotes the development of toxic secondary effects and probably leads to the triggering of regulatory feedback loops. Moreover, most studies used non-specific inhibitors, and very few studies have examined the specific p38 family member involved in a given action. Given the broad spectrum of undesired effects associated with the administration of inhibitors, the results obtained to date are hard to interpret. Further research with transgenic animal models will help to define the complex roles of p38 kinases.

Funding: This research was funded by the following grants to G.S: European Union Seventh Framework Programme (FP7/2007-2013) under grant agreement n° ERC 260464, EFSD/Lilly European Diabetes Research Programme Dr Sabio, 2017 Leonardo Grant for Researchers and Cultural Creators, BBVA Foundation (Investigadores-BBVA-2017) IN [17]_BBM_BAS_0066, MINECO-FEDER SAF2016-79126-R, PID2019-104399RB-I00, EUIN2017-85875, Comunidad de Madrid IMMUNOTHERCAN-CM S2010/BMD-2326 and B2017/BMD-3733, Fundación AECC and intramural grant 12-2016 IGP. R.R-B and A.M.S are fellows of the FPU (FPU17/03847) and FPI Severo Ochoa CNIC program (BES-2016-077635), respectively. C.F has a Sara Borrell contract (CD19/00078). The CNIC is supported by the Instituto de Salud Carlos III (ISCIII), the Ministerio de Ciencia, Innovación y Universidades (MCNU), and the Pro CNIC Foundation and is a Severo Ochoa Center of Excellence (SEV-2015-0505).

Acknowledgments: We thank S. Bartlett for English editing. We thank our collaborators and the students and fellows who contributed to the studies in the Sabio´s group over the years. We regret the inadvertent omission of important references due to space limitations.

Conflicts of Interest: The authors declare no conflict of interest.

References

1. Han, J.; Lee, J.D.; Bibbs, L.; Ulevitch, R.J. A MAP kinase targeted by endotoxin and hyperosmolarity in mammalian cells. *Science* **1994**, *265*, 80888–80911. [CrossRef] [PubMed]
2. Lee, J.C.; Laydon, J.T.; McDonnell, P.C.; Gallagher, T.F.; Kumar, S.; Green, D.; McNulty, D.; Blumenthal, M.J.; Keys, J.R.; Vatter, S.W.L.; et al. A protein kinase involved in the regulation of inflammatory cytokine biosynthesis. *Nature* **1994**, *372*, 73977–74046. [CrossRef] [PubMed]
3. Rouse, J.; Cohen, P.; Trigon, S.; Morange, M.; Alonso-Llamazares, A.; Zamanillo, D.; Hunt, T.; Nebreda, A.R. A novel kinase cascade triggered by stress and heat shock that stimulates MAPKAP kinase-2 and phosphorylation of the small heat shock proteins. *Cell* **1994**, *78*, 10271–11037. [CrossRef]
4. Jiang, Y.; Chen, C.; Li, Z.; Guo, W.; A Gegner, J.; Lin, S.; Han, J. Characterization of the structure and function of a new mitogen-activated protein kinase (p38beta). *J. Biol. Chem.* **1996**, *271*, 17920–17926. [CrossRef] [PubMed]
5. Li, Z.; Jiang, Y.; Ulevitch, R.J.; Han, J. The primary structure of p38 gamma: A new member of p38 group of MAP kinases. *Biochem. Biophys. Res. Commun.* **1996**, *228*, 3340–3343. [CrossRef] [PubMed]

6. Jiang, Y.; Gram, H.; Zhao, M.; New, L.; Gu, J.; Feng, L.; Di Padova, F.; Ulevitch, R.J.; Han, J. Characterization of the structure and function of the fourth member of p38 group mitogen-activated protein kinases, p38delta. *J. Biol. Chem.* **1997**, *272*, 30122–30128. [CrossRef]
7. Li, M.; Liu, J.; Zhang, C. Evolutionary history of the vertebrate mitogen activated protein kinases family. *PLoS ONE* **2011**, *6*, e26999. [CrossRef] [PubMed]
8. Hasegawa, M.; Cuenda, A.; Spillantini, M.G.; Thomas, G.M.; Buée-Scherrer, V.; Cohen, P.; Goedert, M. Stress-activated protein kinase-3 interacts with the PDZ domain of alpha1-syntrophin. A mechanism for specific substrate recognition. *J. Boil. Chem.* **1999**, *274*, 12626–12631. [CrossRef]
9. Sabio, G.; Arthur, J.S.C.; Kuma, Y.; Peggie, M.; Carr, J.; Murray-Tait, V.; Centeno, F.; Goedert, M.; A Morrice, N.; Cuenda, A. p38γ regulates the localisation of SAP97 in the cytoskeleton by modulating its interaction with GKAP. *EMBO J.* **2005**, *24*, 1134–1145. [CrossRef]
10. Sabio, G.; Reuver, S.; Feijoo, C.; Hasegawa, M.; Thomas, G.M.; Centeno, F.; Kuhlendahl, S.; Leal-Ortiz, S.; Goedert, M.; Garner, C.; et al. Stress- and mitogen-induced phosphorylation of the synapse-associated protein SAP90/PSD-95 by activation of SAPK3/p38gamma and ERK1/ERK2. *Biochem. J.* **2004**, *380*, 19–30. [CrossRef]
11. Chang, L.; Karin, M. Mammalian MAP kinase signalling cascades. *Nat. Cell Biol.* **2001**, *410*, 37–40. [CrossRef] [PubMed]
12. Cuadrado, A.; Nebreda, A.R. Mechanisms and functions of p38 MAPK signalling. *Biochem. J.* **2010**, *429*, 403–417. [CrossRef] [PubMed]
13. Arthur, J.S.C.; Ley, S.C. Mitogen-activated protein kinases in innate immunity. *Nat. Rev. Immunol.* **2013**, *13*, 679–692. [CrossRef] [PubMed]
14. Sabio, G.; Davis, R.J. TNF and MAP kinase signalling pathways. *Semin. Immunol.* **2014**, *26*, 237–245. [CrossRef]
15. Salvador, J.M.; Mittelstadt, P.R.; Guszczynski, T.; Copeland, T.D.; Yamaguchi, H.; Appella, E.; Fornace, A.J.; Ashwell, J.D. Alternative p38 activation pathway mediated by T cell receptor–proximal tyrosine kinases. *Nat. Immunol.* **2005**, *6*, 390–395. [CrossRef]
16. Lanna, A.; Henson, S.M.; Escors, D.; Akbar, A.N. The kinase p38 activated by the metabolic regulator AMPK and scaffold TAB1 drives the senescence of human T cells. *Nat. Immunol.* **2014**, *15*, 965–972. [CrossRef]
17. Matesanz, N.; Bernardo, E.; Acín-Pérez, R.; Manieri, E.; Pérez-Sieira, S.; Hernández-Cosido, L.; Montalvo-Romeral, V.; Mora, A.; Rodriguez, É.; Leiva-Vega, L.; et al. MKK6 controls T3-mediated browning of white adipose tissue. *Nat. Commun.* **2017**, *8*, 856. [CrossRef]
18. Ittner, A.; Block, H.; Reichel, C.A.; Varjosalo, M.; Gehart, H.; Sumara, G.; Gstaiger, M.; Krombach, F.; Zarbock, A.; Ricci, R. Regulation of PTEN activity by p38δ-PKD1 signaling in neutrophils confers inflammatory responses in the lung. *J. Exp. Med.* **2012**, *209*, 2229–2246. [CrossRef]
19. Uhlén, M.; Fagerberg, L.; Hallström, B.M.; Lindskog, C.; Oksvold, P.; Mardinoglu, A.; Sivertsson, Å.; Kampf, C.; Sjöstedt, E.; Asplund, A.; et al. Tissue-based map of the human proteome. *Science* **2015**, *347*, 1260419. [CrossRef]
20. Allen, M.; Svensson, L.; Roach, M.; Hambor, J.; McNeish, J.; Gabel, C.A. Deficiency of the Stress Kinase P38α Results in Embryonic Lethality. *J. Exp. Med.* **2000**, *191*, 859–870. [CrossRef]
21. Adams, R.H.; Porras, A.; Alonso, G.; Jones, M.; Vintersten, K.; Panelli, S.; Valladares, A.; Perez, L.; Klein, R.; Nebreda, A.R. Essential role of p38alpha MAP kinase in placental but not embryonic cardiovascular development. *Mol. Cell* **2000**, *6*, 109–116. [CrossRef]
22. Mudgett, J.S.; Ding, J.; Guh-Siesel, L.; Chartrain, N.A.; Yang, L.; Gopal, S.; Shen, M.M. Essential role for p38alpha mitogen-activated protein kinase in placental angiogenesis. *Proc. Natl. Acad. Sci. USA* **2000**, *97*, 10454–10459. [CrossRef] [PubMed]
23. Keren, A.; Tamir, Y.; Bengal, E. The p38 MAPK signaling pathway: A major regulator of skeletal muscle development. *Mol. Cell. Endocrinol.* **2006**, *252*, 224–230. [CrossRef] [PubMed]
24. Eriksson, M.; Leppä, S. Mitogen-activated Protein Kinases and Activator Protein 1 Are Required for Proliferation and Cardiomyocyte Differentiation of P19 Embryonal Carcinoma Cells. *J. Biol. Chem.* **2002**, *277*, 15992–16001. [CrossRef]
25. Aouadi, M.; Bost, F.; Caron, L.; Laurent, K.; Brustel, Y.L.M.; Binétruy, B. p38 Mitogen-Activated Protein Kinase Activity Commits Embryonic Stem Cells to Either Neurogenesis or Cardiomyogenesis. *STEM CELLS* **2006**, *24*, 1399–1406. [CrossRef]

26. Gerits, N.; Kostenko, S.; Moens, U. In vivo functions of mitogen-activated protein kinases: Conclusions from knock-in and knock-out mice. *Transgenic Res.* **2007**, *16*, 281–314. [CrossRef]
27. Engel, F.B.; Schebesta, M.; Duong, M.T.; Lu, G.; Ren, S.; Madwed, J.B.; Jiang, H.; Wang, Y.; Keating, M.T. p38 MAP kinase inhibition enables proliferation of adult mammalian cardiomyocytes. *Genes Dev.* **2005**, *19*, 1175–1187. [CrossRef]
28. Engel, F.B.; Hsieh, P.C.H.; Lee, R.T.; Keating, M.T. FGF1/p38 MAP kinase inhibitor therapy induces cardiomyocyte mitosis, reduces scarring, and rescues function after myocardial infarction. *Proc. Natl. Acad. Sci. USA* **2006**, *103*, 15546–15551. [CrossRef]
29. Hernández-Torres, F.; Martínez-Fernández, S.; Zuluaga, S.; Nebreda, Á.; Porras, A.; Aránega, A.E.; Navarro, F. A role for p38α mitogen-activated protein kinase in embryonic cardiac differentiation. *FEBS Lett.* **2008**, *582*, 1025–1031. [CrossRef]
30. Wu, J.; Kubota, J.; Hirayama, J.; Nagai, Y.; Nishina, S.; Yokoi, T.; Asaoka, Y.; Seo, J.; Shimizu, N.; Kajiho, H.; et al. p38 Mitogen-Activated Protein Kinase Controls a Switch Between Cardiomyocyte and Neuronal Commitment of Murine Embryonic Stem Cells by Activating Myocyte Enhancer Factor 2C-Dependent Bone Morphogenetic Protein 2 Transcription. *Stem Cells Dev.* **2010**, *19*, 1723–1734. [CrossRef]
31. Barrantes, I.D.B.; Coya, J.M.; Maina, F.; Arthur, J.S.C.; Nebreda, A.R. Genetic analysis of specific and redundant roles for p38 and p38 MAPKs during mouse development. *Proc. Natl. Acad. Sci. USA* **2011**, *108*, 12764–12769. [CrossRef] [PubMed]
32. Yokota, T.; Li, J.; Huang, J.; Xiong, Z.; Zhang, Q.; Chan, T.W.; Ding, Y.; Rau, C.D.; Sung, K.; Ren, S.; et al. p38 mitogen-activated protein kinase regulates chamber specific perinatal growth in heart. *J. Clin. Investig.* **2020**, 130. [CrossRef]
33. González-Terán, B.; López, J.A.; Rodríguez, E.; Leiva, L.; Martínez-Martínez, S.; Bernal, J.A.; Jiménez-Borreguero, L.J.; Redondo, J.M.; Vazquez, J.; Sabio, G. p38γ and δ promote heart hypertrophy by targeting the mTOR-inhibitory protein DEPTOR for degradation. *Nat. Commun.* **2016**, *7*, 10477. [CrossRef] [PubMed]
34. Wang, Y.; Su, B.; Sah, V.P.; Brown, J.H.; Han, J.; Chien, K.R. Cardiac Hypertrophy Induced by Mitogen-activated Protein Kinase Kinase 7, a Specific Activator for c-Jun NH2-terminal Kinase in Ventricular Muscle Cells. *J. Boil. Chem.* **1998**, *273*, 5423–5426. [CrossRef] [PubMed]
35. Zechner, D.; Thuerauf, D.J.; Hanford, D.S.; McDonough, P.M.; Glembotski, C.C. A Role for the p38 Mitogen-activated Protein Kinase Pathway in Myocardial Cell Growth, Sarcomeric Organization, and Cardiac-specific Gene Expression. *J. Cell Biol.* **1997**, *139*, 115–127. [CrossRef] [PubMed]
36. Wang, Y.; Huang, S.; Sah, V.P.; Ross, J.; Brown, J.H.; Han, J.; Chien, K.R. Cardiac Muscle Cell Hypertrophy and Apoptosis Induced by Distinct Members of the p38 Mitogen-activated Protein Kinase Family. *J. Biol. Chem.* **1998**, *273*, 2161–2168. [CrossRef]
37. Nishida, K.; Yamaguchi, O.; Hirotani, S.; Hikoso, S.; Higuchi, Y.; Watanabe, T.; Takeda, T.; Osuka, S.; Morita, T.; Kondoh, G.; et al. p38α Mitogen-Activated Protein Kinase Plays a Critical Role in Cardiomyocyte Survival but Not in Cardiac Hypertrophic Growth in Response to Pressure Overload. *Mol. Cell. Biol.* **2004**, *24*, 10611–10620. [CrossRef]
38. Braz, J.C.; Bueno, O.F.; Liang, Q.; Wilkins, B.J.; Dai, Y.-S.; Parsons, S.; Braunwart, J.; Glascock, B.J.; Klevitsky, R.; Kimball, T.F.; et al. Targeted inhibition of p38 MAPK promotes hypertrophic cardiomyopathy through upregulation of calcineurin-NFAT signaling. *J. Clin. Investig.* **2003**, *111*, 1475–1486. [CrossRef]
39. Lopez, I.P.; Cariolato, L.; Maric, D.; Gillet, L.; Abriel, H.; Diviani, D. A-kinase anchoring protein Lbc coordinates a p38 activating signaling complex controlling compensatory cardiac hypertrophy. *Mol. Cell Biol* **2013**, *33*, 2903–2917. [CrossRef]
40. Dingar, D.; Merlen, C.; A Grandy, S.; Gillis, M.-A.; Villeneuve, L.R.; Mamarbachi, A.M.; Fiset, C.; Allen, B.G. Effect of pressure overload-induced hypertrophy on the expression and localization of p38 MAP kinase isoforms in the mouse heart. *Cell. Signal.* **2010**, *22*, 1634–1644. [CrossRef]
41. Poss, K.D.; Wilson, L.G.; Keating, M.T. Heart Regeneration in Zebrafish. *Science* **2002**, *298*, 2188–2190. [CrossRef] [PubMed]
42. Jopling, C.; Suñè, G.; Morera, C.; Belmonte, J.C.I. p38α MAPK regulates myocardial regeneration in zebrafish. *Cell Cycle* **2012**, *11*, 1195–1201. [CrossRef] [PubMed]

43. Ruiz-Bonilla, V.; Perdiguero, E.; Gresh, L.; Serrano, A.L.; Zamora, M.; Sousa-Victor, P.; Jardí, M.; Wagner, E.F.; Muñoz-Cánoves, P. Efficient adult skeletal muscle regeneration in mice deficient in p38β, p38γ and p38δ MAP kinases. *Cell Cycle* **2008**, *7*, 2208–2214. [CrossRef] [PubMed]
44. Gillespie, M.A.; Le Grand, F.; Scimè, A.; Kuang, S.; Von Maltzahn, J.; Seale, V.; Cuenda, A.; Ranish, J.A.; Rudnicki, M.A. p38-γ–dependent gene silencing restricts entry into the myogenic differentiation program. *J. Cell Biol.* **2009**, *187*, 991–1005. [CrossRef]
45. Siles, L.; Ninfali, C.; Cortés, M.; Darling, D.S.; Postigo, A. ZEB1 protects skeletal muscle from damage and is required for its regeneration. *Nat. Commun.* **2019**, *10*, 1364. [CrossRef]
46. Brien, P.; Pugazhendhi, D.; Woodhouse, S.; Oxley, D.; Pell, J.M. p38α MAPK Regulates Adult Muscle Stem Cell Fate by Restricting Progenitor Proliferation During Postnatal Growth and Repair. *STEM CELLS* **2013**, *31*, 1597–1610. [CrossRef] [PubMed]
47. Sousa-Victor, P.; García-Prat, L.; Serrano, A.L.; Perdiguero, E.; Muñoz-Cánoves, P. Muscle stem cell aging: Regulation and rejuvenation. *Trends Endocrinol. Metab.* **2015**, *26*, 287–296. [CrossRef]
48. Bernet, J.D.; Doles, J.D.; Hall, J.K.; Tanaka, K.K.; Carter, T.A.; Olwin, B.B. p38 MAPK signaling underlies a cell-autonomous loss of stem cell self-renewal in skeletal muscle of aged mice. *Nat. Med.* **2014**, *20*, 265–271. [CrossRef]
49. Ge, J.; Liu, K.; Niu, W.; Chen, M.; Wang, M.; Xue, Y.; Gao, C.; Ma, P.X.; Lei, B. Gold and gold-silver alloy nanoparticles enhance the myogenic differentiation of myoblasts through p38 MAPK signaling pathway and promote in vivo skeletal muscle regeneration. *Biomater.* **2018**, *175*, 19–29. [CrossRef]
50. Jin, J.-K.; Blackwood, E.A.; Azizi, K.M.; Thuerauf, N.J.; Fahem, A.G.; Hofmann, C.; Kaufman, R.J.; Doroudgar, S.; Glembotski, C.C. ATF6 Decreases Myocardial Ischemia/Reperfusion Damage and Links ER Stress and Oxidative Stress Signaling Pathways in the Heart. *Circ. Res.* **2016**, *120*, 862–875. [CrossRef]
51. Wallert, M.; Ziegler, M.; Wang, X.; Maluenda, A.; Xu, X.; Yap, M.L.; Witt, R.; Giles, C.; Kluge, S.; Hortmann, M.; et al. α-Tocopherol preserves cardiac function by reducing oxidative stress and inflammation in ischemia/reperfusion injury. *Redox Biol.* **2019**, *26*, 101292. [CrossRef] [PubMed]
52. Tibaut, M.; Mekis, D.; Petrovic, D. Pathophysiology of Myocardial Infarction and Acute Management Strategies. *Cardiovasc. Hematol. Agents Med. Chem.* **2017**, *14*, 150–159. [CrossRef] [PubMed]
53. Marais, E.; Genade, S.; Huisamen, B.; Strijdom, H.; Moolman, J.; Lochner, A. Activation of p38 MAPK Induced by a Multi-cycle Ischaemic Preconditioning Protocol is Associated with Attenuated p38 MAPK Activity During Sustained Ischaemia and Reperfusion. *J. Mol. Cell. Cardiol.* **2001**, *33*, 769–778. [CrossRef] [PubMed]
54. Ballard-Croft, C.; Kristo, G.; Yoshimura, Y.; Reid, E.; Keith, B.J.; Mentzer, R.M.; Lasley, R.D. Acute adenosine preconditioning is mediated by p38 MAPK activation in discrete subcellular compartments. *Am. J. Physiol. Circ. Physiol.* **2005**, *288*, H1359–H1366. [CrossRef] [PubMed]
55. Conrad, P.W.; Rust, R.T.; Han, J.; Millhorn, D.E.; Beitner-Johnson, D. Selective Activation of p38α and p38γ by Hypoxia. *J. Biol. Chem.* **1999**, *274*, 23570–23576. [CrossRef]
56. Saurin, A.T.; Martin, J.L.; Heads, R.; Foley, C.; Mockridge, J.W.; Wright, M.J.; Wang, Y.; Marber, M.S. The role of differential activation of p38-mitogen-activated protein kinase in preconditioned ventricular myocytes. *FASEB J.* **2000**, *14*, 2237–2246. [CrossRef]
57. Yoshimura, Y.; Kristo, G.; Keith, B.J.; Jahania, S.A.; Mentzer, R.M.; Lasley, R.D. The p38 MAPK Inhibitor SB203580 Blocks Adenosine A1 Receptor-Induced Attenuation of In Vivo Myocardial Stunning. *Cardiovasc. Drugs Ther.* **2004**, *18*, 433–440. [CrossRef]
58. Zhao, T.C.; Hines, D.S.; Kukreja, R.C. Adenosine-induced late preconditioning in mouse hearts: Role of p38 MAP kinase and mitochondrial KATP channels. *Am. J. Physiol. Circ. Physiol.* **2001**, *280*, H1278–H1285. [CrossRef]
59. Dana, A.; Skarli, M.; Papakrivopoulou, J.; Yellon, D.M. Adenosine A1Receptor Induced Delayed Preconditioning in Rabbits. *Circ. Res.* **2000**, *86*, 989–997. [CrossRef]
60. Emerling, B.M.; Platanias, L.C.; Black, E.; Nebreda, A.R.; Davis, R.J.; Chandel, N.S. Mitochondrial Reactive Oxygen Species Activation of p38 Mitogen-Activated Protein Kinase Is Required for Hypoxia Signaling. *Mol. Cell. Biol.* **2005**, *25*, 4853–4862. [CrossRef]
61. Barajas-Espinosa, A.; Basye, A.; Angelos, M.G.; Chen, C.-A. Modulation of p38 kinase by DUSP4 is important in regulating cardiovascular function under oxidative stress. *Free. Radic. Boil. Med.* **2015**, *89*, 170–181. [CrossRef] [PubMed]

62. Kurian, G.A.; Paddikkala, J. N-acetylcysteine and magnesium improve biochemical abnormalities associated with myocardial ischaemic reperfusion in South Indian patients undergoing coronary artery bypass grafting: A comparative analysis. *Singap. Med J.* **2010**, *51*, 3813–3888.
63. Kurian, G.A.; Paddikkala, J. Administration of aqueous extract of Desmodium gangeticum (L) root protects rat heart against ischemic reperfusion injury induced oxidative stress. *Indian J. Exp. Boil.* **2009**, *47*, 1291–1335.
64. Kurian, G.A.; Suryanarayanan, S.; Raman, A.; Padikkala, J. Antioxidant effects of ethyl acetate extract of Desmodium gangeticum root on myocardial ischemia reperfusion injury in rat hearts. *Chin. Med.* **2010**, *5*, 3. [CrossRef] [PubMed]
65. Ashraf, M.I.; Ebner, M.; Wallner, C.; Haller, M.; Khalid, S.; Schwelberger, H.G.; Koziel, K.; Enthammer, M.; Hermann, M.; Sickinger, S.; et al. A p38MAPK/MK2 signaling pathway leading to redox stress, cell death and ischemia/reperfusion injury. *Cell Commun. Signal.* **2014**, *12*, 6. [CrossRef]
66. Hu, J.; Li, Z.; Xu, L.-T.; Sun, A.-J.; Fu, X.-Y.; Zhang, L.; Jing, L.-L.; Lu, A.-D.; Dong, Y.-F.; Jia, Z.-P. Protective Effect of Apigenin on Ischemia/Reperfusion Injury of the Isolated Rat Heart. *Cardiovasc. Toxicol.* **2014**, *15*, 241–249. [CrossRef]
67. Yang, X.; Yang, J.; Hu, J.; Li, X.; Zhang, X.; Li, Z. Apigenin attenuates myocardial ischemia/reperfusion injury via the inactivation of p38 mitogen-activated protein kinase. *Mol. Med. Rep.* **2015**, *12*, 6873–6878. [CrossRef]
68. Guo, W.; Liu, X.; Li, J.; Shen, Y.; Zhou, Z.; Wang, M.; Xie, Y.; Feng, X.; Wang, L.; Wu, X. Prdx1 alleviates cardiomyocyte apoptosis through ROS-activated MAPK pathway during myocardial ischemia/reperfusion injury. *Int. J. Biol. Macromol.* **2018**, *112*, 608–615. [CrossRef]
69. Bassi, R.; Burgoyne, J.R.; DeNicola, G.F.; Rudyk, O.; DeSantis, V.; Charles, R.L.; Eaton, P.; Marber, M.S. Redox-dependent dimerization of p38α mitogen-activated protein kinase with mitogen-activated protein kinase kinase. *J. Biol. Chem.* **2017**, *292*, 16161–16173. [CrossRef]
70. Säkkinen, H.; Aro, J.; Kaikkonen, L.; Ohukainen, P.; Näpänkangas, J.; Tokola, H.; Ruskoaho, H.; Rysä, J. Mitogen-activated protein kinase p38 target regenerating islet-derived 3γ expression is upregulated in cardiac inflammatory response in the rat heart. *Physiol. Rep.* **2016**, *4*, e12996. [CrossRef]
71. Na, D.; Aijie, H.; Bo, L.; Zhilin, M.; Long, Y. Gambogic acid exerts cardioprotective effects in a rat model of acute myocardial infarction through inhibition of inflammation, iNOS and NF-κB/p38 pathway. *Exp. Ther. Med.* **2017**, *15*, 1742–1748. [CrossRef] [PubMed]
72. Ahmad, F.; Tomar, D.; A C, S.A.; Elmoselhi, A.B.; Thomas, M.; Elrod, J.W.; Tilley, D.G.; Force, T. Nicotinamide riboside kinase-2 alleviates ischemia-induced heart failure through P38 signaling. *Biochim. et Biophys. Acta (BBA) - Mol. Basis Dis.* **2020**, *1866*, 165609. [CrossRef] [PubMed]
73. Xu, L.; Yates, C.C.; Lockyer, P.; Xie, L.; Bevilacqua, A.; He, J.; Lander, C.; Patterson, C.; Willis, M.S. MMI-0100 inhibits cardiac fibrosis in myocardial infarction by direct actions on cardiomyocytes and fibroblasts via MK2 inhibition. *J. Mol. Cell. Cardiol.* **2014**, *77*, 86–101. [CrossRef] [PubMed]
74. Molkentin, J.D.; Bugg, D.; Ghearing, N.; Dorn, L.E.; Kim, P.; Sargent, M.A.; Gunaje, J.; Otsu, K.; Davis, J. Fibroblast-Specific Genetic Manipulation of p38 Mitogen-Activated Protein Kinase In Vivo Reveals Its Central Regulatory Role in Fibrosis. *Circulation* **2017**, *136*, 549–561. [CrossRef] [PubMed]
75. Wang, M.; Li, Z.; Zhang, X.; Xie, X.; Zhang, Y.; Wang, X.; Hou, Y. Rosuvastatin Attenuates Atrial Structural Remodelling in Rats with Myocardial Infarction through the Inhibition of the p38 MAPK Signalling Pathway. *Hear. Lung Circ.* **2015**, *24*, 386–394. [CrossRef]
76. Li, M.; Liu, F.; Sang, M.; Sun, X.; Li, L.; Wang, X. Effects of atorvastatin on p38 phosphorylation and cardiac remodeling after myocardial infarction in rats. *Exp. Ther. Med.* **2018**, *16*, 751–757. [CrossRef]
77. Prompunt, E.; Sanit, J.; Barrère-Lemaire, S.; Nargeot, J.; Noordali, H.; Madhani, M.; Kumphune, S. The cardioprotective effects of secretory leukocyte protease inhibitor against myocardial ischemia/reperfusion injury. *Exp. Ther. Med.* **2018**, *15*, 5231–5242. [CrossRef]
78. Kumphune, S.; Surinkaew, S.; Chattipakorn, S.C.; Chattipakorn, N. Inhibition of p38 MAPK activation protects cardiac mitochondria from ischemia/reperfusion injury. *Pharm. Biol.* **2015**, *53*, 1831–1841. [CrossRef]
79. Song, N.; Ma, J.; Meng, X.-W.; Liu, H.; Wang, H.; Song, S.-Y.; Chen, Q.-C.; Liu, H.-Y.; Zhang, J.; Peng, K.; et al. Heat Shock Protein 70 Protects the Heart from Ischemia/Reperfusion Injury through Inhibition of p38 MAPK Signaling. *Oxidative Med. Cell. Longev.* **2020**, *2020*, 1–19. [CrossRef]
80. Zhu, S.; Xu, T.; Luo, Y.; Zhang, Y.; Xuan, H.; Ma, Y.; Pan, D.; Li, N.; Zhu, H. Luteolin Enhances Sarcoplasmic Reticulum Ca2+-ATPase Activity through p38 MAPK Signaling thus Improving Rat Cardiac Function after Ischemia/Reperfusion. *Cell. Physiol. Biochem.* **2017**, *41*, 999–1010. [CrossRef]

81. Lee, M.-L.; Sulistyowati, E.; Hsu, J.-H.; Huang, B.-Y.; Dai, Z.-K.; Wu, B.-N.; Chao, Y.-Y.; Yeh, J.-L. KMUP-1 Ameliorates Ischemia-Induced Cardiomyocyte Apoptosis through the NO⁻ cGMP⁻MAPK Signaling Pathways. *Molecules* **2019**, *24*, 1376. [CrossRef] [PubMed]
82. Sanit, J.; Prompunt, E.; Adulyaritthikul, P.; Nokkaew, N.; Mongkolpathumrat, P.; Kongpol, K.; Kijtawornrat, A.; Petchdee, S.; Kumphune, S. Combination of metformin and p38 MAPK inhibitor, SB203580, reduced myocardial ischemia/reperfusion injury in non-obese type 2 diabetic Goto-Kakizaki rats. *Exp. Ther. Med.* **2019**, *18*, 1701–1714. [CrossRef] [PubMed]
83. Surinkaew, S.; Kumphune, S.; Chattipakorn, S.; Chattipakorn, N. Inhibition of p38 MAPK During Ischemia, But Not Reperfusion, Effectively Attenuates Fatal Arrhythmia in Ischemia/Reperfusion Heart. *J. Cardiovasc. Pharmacol.* **2013**, *61*, 133–141. [CrossRef] [PubMed]
84. Meijles, D.N.; Cull, J.J.; Markou, T.; Cooper, S.T.; Haines, Z.H.; Fuller, S.J.; O'Gara, P.; Sheppard, M.; Harding, S.E.; Sugden, P.H.; et al. Redox Regulation of Cardiac ASK1 (Apoptosis Signal-Regulating Kinase 1) Controls p38-MAPK (Mitogen-Activated Protein Kinase) and Orchestrates Cardiac Remodeling to Hypertension. *Hypertension* **2020**, *76*, 1208–1218. [CrossRef]
85. Ziaeian, B.; Fonarow, G.C. Epidemiology and aetiology of heart failure. *Nat. Rev. Cardiol.* **2016**, *13*, 368–378. [CrossRef]
86. Savarese, G.; Lund, L.H. Global Public Health Burden of Heart Failure. *Card. Fail. Rev.* **2017**, *3*, 7–11. [CrossRef]
87. Cardin, S.; Li, D.; Thorin-Trescases, N.; Leung, T.-K.; Thorin, E.; Nattel, S. Evolution of the atrial fibrillation substrate in experimental congestive heart failure: Angiotensin-dependent and -independent pathways. *Cardiovasc. Res.* **2003**, *60*, 315–325. [CrossRef]
88. Li, D.; Shinagawa, K.; Pang, L.; Leung, T.K.; Cardin, S.; Wang, Z.; Nattel, S. Effects of angiotensin-converting enzyme inhibition on the development of the atrial fibrillation substrate in dogs with ventricular tachypacing-induced congestive heart failure. *Circulation* **2001**, *104*, 2608–2614. [CrossRef]
89. Li, M. p38 MAP Kinase Mediates Inflammatory Cytokine Induction in Cardiomyocytes and Extracellular Matrix Remodeling in Heart. *Circulation* **2005**, *111*, 2494–2502. [CrossRef]
90. Kyoi, S.; Otani, H.; Matsuhisa, S.; Akita, Y.; Tatsumi, K.; Enoki, C.; Fujiwara, H.; Imamura, H.; Kamihata, H.; Iwasaka, T. Opposing effect of p38 MAP kinase and JNK inhibitors on the development of heart failure in the cardiomyopathic hamster. *Cardiovasc. Res.* **2006**, *69*, 888–898. [CrossRef]
91. Khan, R.; Sheppard, R.J.M.F. Fibrosis in heart disease: Understanding the role of transforming growth factor-β1 in cardiomyopathy, valvular disease and arrhythmia. *Immunology* **2006**, *118*, 10–24. [CrossRef] [PubMed]
92. Gonzalez, A.; Schelbert, E.B.; Díez, J.; Butler, J. Myocardial Interstitial Fibrosis in Heart Failure. *J. Am. Coll. Cardiol.* **2018**, *71*, 1696–1706. [CrossRef] [PubMed]
93. Segura, A.M.; Frazier, O.H.; Buja, L.M. Fibrosis and heart failure. *Hear. Fail. Rev.* **2012**, *19*, 173–185. [CrossRef] [PubMed]
94. Liao, P.; Georgakopoulos, D.; Kovacs, A.; Zheng, M.; Lerner, D.; Pu, H.; Saffitz, J.; Chien, K.; Xiao, R.-P.; Kass, D.A.; et al. The in vivo role of p38 MAP kinases in cardiac remodeling and restrictive cardiomyopathy. *Proc. Natl. Acad. Sci. USA* **2001**, *98*, 12283–12288. [CrossRef] [PubMed]
95. Shirazi, L.F.; Bissett, J.; Romeo, F.; Mehta, J.L. Role of Inflammation in Heart Failure. *Curr. Atheroscler. Rep.* **2017**, *19*, 49. [CrossRef]
96. Porter, K.E.; Turner, N.A. Cardiac fibroblasts: At the heart of myocardial remodeling. *Pharmacol. Ther.* **2009**, *123*, 255–278. [CrossRef]
97. Li, Y.; Li, Z.; Zhang, C.; Li, P.; Wu, Y.; Wang, C.; Lau, W.B.; Ma, X.L.; Du, J. Cardiac Fibroblast–Specific Activating Transcription Factor 3 Protects Against Heart Failure by Suppressing MAP2K3-p38 Signaling. *Circulation* **2017**, *135*, 2041–2057. [CrossRef]
98. Bujak, M.; Frangogiannis, N.G. The role of TGF-β signaling in myocardial infarction and cardiac remodeling. *Cardiovasc. Res.* **2007**, *74*, 184–195. [CrossRef]
99. Lijnen, P.; Petrov, V.; Fagard, R. Induction of Cardiac Fibrosis by Transforming Growth Factor-β1. *Mol. Genet. Metab.* **2000**, *71*, 418–435. [CrossRef]
100. Dorn, G.W.; Molkentin, J.D. Manipulating Cardiac Contractility in Heart Failure. *Circulation* **2004**, *109*, 150–158. [CrossRef]

101. Chen, Y.; Rajashree, R.; Liu, Q.; Hofmann, P. Acute p38 MAPK activation decreases force development in ventricular myocytes. *Am. J. Physiol. Circ. Physiol.* **2003**, *285*, H2578–H2586. [CrossRef] [PubMed]
102. Liao, P.; Wang, S.-Q.; Wang, S.; Zheng, M.; Zheng, M.; Zhang, S.-J.; Cheng, H.; Wang, Y.; Xiao, R.-P. p38 Mitogen-Activated Protein Kinase Mediates a Negative Inotropic Effect in Cardiac Myocytes. *Circ. Res.* **2002**, *90*, 190–196. [CrossRef] [PubMed]
103. Szokodi, I.; Kerkelä, R.; Kubin, A.-M.; Sármán, B.; Pikkarainen, S.; Kónyi, A.; Horváth, I.G.; Papp, L.; Tóth, M.; Skoumal, R.; et al. Functionally Opposing Roles of Extracellular Signal-Regulated Kinase 1/2 and p38 Mitogen-Activated Protein Kinase in the Regulation of Cardiac Contractility. *Circulation* **2008**, *118*, 1651–1658. [CrossRef] [PubMed]
104. Zheng, M.; Zhang, S.-J.; Zhu, W.-Z.; Ziman, B.; Kobilka, B.K.; Xiao, R.-P. β2-Adrenergic Receptor-induced p38 MAPK Activation Is Mediated by Protein Kinase A Rather than by Gior Gβγ in Adult Mouse Cardiomyocytes. *J. Boil. Chem.* **2000**, *275*, 40635–40640. [CrossRef]
105. Palomeque, J.; Sapia, L.; Hajjar, R.J.; Mattiazzi, A.; Petroff, M.V. Angiotensin II-induced negative inotropy in rat ventricular myocytes: Role of reactive oxygen species and p38 MAPK. *Am. J. Physiol. Circ. Physiol.* **2006**, *290*, H96–H106. [CrossRef]
106. Vahebi, S.; Ota, A.; Li, M.; Warren, C.M.; De Tombe, P.P.; Wang, Y.; Solaro, R.J. p38-MAPK Induced Dephosphorylation of α-Tropomyosin Is Associated With Depression of Myocardial Sarcomeric Tension and ATPase Activity. *Circ. Res.* **2007**, *100*, 408–415. [CrossRef]
107. Espejo, M.S.; Aiello, E.A.; Sepúlveda, M.; Petroff, M.G.V.; Aiello, E.A.; De Giusti, V.C. The reduced myofilament responsiveness to calcium contributes to the negative force-frequency relationship in rat cardiomyocytes: Role of reactive oxygen species and p-38 map kinase. *Pflügers Archiv. Eur. J. Physiol.* **2017**, *469*, 1663–1673. [CrossRef]
108. Kaikkonen, L.; Magga, J.; Ronkainen, V.-P.; Koivisto, E.; Perjés, Á.; Chuprun, J.K.; Vinge, L.E.; Kilpiö, T.; Aro, J.; Ulvila, J.; et al. p38α regulates SERCA2a function. *J. Mol. Cell. Cardiol.* **2014**, *67*, 86–93. [CrossRef]
109. Bers, D.M. Calcium fluxes involved in control of cardiac myocyte contraction. *Circ. Res.* **2000**, *87*, 275–281. [CrossRef]
110. Luo, M.; Anderson, M.E. Mechanisms of altered Ca^{2+} handling in heart failure. *Circ. Res.* **2013**, *113*, 690–708. [CrossRef]
111. Heidkamp, M.C.; Scully, B.T.; Vijayan, K.; Engman, S.J.; Szotek, E.L.; Samarel, A.M. PYK2 regulates SERCA2 gene expression in neonatal rat ventricular myocytes. *Am. J. Physiol. Physiol.* **2005**, *289*, C471–C482. [CrossRef] [PubMed]
112. Andrews, C.; Ho, P.D.; Dillmann, W.H.; Glembotski, C.C.; McDonough, P.M. The MKK6-p38 MAPK pathway prolongs the cardiac contractile calcium transient, downregulates SERCA2, and activates NF-AT. *Cardiovasc. Res.* **2003**, *59*, 46–56. [CrossRef]
113. Scharf, M.; Neef, S.; Freund, R.; Geers-Knörr, C.; Franz-Wachtel, M.; Brandis, A.; Krone, D.; Schneider, H.; Groos, S.; Menon, M.B.; et al. Mitogen-Activated Protein Kinase-Activated Protein Kinases 2 and 3 Regulate SERCA2a Expression and Fiber Type Composition To Modulate Skeletal Muscle and Cardiomyocyte Function. *Mol. Cell. Biol.* **2013**, *33*, 2586–2602. [CrossRef] [PubMed]
114. Clerk, A.; Michael, A.; Sugden, P.H. Stimulation of the p38 Mitogen-activated Protein Kinase Pathway in Neonatal Rat Ventricular Myocytes by the G Protein–coupled Receptor Agonists, Endothelin-1 and Phenylephrine: A Role in Cardiac Myocyte Hypertrophy? *J. Cell Biol.* **1998**, *142*, 523–535. [CrossRef]
115. Kranias, E.G.; Hajjar, R.J. Modulation of Cardiac Contractility by the Phopholamban/SERCA2a Regulatome. *Circ. Res.* **2012**, *110*, 1646–1660. [CrossRef]
116. Akaike, T.; Du, N.; Lu, G.; Minamisawa, S.; Wang, Y.; Ruan, H. A Sarcoplasmic Reticulum Localized Protein Phosphatase Regulates Phospholamban Phosphorylation and Promotes Ischemia Reperfusion Injury in the Heart. *JACC. Basic Transl. Sci.* **2017**, *2*, 160–180. [CrossRef]
117. Lei, M.; Wang, X.; Ke, Y.; Solaro, R.J. Regulation of Ca2+ transient by PP2A in normal and failing heart. *Front. Physiol.* **2015**, *6*, 13. [CrossRef]
118. Liu, Q.; Hofmann, P.A. Modulation of protein phosphatase 2a by adenosine A1 receptors in cardiomyocytes: Role for p38 MAPK. *Am. J. Physiol. Circ. Physiol.* **2003**, *285*, H97–H103. [CrossRef]
119. Xu, L.; Kappler, C.; Menick, D. The role of p38 in the regulation of Na?Ca exchanger expression in adult cardiomyocytes. *J. Mol. Cell. Cardiol.* **2005**, *38*, 735–743. [CrossRef]

120. Menick, D.R.; Renaud, L.; Buchholz, A.; Müller, J.G.; Zhou, H.; Kappler, C.S.; Kubalak, S.W.; Conway, S.J.; Xu, L. Regulation of Ncx1 gene expression in the normal and hypertrophic heart. *Ann. New York Acad. Sci.* **2007**, *1099*, 195–203. [CrossRef]
121. Xu, L.; Kappler, C.S.; Mani, S.K.; Shepherd, N.R.; Renaud, L.; Snider, P.; Conway, S.J.; Menick, D.R. Chronic Administration of KB-R7943 Induces Up-regulation of Cardiac NCX1. *J. Biol. Chem.* **2009**, *284*, 27265–27272. [CrossRef] [PubMed]
122. Pogwizd, S.M.; Schlotthauer, K.; Li, L.; Yuan, W.; Bers, D.M. Arrhythmogenesis and Contractile Dysfunction in Heart Failure. *Circ. Res.* **2001**, *88*, 1159–1167. [CrossRef] [PubMed]
123. Lu, Y.-Y.; Chen, Y.-C.; Kao, Y.-H.; Chen, S.-A.; Chen, Y.-J. Extracellular matrix of collagen modulates arrhythmogenic activity of pulmonary veins through p38 MAPK activation. *J. Mol. Cell. Cardiol.* **2013**, *59*, 159–166. [CrossRef]
124. Cheng, W.; Zhu, Y.; Wang, H. The MAPK pathway is involved in the regulation of rapid pacing-induced ionic channel remodeling in rat atrial myocytes. *Mol. Med. Rep.* **2016**, *13*, 2677–2682. [CrossRef] [PubMed]
125. Mollenhauer, M.; Friedrichs, K.; Lange, M.; Gesenberg, J.; Remane, L.; Kerkenpaß, C.; Krause, J.; Schneider, J.; Ravekes, T.; Maass, M.; et al. Myeloperoxidase Mediates Postischemic Arrhythmogenic Ventricular Remodeling. *Circ. Res.* **2017**, *121*, 56–70. [CrossRef]
126. Desplantez, T. Cardiac Cx43, Cx40 and Cx45 co-assembling: Involvement of connexins epitopes in formation of hemichannels and Gap junction channels. *BMC Cell Biol.* **2017**, *18*, 3. [CrossRef]
127. Marian, A.J.; Asatryan, B.; Wehrens, X.H. Genetic basis and molecular biology of cardiac arrhythmias in cardiomyopathies. *Cardiovasc. Res.* **2020**, 116. [CrossRef]
128. Solan, J.L.; Lampe, P.D. Connexin43 phosphorylation: Structural changes and biological effects. *Biochem. J.* **2009**, *419*, 261–272. [CrossRef]
129. Polontchouk, L.; Ebelt, B.; Jackels, M.; Dhein, S. Chronic effects of endothelin-1 and angiotensin-II on gap junctions and intercellular communication in cardiac cells. *FASEB J.* **2001**, *16*, 1–15. [CrossRef]
130. Salameh, A.; Schneider, P.; Mühlberg, K.; Hagendorff, A.; Dhein, S.; Pfeiffer, D. Chronic regulation of the expression of gap junction proteins connexin40, connexin43, and connexin45 in neonatal rat cardiomyocytes. *Eur. J. Pharmacol.* **2004**, *503*, 9–16. [CrossRef]
131. Inoue, N.; Ohkusa, T.; Nao, T.; Lee, J.-K.; Matsumoto, T.; Hisamatsu, Y.; Satoh, T.; Yano, M.; Yasui, K.; Kodama, I.; et al. Rapid electrical stimulation of contraction modulates gap junction protein in neonatal rat cultured cardiomyocytes. *J. Am. Coll. Cardiol.* **2004**, *44*, 914–922. [CrossRef] [PubMed]
132. Yao, J.; Ke, J.; Zhou, Z.; Tan, G.; Yin, Y.; Liu, M.; Chen, J.; Wu, W. Combination of HGF and IGF-1 promotes connexin 43 expression and improves ventricular arrhythmia after myocardial infarction through activating the MAPK/ERK and MAPK/p38 signaling pathways in a rat model. *Cardiovasc. Diagn. Ther.* **2019**, *9*, 346–354. [CrossRef] [PubMed]
133. Salameh, A.; Krautblatter, S.; Baeβler, S.; Karl, S.; Gomez, D.R.; Dhein, S.; Pfeiffer, D.; Baessler, S. Signal Transduction and Transcriptional Control of Cardiac Connexin43 Up-Regulation after α1-Adrenoceptor Stimulation. *J. Pharmacol. Exp. Ther.* **2008**, *326*, 315–322. [CrossRef] [PubMed]
134. Wang, C.-Y.; Liu, H.-J.; Chen, H.-J.; Lin, Y.-C.; Wang, H.-H.; Hung, T.-C.; Yeh, H.-I. AGE-BSA down-regulates endothelial connexin43 gap junctions. *BMC Cell Biol.* **2011**, *12*, 19. [CrossRef] [PubMed]
135. Schulz, R.; Gres, P.; Skyschally, A.; Duschin, A.; Belosjorow, S.; Konietzka, I.; Heusch, G. Ischemic preconditioning preserves connexin 43 phosphorylation during sustained ischemia in pig hearts in vivo. *FASEB J.* **2003**, *17*, 1355–1357. [CrossRef]
136. Xue, J.; Yan, X.; Yang, Y.; Chen, M.; Wu, L.; Gou, Z.; Sun, Z.; Talabieke, S.; Zheng, Y.; Luo, D. Connexin 43 dephosphorylation contributes to arrhythmias and cardiomyocyte apoptosis in ischemia/reperfusion hearts. *Basic Res. Cardiol.* **2019**, *114*, 40. [CrossRef]
137. Ogawa, T.; Hayashi, T.; Kyoizumi, S.; Kusunoki, Y.; Nakachi, K.; Macphee, D.G.; Trosko, J.E.; Kataoka, K.; Yorioka, N. Anisomycin downregulates gap-junctional intercellular communication via the p38 MAP-kinase pathway. *J. Cell Sci.* **2004**, *117*, 2087–2096. [CrossRef]

138. Zhong, C.; Chang, H.; Wu, Y.; Zhou, L.; Wang, Y.; Wang, M.; Wu, P.; Qi, Z.; Zou, J. Up-regulated Cx43 phosphorylation at Ser368 prolongs QRS duration in myocarditis. *J. Cell. Mol. Med.* **2018**, *22*, 3537–3547. [CrossRef]
139. Newby, L.K.; Marber, M.S.; Melloni, C.; Sarov-Blat, L.; Aberle, L.H.; E Aylward, P.; Cai, G.; De Winter, R.J.; Hamm, C.W.; Heitner, J.F.; et al. Losmapimod, a novel p38 mitogen-activated protein kinase inhibitor, in non-ST-segment elevation myocardial infarction: A randomised phase 2 trial. *Lancet* **2014**, *384*, 1187–1195. [CrossRef]
140. O'Donoghue, M.L.; Glaser, R.; Cavender, M.A.; Aylward, P.E.; Bonaca, M.P.; Budaj, A.; Davies, R.Y.; Dellborg, M.; Fox, K.A.A.; Gutierrez, J.A.T.; et al. Effect of Losmapimod on Cardiovascular Outcomes in Patients Hospitalized With Acute Myocardial Infarction. *JAMA* **2016**, *315*, 1591. [CrossRef]
141. Dulos, J.; Wijnands, F.P.G.; Alebeek, J.A.J.V.D.H.-V.; Van Vugt, M.J.H.; Rullmann, J.A.C.; Schot, J.-J.G.; De Groot, M.W.G.D.M.; Wagenaars, J.L.; Os, R.V.R.-V.; Smets, R.L.; et al. p38 inhibition and not MK2 inhibition enhances the secretion of chemokines from TNF-? activated rheumatoid arthritis fibroblast-like synoviocytes. *Clin. Exp. Rheumatol.* **2013**, *31*, 515–525. [PubMed]
142. Shiroto, K.; Otani, H.; Yamamoto, F.; Huang, C.-K.; Maulik, N.; Das, D.K. MK2 gene knockout mouse hearts carry anti-apoptotic signal and are resistant to ischemia reperfusion injury. *J. Mol. Cell. Cardiol.* **2005**, *38*, 93–97. [CrossRef] [PubMed]

International Journal of
Molecular Sciences

Review

p38 MAPK in Glucose Metabolism of Skeletal Muscle: Beneficial or Harmful?

Eyal Bengal [1,*], Sharon Aviram [1] and Tony Hayek [2]

[1] Department of Biochemistry, Rappaport Faculty of Medicine, Technion-Israel Institute of Technology, P.O. Box 9649, Haifa 31096, Israel; avirams@technion.ac.il
[2] Department of Internal Medicine E, Rambam Medical Center, Haifa 31096, Israel; t_hayek@rambam.health.gov.il
* Correspondence: bengal@technion.ac.il; Tel.: +972-4-8295-287; Fax: +972-4-8553-299

Received: 30 July 2020; Accepted: 2 September 2020; Published: 4 September 2020

Abstract: Skeletal muscles respond to environmental and physiological changes by varying their size, fiber type, and metabolic properties. P38 mitogen-activated protein kinase (MAPK) is one of several signaling pathways that drive the metabolic adaptation of skeletal muscle to exercise. p38 MAPK also participates in the development of pathological traits resulting from excessive caloric intake and obesity that cause metabolic syndrome and type 2 diabetes (T2D). Whereas p38 MAPK increases insulin-independent glucose uptake and oxidative metabolism in muscles during exercise, it contrastingly mediates insulin resistance and glucose intolerance during metabolic syndrome development. This article provides an overview of the apparent contradicting roles of p38 MAPK in the adaptation of skeletal muscles to exercise and to pathological conditions leading to glucose intolerance and T2D. Here, we focus on the involvement of p38 MAPK in glucose metabolism of skeletal muscle, and discuss the possibility of targeting this pathway to prevent the development of T2D.

Keywords: skeletal muscle; energy metabolism; signal transduction; p38 MAPK; exercise; type 2 diabetes

---

1. Introduction

1.1. Skeletal Muscle Energy Metabolism

Skeletal muscle comprises about 40% of the total body mass and accounts for around 30% of resting metabolic rate [1]. Due to its role in locomotion, this tissue is a major energy consumer, especially during exercise. Thus, it is the primary site of glucose disposal and the major glycogen storage organ and is, therefore, absolutely critical for glycemic control and the metabolic homeostasis of the body [2]. Since ATP's resting intramuscular stores are small, energy needs to tightly regulate the metabolic pathways that generate ATP to maintain ATP at constant levels. ATP is generated in the muscle by the oxidation of carbohydrates and lipids. The primary carbohydrate energy source in skeletal muscle is glucose that is processed from internal glycogen stores (glycogenolysis) or is extracted from the blood.

1.1.1. Glucose Uptake

Glucose is transported by the Glut family of membrane transporters, with Glut1 and Glut4 being the major transporters in skeletal muscle. Whereas Glut1 is responsible for the basal uptake of glucose, Glut4 is an inducible transporter that facilitates glucose uptake by insulin or muscle contraction. In the basal state, Glut4 is mainly associated with intracellular vesicles lying adjacent to the plasma membrane of the muscle fiber (sarcolemma). Insulin and muscle contraction facilitate the translocation of vesicular Glut4 to the sarcolemma and the T tubular system where Glut4 transports glucose from the blood into the cytoplasm of muscle cells. Distinct molecular mechanisms are responsible for Glut4 translocation into the sarcolemma by contraction and insulin [3]. However, the two pathways converge at the GTPase activating proteins (GAPs), TBC1D1, and TBC1D4 (AS160). Phosphorylation of TBC1D1/TBC1D4 at particular residues by either AMP-activated protein kinase (AMPK) in contracting muscle or by insulin-activated Akt inhibits their GTPase-activating domain and enables the exchange of GDP to GTP–bound state of Rab. Once activated, Rab mediates the translocation of vesicular Glut4 to the plasma membrane and facilitates glucose uptake [3]. Hereafter, we will describe the role of p38 mitogen-activated protein kinase (MAPK) in glucose uptake under different physiological conditions.

1.1.2. Fiber Type and Glucose Metabolism

Skeletal muscle fibers are heterogeneous with respect to their contractile apparatus and metabolism [4]. Three types of motor units, called slow twitch, fast fatigable, and fast fatigue-resistant, are composed of type 1, 2A, 2B, and 2X fibers. These fibers are distinguished by their myosin heavy chain (MyHC) isoforms and their oxidative/glycolytic metabolism. Whereas type 1 and 2A fibers are rich in mitochondria and derive their energy mostly from oxidative metabolism, the mitochondria-poor type 2B and 2X fibers primarily utilize anaerobic glycolysis for ATP production. Different muscles contain particular ratios of fast and slow-twitch fibers. The skeletal muscle is a highly plastic organ that adapts to the changes in metabolic needs. Whereas physical inactivity and obesity induce a phenotypic shift from oxidative type 1 and 2A fibers to glycolytic type 2B fibers [5], endurance exercise causes a switch from glycolytic type 2B fibers towards more oxidative type 1 and 2A fibers [6].

1.2. P38 MAPK Is Involved in Various Aspects of Whole-Body Energy Metabolism

Four isoforms of p38 MAPK, alpha, beta, gamma, and delta, are encoded by four different genes and have different tissue expression patterns [7]. These kinases are members of a larger family that includes at least three additional kinases; the extracellular signal-regulated kinases (ERK1/2), ERK5 (also known as BMK1), and the Jun amino-terminal kinases (JNK1-3). P38 and JNK are considered stress-activated protein kinases (SAPKs) that participate in the cellular response to metabolic and other stress conditions. In these roles, they are likely to contribute to metabolism-related pathogenesis. The four isoforms of p38 are activated by environmental and genotoxic stress conditions, inflammatory cytokines, and hormones that drive dual phosphorylation in the activation loop sequence Thr-Gly-Tyr. The primary activating kinases (MAPKK) of the four-p38 isoforms are MKK3 and MKK6. In turn, p38 isoforms phosphorylate and activate many substrates, including proteins and enzymes of glucose and lipid metabolism. Below is a list of p38 MAPK targets, mostly transcription factors involved in the energy metabolism of different tissues.

1.2.1. PPAR Gamma Activator-1 Alpha (PGC1α)

PGC1α is a master metabolic co-factor of mitochondrial biogenesis. It is involved in respiration and oxidative phosphorylation, Glut4 expression, conversion of type 2B muscle fibers to type 2A and 1 fibers and in gluconeogenesis in the liver. PGC1α is a transcriptional coactivator whose phosphorylation by p38 prevents the binding of the repressor p160MBP and enables its activity in adipose tissue, muscle, and liver [8]. p38 MAPK also phosphorylates and activates the ATF2 and Mef2 transcription factors, which in turn upregulate the expression of PGC1α. Hence, p38 MAPK positively affects both the transcription and activity of PGC1α.

1.2.2. Activating Transcription Factor 2 (ATF2)

ATF2 is a transcription factor, a member of the leucine zipper (bZip) family, binds to a cAMP-responsive element (CRE) as homodimer or heterodimer with c-Jun. ATF2 is usually phosphorylated and activated by SAPKs, p38, and JNK [9]. P38 MAPK regulates brown adipose tissue differentiation by increasing the expression levels of uncoupling protein 1 (UCP1) through the phosphorylation of ATF2 and PGC1α [10–12]. Inhibition of p38 MAPK in high-fat diet (HFD)-fed mice prevented obesity and alleviated insulin resistance [11].

1.2.3. CCAAT/Enhancer-Binding Protein α (C/EBPα)

This bZip transcription factor binds to DNA as a homodimer, or as a heterodimer with the related proteins C/EBPβ and C/EBPγ, as well as with distinct transcription factors such as c-Jun. It is a crucial regulator of adipogenesis, accumulation of lipids in those cells, and the metabolism of glucose and lipids in the liver [13]. P38 MAPK phosphorylates Ser21 of C/EBPα, which regulates whole-body glucose homeostasis [14]. Serine phosphorylation by p38 enhances C/EBPα activity necessary for the transcription of phosphoenolpyruvate carboxykinase (PEPCK), a rate-limiting enzyme in liver gluconeogenesis [15]. P38 MAPK is also involved in adipocyte differentiation through the phosphorylation of C/EBPβ, which induces peroxisome proliferator-activated receptor gamma (PPARγ).

1.2.4. cAMP Response Element-Binding Protein (CREB)

Phosphorylation of CREB at Ser133 by several kinases induces its transcriptional activity [16]. P38 MAPK does not directly phosphorylate CREB, but rather induces CREB phosphorylation via MSK1 (mitogen-and stress-activated kinase 1). P38 MAPK augments liver gluconeogenesis via the CREB protein that activates the transcription of *Pepck*, *Glucose 6-phosphatase*, and *Pgc1α* genes [10,17,18].

1.2.5. Glycogen Synthase (GS)

It is the rate-limiting enzyme in the glycogen biosynthesis pathway. The β isoform of p38 has been shown to phosphorylate GS and inhibit its activity and the conversion of glucose to glycogen in the liver [19,20].

1.2.6. Peroxisome Proliferator-Activated Receptor α (PPARα)

PPARα belongs to a group of nuclear receptor proteins that function as transcription factors as heterodimers with the retinoid X receptor (RXR). P38 MAPK promotes the β oxidation of fatty acids in the liver and cardiomyocytes through the phosphorylation of PPARα [21,22]. p38 MAPK also inhibits hepatic lipogenesis [23], and therefore, when activated, it prevents lipid accumulation in the liver.

1.2.7. X-Box Binding Protein 1 (Xbp1)

This member of the CREB/ATF family of transcription factors which is activated by splicing under conditions of endoplasmic reticulum (ER) stress, is a master regulator of unfolded protein response (UPR) folding capacity. Obesity and T2D increase ER stress in the liver, and expression of Xbp1 significantly decreases ER stress, restores glucose tolerance, and reduces blood glucose levels [24]. By phosphorylating Thr48 and Ser61 residues of the spliced form of Xbp1, p38 MAPK enhances its nuclear translocation and activity. P38 MAPK activity is diminished in the livers of obese mice compared with lean mice. Conversely, its activation by the expression of constitutively active MKK6 kinase (MKK6(E)) reduces ER stress and establishes euglycemia in obese diabetic mice [25].

2. Regulation of Glucose Metabolism by p38 MAPK in the Adaptation of Skeletal Muscle to Exercise

It is well established that physical activity and exercise training are beneficial to health and can prevent insulin resistance, type 2 diabetes (T2D) [26], and sarcopenia [27]. Muscle adaptation to exercise includes changes in contractile proteins, mitochondrial function, metabolic regulation, and specific signaling pathways that regulate gene expression. Muscle adaptation is necessary, since maximal exercise induces a 20-fold increase in whole-body metabolism and up to a 100-fold increase in ATP consumption relative to resting skeletal muscle [28]. Exercise affects aerobic (endurance) and anaerobic (resistance) metabolism, with different benefits to each modality, as is reviewed in detail elsewhere [28]. The combination of both types of exercise is more effective than each in preventing obesity, reducing insulin resistance, and improving glycemic control in T2D [29]. Here, we focus on p38 MAPK in the adaptation of skeletal muscle carbohydrate metabolism to exercise. Three of the four isoforms of p38, α, β, and γ are expressed in skeletal muscle [30]. Exercise induces their activation, whose level and duration depend on the mode of exercise [31–33]. P38 MAPK affects the crucial processes necessary for the adaptation to the metabolic demands and energy needs of the exercising skeletal muscle. This is done by increasing glucose transport into the tissue, elevating glycolytic and citric acid cycle flux, and raising the mitochondria's number and functional quality. P38 MAPK mediates these positive effects by phosphorylating diverse transcription factors and coactivators involved in carbohydrate metabolism.

2.1. Glucose Uptake

Exercise induces a dramatic increase in glucose uptake by the Glut4 transporter. The primary sensor of energy demand in contracting muscle is AMP-activated protein kinase (AMPK), a serine/threonine kinase activated by low energy reservoirs and muscle contraction. Low intracellular energy levels are reflected by increased AMP: ATP ratio and muscle contraction by increasing the concentration of cytoplasmic Ca^{2+}. The alpha subunit of the heterotrimeric AMPK is phosphorylated at Thr172 by liver kinase B1 (LKB1), by calcium/calmodulin-dependent protein kinase β (CaMKβ), and by the transforming growth factor-beta activated kinase 1 (TAK1) that dramatically increases its kinase activity towards metabolic substrates [34]. Among these substrates are the Rab GTPase-activating proteins (GAPs), TBC1D1 and TBC1D4 (AS160) that are inactivated by phosphorylation, a step necessary for Glut4 translocation to the plasma membrane. A possible role of p38 MAPK in glucose uptake was deduced from the findings that its activity is positively correlated with and depends on AMPK [35–38]. Activation of both kinases, AMPK and p38 MAPK by TAK1 may explain their synchronization. Moreover, the Mef2 transcription factors involved in the expression of Glut4 in the exercised muscle are substrates of p38 MAPK [39,40]. Endurance exercise increases *Glut4* gene expression by the binding of Mef2 to its cognate site at the *Glut4* promoter [41]. Exercise also increases the kinase activity of p38α/β, which associates with and phosphorylates Mef2 to enhance its transcriptional activity. Indeed, Chambers and colleagues showed that stretch-stimulated glucose uptake was diminished by inhibitors of p38α and β [35,42]. However, other studies investigating adipocytes and myotubes revealed an unexpected off-target activity of the p38α, β inhibitor SB203580, that directly bound to Glut4 transporter and interfered with its activity. The authors concluded that this inhibitor could diminish glucose uptake through the inhibition of Glut4 and not necessarily through p38 MAPK [43,44]. Nevertheless, the addition of other p38α, β inhibitors that are chemically unrelated to SB203580, and the expression of a dominant-negative p38α mutant (p38AGF) reduced insulin-stimulated glucose uptake [45]. In another study, activation of p38 MAPK by expressing constitutively active MKK6 (MKK6E) upregulated Glut1 and down-regulated Glut4, thereby increased basal glucose uptake, but diminished insulin-mediated glucose uptake in L6 myotubes [46]. Studies investigating intracellular and extracellular agonists of p38 MAPK implicate this kinase in insulin-independent glucose uptake by muscles. For example, the drug anisomycin, which activates the stress kinases JNK and p38 MAPKs, increased glucose uptake in resting muscles [47]. Homocysteine sulfonic acid, a glutamate receptor agonist, was shown to stimulate glucose uptake in myotubes through an AMPK-p38 -dependent pathway [48]. Rac1 GTPase, one of the known effectors of p38, facilitates exercise mediated translocation of Glut4 to the plasma membrane and glucose uptake [49]. Furthermore, the accumulation of cytosolic reactive oxygen species (ROS) produced by NADPH oxidase 2 (NOX2), entailed phosphorylation of p38 MAPK, and stimulated Glut4-dependent glucose uptake in muscle during exercise [50]. However, p38 MAPK phosphorylation levels were not always consistent with the reduction in glucose uptake in muscles of mice devoid of NOX2 activity, indicative that p38 may not be the mediator of glucose uptake, due to the accumulation of cytoplasmic ROS. Investigation of the role skeletal muscle's most abundant isoform, p38γ, in glucose transport revealed that its overexpression in mature muscle fibers reduced the expression of Glut4 and decreased contraction-induced glucose uptake [51]. Overall, most studies demonstrated the involvement of p38 MAPK activity in exercise-mediated glucose uptake. Yet, it remains unclear whether this family of kinases mediates glucose uptake by inducing the expression of Glut4, Glut1, or both, and the particular roles of the different isoforms in basal and stimulated glucose uptake.

2.2. Mitochondrial Activities

The transcriptional coactivator PGC1α is a crucial regulator of mitochondrial function, oxidative metabolism, and energy homeostasis in a variety of tissues [52]. The expression of PGC1α is induced in muscle following endurance or resistance exercise [53]. Forced expression of PGC1α in mice's skeletal muscle is sufficient to improve performance capacity during exercise [54] and protect skeletal muscle from sarcopenia and aging-related metabolic diseases [55]. The core function of PGC1α in skeletal muscle is to increase the number and capacity of mitochondria and enhance oxidative metabolism. It orchestrates these functions by co-activing the estrogen-related receptor α (ERRα), the nuclear respiratory factor 1 and 2 (NRF1 and 2), and subsequently increasing the level of the mitochondrial transcription factor A (TFAM). P38 MAPK plays a central role in the expression and activity of PGC1α. Several inducible transcription factors rapidly and robustly increase PGC1α gene expression; CREB and ATF2 [56], that bind to cAMP-response elements (CRE), and Mef2C and D [57], that bind to the YTA(A/T)$_4$TAR sequence found within regulatory sequences of the *Pgc-1α* gene. These transcription factors are phosphorylated and activated by p38 MAPK and by Ca^{2+} signaling mediated by calcineurin and CaMK pathways [58,59]. Therefore, several signaling pathways, including p38 MAPK, converge onto *Pgc1α* gene expression and muscle adaptation following exercise. Moreover, transgenic mice with skeletal muscle-specific expression of active MKK6 express higher levels of PGC1α and other markers of mitochondrial biogenesis in fast-twitch muscle [56]. At another level, p38 MAPK directly phosphorylates the PGC-1α protein to promote its stability and activity [8]. PGC1α contains a negative regulatory domain that inhibits the function of the transcriptional activation domain [60]. Transcription suppression is relieved when p38 MAPK phosphorylates residues Thr262, Ser265, and Thr298 within the negative regulatory domain of PGC1α [61,62]. These phosphorylation events stabilize PGC1α and drive its dissociation from the p160MBP repressor [8]. Forced expression of PGC1α in skeletal muscle adapting to exercise also leads to the conversion of muscle fibers from type 2 (fast-twitch) to type 1 (slow-twitch) [63]. Moreover, an autoregulatory loop established between Mef2 and PGC1α maintains type 1 fibers in the contracting muscle [64]. Although there is no direct supporting evidence, it is reasonable to conjecture that p38 MAPK is also involved in fiber type adaptation by activating Mef2 and PGC1α. However, some results suggest otherwise; muscle-specific deletions of each p38 MAPK isoform (α, β, γ) indicated that none of the isoforms were required for exercise-induced slow fiber type transformation [65]. Still, p38γ MAPK was the only isoform required for endurance exercise-induced mitochondrial biogenesis. Overexpression of a dominant-negative form of p38γ MAPK, but not of p38α MAPK or p38β MAPK, blocked contraction induced *Pcg 1α* transcription [65]. It was, therefore, concluded that p38γ MAPK was involved in metabolic adaptation, but not in the adaptation of skeletal muscle contractile machinery to exercise. Interestingly, the results of another study indicate that p38γ MAPK is phosphorylated and activated in slow (soleus) but not in fast (gastrocnemius) muscles. Loss of p38γ MAPK reduced slow myosin-expressing fibers and increased the number of fast myosin-expressing fibers in the soleus [66]. Overall, these studies suggest that the p38 MAPK pathway is involved in the adaptation of mitochondria and oxidative metabolism to exercise. Still, its involvement in the adaptation of the contractile machinery calls for further inquiry. Future studies should also look for possible contributions of different p38 MAPK isoforms to the metabolic adaptation of muscle to exercise.

3. P38 MAPK in the Development of Insulin Resistance and Type 2 Diabetes (T2D)

3.1. Insulin Resistance

Insulin resistance is defined as the reduction in insulin's ability to stimulate glucose uptake by the body's peripheral tissues. Insulin is known to activate the canonical IRS-PI3K-Akt pathway and phosphorylate and inactivate Akt substrate of 160 kDa (AS160, TBC1D4). By inactivating AS160, insulin promotes Glut4-intracellular vesicles' transport and their fusion with the plasma membrane, consequently increasing glucose uptake by tissues [3]. Obesity and genetic factors induce chronic metabolic syndrome that involves dyslipidemia, inflammation, hypertension, and insulin resistance. Systemic insulin resistance triggers chronic hyperglycemia, which causes pancreatic β cells to secrete more insulin. At later stages of the syndrome, increased insulin secretion induces ER stress, β cell death, insulin deficiency, and diabetes. Skeletal muscle accounts for ~30% of the resting metabolic rate in humans [1,67], and up to 80% of glucose disposal under insulin-stimulated conditions, and it is thus the primary organ regulating glucose balance [1,2]. One model explaining the development of insulin resistance in skeletal muscle argues that the excess of lipids stored in the adipose tissue are released into the circulation as fatty acids that are taken up and accumulate in organs, such as the liver and the skeletal muscles [68]. Intramuscular lipid metabolites damage mitochondrial activity and consequently increase the production of reactive oxygen radicals. Another model predicts that calorie surplus imposes maximal mitochondrial respiration and leakage of electrons from the electron transport chain, forming oxygen radicals [69]. Oxidative stress inhibits insulin action via activation of serine/threonine kinases, including p38 MAPK that phosphorylate and neutralize insulin receptor substrate 1 (IRS1) [70,71]. Excessive oxygen radicals also inflict cellular damage by oxidizing proteins, fatty acids, and carbohydrates.

3.2. P38 MAPK in Skeletal Muscle Insulin Resistance

The role of the p38 MAPK pathway in the development of insulin resistance remains controversial. Likely sources of the dispute are different experimental-settings and systems, some of which investigated insulin resistance in muscle cell cultures and others in animal models. The general concept for the involvement of p38 MAPK in insulin resistance is that its activity inhibits insulin signaling by enhancing inhibitory phosphorylation of IRS1 at Ser307 and other related residues [72]. IRS1 may not be a direct target of p38 MAPK, but of other associated kinases, including IkappaB kinase (IKK), JNK, and some novel atypical PKC isoforms. The treatment of primary myotubes with tumor necrosis factor (TNF)α produced insulin resistance associated with Ser307 phosphorylation, which was mediated by p38 α,β MAPK [73]. Analysis of oxidative stress-induced insulin resistance in soleus muscle strips revealed an exciting insight into the possible role of p38 MAPK [74]. Oxidative stress-induced chronic activity of p38 MAPK increased basal glucose transport activity (in the absence of insulin), but blocked, at the same time, insulin signaling and, consequently, insulin-induced glucose transport [75]. In another study, forced expression of constitutive active MKK6/3 in L6 myotubes increased the expression of Glut1 while decreasing that of Glut4, thereby enhancing basal glucose transport and diminishing insulin-induced transport [46]. In this respect, it is interesting to note that the basal phosphorylation of p38 MAPK was higher in the skeletal muscle of type 2 diabetic human subjects than in healthy controls. Insulin treatment transiently increased phosphorylation of p38 MAPK in skeletal muscles of non-diabetic individuals, but not of T2D patients [76]. Ex vivo treatment of isolated soleus muscle with oxidative radicals entailed a significant decrease in insulin-stimulated glucose transport activity associated with selective loss of IRS1 and IRS2 proteins and augmented phosphorylation of IRS1 at Ser307. Inhibition of p38α/β MAPK partially restored the impaired insulin-stimulated glucose transport activity [77]. Together, these findings suggest a role of p38 MAPK in the mediation of basal glucose transport and in the negative regulation of insulin-stimulated glucose transport activities that may be regulated by different isoforms of p38 (α, β or γ) MAPK. At least one study pointed at p38γ MAPK as the isoform mediating both activities, inducing basal glucose uptake, and inhibiting muscle contraction-stimulated glucose uptake [51]. An additional aspect of the involvement p38 MAPK in insulin resistance is its suppression of PGC1α α and PGC1β expression that entails the accumulation of lipid metabolites [78]. This result is unexpected given the reported role of p38 MAPK in increasing PGC1α expression/function in contracting muscles. Insulin resistance associated with the elevated phosphorylation of JNK and p38 MAPKs was also observed in rat skeletal muscles after 6 h of hind limb immobilization [79]. Inflammatory cytokines such as TNFα and Il-6 also cause insulin resistance, and p38 MAPK itself is a crucial mediator of the expression of genes encoding for proinflammatory cytokines in skeletal muscles. Indeed, inhibition of p38 MAPK in myotubes derived from T2D subjects prevented proinflammatory cytokines' secretion, but did not improve insulin resistance [80]. In another study, the treatment of primary myotubes with TNFα produced insulin resistance dependent on the elevated activity of p38 MAPK [73]. TNFα-P38 MAPK pathway induced serine phosphorylation of insulin receptor (IR) and IRS1, and reduced tyrosine phosphorylation of the same molecules. P38 MAPK-dependent insulin resistance was also reported in myotubes that were cultured in conditioned medium of macrophages treated with the saturated fatty acid, palmitate, and contained increased levels of proinflammatory cytokines [81]. In conclusion, different metabolites known to be involved in the development of insulin resistance (saturated fatty acids, oxidative radicals, and inflammatory cytokines), induce the chronically elevated p38 MAPK in skeletal muscles, that is also likely to mediate insulin resistance.

Despite the above studies that establish a role for p38 MAPK in insulin resistance, its requirement for oxidative metabolism and its possible function in the transition of fast to slow-twitch fibers during exercise is in line with the idea that elevated p38 MAPK activity might prevent the development of obesity and insulin resistance. Indeed, an accepted perception is that insulin-stimulated glucose transport is higher in muscles enriched with slow-twitched oxidative fibers [82]. Several animal and cell culture studies support a role for p38 MAPK in preventing insulin resistance development. By manipulating the expression of MAPK phosphatase-1 (MKP-1) that inactivates both JNK and p38 MAPK in mouse skeletal muscle, Bennett and colleagues suggested that the activities of p38 MAPK and JNK were required for whole-body energy expenditure [22,83,84]. High-fat diet (HFD) nutrition of mice that upregulated the expression of MKP-1 in skeletal muscle entailed obesity and insulin resistance through the inactivation of p38/JNK MAPKs. Conversely, knockout of MKP-1 in skeletal muscles of mice fed on HFD prevented the development of insulin resistance [83]. These authors suggested that increased p38/JNK activity in skeletal muscle prevents obesity and insulin resistance by augmenting oxidative metabolism. However, all three MAPKs, ERK, JNK, and p38 targeted by MKP-1, can affect metabolic regulation. For example, mice lacking ERK1 display resistance to diet-induced obesity, coupled with protection from insulin resistance [85]. Thus, muscle expression of MKP-1 probably reflects complex changes in the activities of all three major classes of MAPK. Adiponectin, the most abundant peptide secreted from adipocytes, is also secreted by additional tissues, including skeletal muscle. Adiponectin has many metabolic benefits and improves glucose uptake, utilization, and fatty acid oxidation in myotubes [86]. The peptide binds to its receptors AdipoR1, and AdipoR2 found on membranes of skeletal muscle cells, and induces AMPK and p38 MAPK [87]. Activation of p38 MAPK and AMPK is essential for adiponectin–induced glucose uptake and fatty acid oxidation. These two kinases mediate metabolic changes by stimulating the transcriptional activity of PPARα [88]. Moreover, the forced expression of APPL1, an intracellular adaptor protein of adiponectin signaling, enhances phosphorylation of AMPK and p38 MAPK in myotubes, leading to the translocation of Glut4 to the plasma membrane [89]. By activating the two kinases, AMPK and p38 MAPK, a conserved 13 residue-long peptide derived from the adiponectin collagen domain (ADP-1) was sufficient to induce glucose uptake in TNFα-treated insulin-resistant myotubes [90].

Although the common idea that p38 MAPK is involved in insulin-independent glucose uptake, several studies also suggest its involvement in insulin-dependent glucose uptake. First, p38 MAPK was transiently phosphorylated, not only by muscle contraction, but also by the administration of insulin [91–93]. Moreover, exercise increased p38 MAPK phosphorylation, and insulin administration following exercise induced even higher p38 phosphorylation levels that were correlated with enhanced insulin sensitivity in human skeletal muscles [93]. A more recent study that tested the hypothesis that exercise increases insulin sensitivity revealed that anisomycin-mediated activation of p38 in skeletal muscles increased insulin-mediated glucose transport. However, the same study also showed that p38 was not necessary to increase insulin sensitivity following muscle contraction [47]. In this respect, it is essential to mention the early report by Klip and colleagues that SB203580, an inhibitor of p38 α and β attenuated insulin-stimulated glucose uptake [42]. However, later studies by the same group ruled out the involvement of p38 in the process [43]. Moreover, SB203580 reduced insulin-mediated glucose transport through the inhibition of Glut4 transporter activity in a mechanism that might involve direct interaction of the inhibitor with Glut4 that maintained the transporter in an inactive conformation, or through the removal of an inhibitory protein [44,94,95].

In summary, under conditions that affect insulin resistance, chronically activated p38 MAPK is involved with additional kinases, in the inactivation of IRS molecules. Interestingly, despite its role in preventing insulin-mediated glucose uptake, p38 MAPK may increase basal and insulin-independent glucose uptake by skeletal muscle. Different treatments/conditions that induced a burst of p38 MAPK activity prevented obesity and insulin resistance. These conditions include muscle contraction and hormones like adiponectin. Therefore, whereas p38 MAPK prevents insulin-mediated glucose transport, it participates with other signaling molecules in insulin-independent glucose uptake. Acute activation of p38 MAPK may also play a role in insulin-dependent glucose uptake. The intensities and duration of p38 MAPK signals, the involvement of specific p38 MAPK isoforms, and the molecular mechanisms of modulation of glucose uptake await future investigation.

3.3. *P38 MAPK as a Potential Target for Prevention of Obesity-Induced T2DM*

The double edged-sword activity of p38 MAPK in the metabolism of skeletal muscle raises the question of what intervention with this kinase or pathway could beneficially prevent the development of T2DM. On the one hand, elevation of p38 MAPK activity is expected to increase skeletal muscle oxidative metabolism and glucose uptake. To date, regular exercise is more effective than pharmacological intervention in the treatment and prevention of T2DM [26] and sarcopenia [27]. P38 MAPK mediates stimulation of glucose uptake during exercise by elevating the expression levels of Glut4 and increasing PGC1α activity that entails mitochondrial biogenesis and oxidative metabolism. These positive effects of p38 MAPK are insulin-independent, and therefore can bypass insulin resistance. For decades, the most important first-line drug for T2D patients' treatment is metformin, which induces the activation of AMPK [96]. AMPK facilitates the translocation of Glut4 to the plasma membrane and increases PGC1α-mediated mitochondrial biogenesis and respiration. The concurrent activation of AMPK and p38 MAPK signaling during exercise indicates that the two pathways cross-react and synergize in increasing glucose uptake and oxidative metabolism in skeletal muscle. Interestingly, forced activation of p38 MAPK in livers of obese mice reduces endoplasmic reticulum (ER) stress and establishes euglycemia [25]. Therefore, the application of pharmacoactive agents that augment p38 MAPK activity together with AMPK agonists may synergistically improve glucose import and oxidative metabolism of skeletal muscle in an insulin-independent fashion. However, activation of p38 MAPK to treat obesity and T2D holds a risk due to its possible deleterious effects on whole-body metabolism. For example, the elevated chronic activity of p38α MAPK and p38β MAPK that result from inactivity, denervation, aging (sarcopenia), or cancer (cachexia) participates in catabolism and muscle atrophy and weakness. Under these conditions, p38 MAPK upregulates E3 ligases like MAFbx/Atrogin1 and Murf1 targeting proteins for proteasome degradation [97–101]. Inhibition of p38 MAPK activity protects muscle from the oxidative damage and prevents proteolysis, and was, therefore, suggested as a potential target for the treatment of muscle atrophy [102–104]. The continuous activity of p38 MAPK is also involved in the etiology of inflammatory diseases and specifically bowel diseases, like ulcerative colitis and Crohn's disease [105]. P38 induces the expression of proinflammatory cytokines like TNFα and Il1-β, and the inhibition of p38 MAPK can effectively suppress the expression of circulating inflammatory mediators.

An alternative option for preventing insulin resistance is the inhibition of p38 MAPK activity. This approach's primary rationale is prevention of the inhibitory phosphorylation of IRS molecules that p38 MAPK mediates and ensuing the restoration of insulin signaling. Moreover, inhibition of p38 MAPK should block hepatic gluconeogenesis and to consequently reduce hepatic glucose release and lower blood glucose levels [10,23]. Interestingly, the deletion of the *mapk14* gene encoding for p38α MAPK in the liver is associated with elevated levels of active AMPK, which is known to suppress gluconeogenesis [106]. In fact, metformin's significant therapeutic effect is the suppression of liver gluconeogenesis, likely by the activation of AMPK [96]. Therefore, some of the beneficial effects of p38 MAPK inhibition in reducing blood glucose levels may be mediated by liver AMPK. However, as mentioned before, since p38 MAPK activity is vital for mitochondrial function and oxidative metabolism, its inhibition may have adverse effects on muscle glucose metabolism [8,56,62]. Inhibition of p38 MAPK is anticipated to reduce skeletal muscle oxidative phosphorylation and compensate for the lost energy by increasing anaerobic glycolysis [102]. The prevailing model is that excessive calorie intake cause mitochondrial electron leak, accumulation of ROS, mitochondrial dysfunction, and insulin resistance [68,107,108]. Studies in recent years showed that insulin resistance occurred before the deterioration of mitochondrial function [109,110]. Therefore, the loss of mitochondrial oxidative metabolism may not be the cause of insulin resistance. One approach for preventing insulin resistance and improving glucose/fat metabolism is to increase slow/oxidative muscles [111,112]. An equally effective different approach is to increase the relative proportion of glycolytic muscle fibers, which were proven to reduce fat mass and increase insulin sensitivity [113–116]. By increasing the glucose uptake rate from the plasma and their glycolytic flux, fast-glycolytic muscles maintain energy levels similar to those of slow-oxidative muscles. In fact, the commonly prescribed drugs to treat T2D, biguanides such as metformin, inhibit complex I of the respiratory chain and thereby impair both mitochondrial function and cell respiration [117,118]. Therefore, increasing glycolytic metabolism in skeletal muscle by inhibiting p38 MAPK may help restore glucose tolerance in diabetic patients.

4. Conclusions

Despite the substantial progress in understanding p38 MAPK involvement in the regulation of glucose metabolism in skeletal muscle, there remain open questions concerning the identities and involvement of p38 MAPK isoforms in different physiologic states and their modes of affecting their metabolic targets. Due to its massive energy consumption, skeletal muscle functionality is critical for maintaining whole-body glucose homeostasis. Insulin resistance and sarcopenia develop mainly in obese and old immobile individuals due to dysregulated energy metabolism of skeletal muscles, and p38 MAPK is involved in the development of these maladies. Hence, depending on the physiological context, p38 MAPK activity may lead to harmful consequences of sarcopenia and insulin resistance in the immobile muscle or beneficial effects of increased glucose sensitivity and metabolism in the contracting muscle. This review summarized the involvement of p38 MAPK in glucose metabolism of skeletal muscle under the two extreme conditions of exercise (health) and obesity (disease), as schemed in Figure 1. In exercise training, muscle contraction utilizes an enormous amount of energy, supplied by the adaptation of muscles to glycolytic (resistance) or oxidative (endurance) glucose metabolism. Under these conditions, p38 MAPK facilitates glucose transport in an insulin-independent manner and improves insulin-dependent glucose transport by inducing transcription of genes encoding the glucose transporters, Glut1, and Glut4. P38 MAPK also increases oxidative metabolism by inducing the transcription and the activity of PGC1α, the "master coactivator" of mitochondrial biogenesis and activity. Interestingly, in the contracting muscle, p38 MAPK is co-expressed with AMPK, the "energy sensor" of the cell. The activities of these two kinases synergistically increase glucose import in a pathway alternative to that of insulin. They also facilitate together mitochondrial oxidative metabolism to provide the necessary amount of energy for muscle contraction.

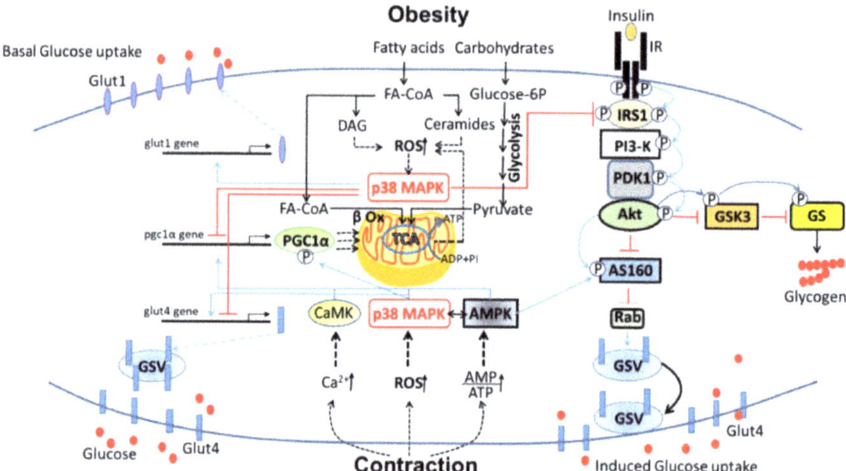

Figure 1. The involvement of p38 mitogen-activated protein kinase (MAPK) in glucose metabolism of skeletal muscle in health and disease. The upper half of the scheme describes the involvement of p38 MAPK in glucose metabolism of skeletal muscle in obesity: excessive intake of fatty acids and carbohydrates cause mitochondrial electron leak from the electron transport chain (ETC), accumulation of ROS, and mitochondrial dysfunction. Intramuscular fat metabolites (ceramide and DAG) reduce mitochondrial oxidative capacity and increase in the generation of mitochondrial reactive oxygen species (ROS) that induce p38 MAPK activity. Activated p38 MAPK inhibits IRS1 of insulin signaling through inhibitory phosphorylation. It also inhibits the transcription of *Pgc1α*, *Glut4* genes, and activates that of the *Glut1* gene. As a result, insulin-dependent glucose uptake is blocked, while insulin-independent glucose uptake is elevated. The lower half of the scheme describes the involvement of p38 MAPK in glucose metabolism of skeletal muscle in exercise: transient elevation in ROS induces the activity of p38 MAPK, which in turn stimulates the transcription of *Pgc1α* and *Glut4* genes. Besides, p38 phosphorylates PGC1α and augments its activity needed for mitochondrial integrity and function. P38 MAPK synergizes with AMPK in glucose uptake; the first increases the levels of Glut4 and the second drives the transport of vesicular Glut4 to the plasma membrane. Abbreviations: GSV, Glut4 storage vesicles; DAG, diacylglycerol; PI3-K, Phosphoinositide 3-kinase; PDK1, Phosphoinositide-dependent protein kinase-1; GSK3, Glycogen synthase kinase 3; GS, Glycogen synthase; IR, Insulin receptor; FA, Fatty acid.

In obesity, however, the activity of p38 MAPK contributes to the development of insulin resistance. The surplus in skeletal muscle calorie intake and the elevated levels of circulating fatty acids that penetrate muscle cells, increase intramuscular fat metabolites that reduce mitochondrial oxidative capacity and high electron leakage that generate ROS. ROS and lipid metabolites activate some kinases, including p38 MAPK, that impair the insulin-signaling pathway by phosphorylating serine/threonine residues of IRS1 and preventing tyrosine phosphorylation and activation of IRS1 by insulin. Consequently, insulin signaling is impaired and Glut4-mediated glucose uptake is prevented. The chronic activity of p38 MAPK also decreases the expression of PGC1 proteins and, consequently, diminishes glucose utilization by the mitochondria. Under these pathological conditions, p38 MAPK activates the *Glut1* gene expression and increases the basal diffusion of glucose independently of insulin signaling.

In sum, the p38 MAPK pathway is a double-edged sword that increases insulin-independent glucose uptake and mitochondrial oxidative phosphorylation in a healthy lifestyle while inhibiting, in unhealthy lifestyles, the same processes mediated by insulin signaling, leading to metabolic syndrome.

Author Contributions: E.B. and T.H. collected and analyzed the material, and wrote the paper. S.A. critically reviewed, commented and added changes to the paper. All authors have read and agreed to the published version of the manuscript.

Funding: This work was supported by a grant from "Rappaport Family Institute for Research in the Medical Sciences".

Acknowledgments: We thank Michael Fry for critical reading of the manuscript.

Conflicts of Interest: The authors declare that they have no conflict of interest.

Abbreviations

AMPK	AMP-activated protein kinase
MAPK	Mitogen-Activated Protein Kinase
ROS	reactive oxygen species
PGC-1α	Peroxisome proliferator-activated receptor gamma coactivator 1-alpha;
SAPK	Stress-activated protein kinase
T2D	Type 2 Diabetes
AS160	Akt substrate of 160 kDa
IRS	Insulin receptor substrate
Glut	Glucose transporter

References

1. Zurlo, F.; Larson, K.; Bogardus, C.; Ravussin, E. Skeletal muscle metabolism is a major determinant of resting energy expenditure. *J. Clin. Investig.* **1990**, *86*, 1423–1427. [CrossRef]
2. DeFronzo, R.A.; Ferrannini, E.; Sato, Y.; Felig, P.; Wahren, J. Synergistic interaction between exercise and insulin on peripheral glucose uptake. *J. Clin. Investig.* **1981**, *68*, 1468–1474. [CrossRef] [PubMed]
3. Richter, E.A.; Hargreaves, M. Exercise, GLUT4, and skeletal muscle glucose uptake. *Physiol. Rev.* **2013**, *93*, 993–1017. [CrossRef] [PubMed]
4. Ciciliot, S.; Rossi, A.C.; Dyar, K.A.; Blaauw, B.; Schiaffino, S. Muscle type and fiber type specificity in muscle wasting. *Int. J. Biochem. Cell Biol.* **2013**, *45*, 2191–2199. [CrossRef] [PubMed]
5. Kriketos, A.D.; Baur, L.A.; O'Connor, J.; Carey, D.; King, S.; Caterson, I.D.; Storlien, L.H. Muscle fibre type composition in infant and adult populations and relationships with obesity. *Int. J. Obes. Relat. Metab. Disord. J. Int. Assoc. Study Obes.* **1997**, *21*, 796–801. [CrossRef]
6. Thayer, R.; Collins, J.; Noble, E.G.; Taylor, A.W. A decade of aerobic endurance training: Histological evidence for fibre type transformation. *J. Sports Med. Phys. Fit.* **2000**, *40*, 284–289.
7. Sabio, G.; Davis, R.J. TNF and MAP kinase signalling pathways. *Semin. Immunol.* **2014**, *26*, 237–245. [CrossRef]
8. Fan, M.; Rhee, J.; St-Pierre, J.; Handschin, C.; Puigserver, P.; Lin, J.; Jaeger, S.; Erdjument-Bromage, H.; Tempst, P.; Spiegelman, B.M. Suppression of mitochondrial respiration through recruitment of p160 myb binding protein to PGC-1alpha: Modulation by p38 MAPK. *Genes Dev.* **2004**, *18*, 278–289. [CrossRef]
9. Gupta, S.; Campbell, D.; Derijard, B.; Davis, R.J. Transcription factor ATF2 regulation by the JNK signal transduction pathway. *Science* **1995**, *267*, 389–393. [CrossRef]
10. Cao, W.; Collins, Q.F.; Becker, T.C.; Robidoux, J.; Lupo, E.G., Jr.; Xiong, Y.; Daniel, K.W.; Floering, L.; Collins, S. p38 Mitogen-activated protein kinase plays a stimulatory role in hepatic gluconeogenesis. *J. Biol. Chem.* **2005**, *280*, 42731–42737. [CrossRef]
11. Maekawa, T.; Jin, W.; Ishii, S. The role of ATF-2 family transcription factors in adipocyte differentiation: Antiobesity effects of p38 inhibitors. *Mol. Cell Biol.* **2010**, *30*, 613–625. [CrossRef] [PubMed]
12. Tseng, Y.H.; Kokkotou, E.; Schulz, T.J.; Huang, T.L.; Winnay, J.N.; Taniguchi, C.M.; Tran, T.T.; Suzuki, R.; Espinoza, D.O.; Yamamoto, Y.; et al. New role of bone morphogenetic protein 7 in brown adipogenesis and energy expenditure. *Nature* **2008**, *454*, 1000–1004. [CrossRef] [PubMed]
13. Olofsson, L.E.; Orho-Melander, M.; William-Olsson, L.; Sjoholm, K.; Sjostrom, L.; Groop, L.; Carlsson, B.; Carlsson, L.M.; Olsson, B. CCAAT/enhancer binding protein alpha (C/EBPalpha) in adipose tissue regulates genes in lipid and glucose metabolism and a genetic variation in C/EBPalpha is associated with serum levels of triglycerides. *J. Clin. Endocrinol. Metab.* **2008**, *93*, 4880–4886. [CrossRef]

14. Yang, J.; Croniger, C.M.; Lekstrom-Himes, J.; Zhang, P.; Fenyus, M.; Tenen, D.G.; Darlington, G.J.; Hanson, R.W. Metabolic response of mice to a postnatal ablation of CCAAT/enhancer-binding protein alpha. *J. Biol. Chem.* **2005**, *280*, 38689–38699. [CrossRef] [PubMed]
15. Qiao, L.; MacDougald, O.A.; Shao, J. CCAAT/enhancer-binding protein alpha mediates induction of hepatic phosphoenolpyruvate carboxykinase by p38 mitogen-activated protein kinase. *J. Biol. Chem.* **2006**, *281*, 24390–24397. [CrossRef]
16. Mayr, B.; Montminy, M. Transcriptional regulation by the phosphorylation-dependent factor CREB. *Nat. Rev. Mol. Cell Biol.* **2001**, *2*, 599–609. [CrossRef]
17. Collins, Q.F.; Xiong, Y.; Lupo, E.G., Jr.; Liu, H.Y.; Cao, W. p38 Mitogen-activated protein kinase mediates free fatty acid-induced gluconeogenesis in hepatocytes. *J. Biol. Chem.* **2006**, *281*, 24336–24344. [CrossRef]
18. Herzig, S.; Long, F.; Jhala, U.S.; Hedrick, S.; Quinn, R.; Bauer, A.; Rudolph, D.; Schutz, G.; Yoon, C.; Puigserver, P.; et al. CREB regulates hepatic gluconeogenesis through the coactivator PGC-1. *Nature* **2001**, *413*, 179–183. [CrossRef]
19. Kuma, Y.; Campbell, D.G.; Cuenda, A. Identification of glycogen synthase as a new substrate for stress-activated protein kinase 2b/p38beta. *Biochem. J.* **2004**, *379*, 133–139. [CrossRef]
20. Lawrence, J.C., Jr.; Roach, P.J. New insights into the role and mechanism of glycogen synthase activation by insulin. *Diabetes* **1997**, *46*, 541–547. [CrossRef]
21. Barger, P.M.; Browning, A.C.; Garner, A.N.; Kelly, D.P. p38 mitogen-activated protein kinase activates peroxisome proliferator-activated receptor alpha: A potential role in the cardiac metabolic stress response. *J. Biol. Chem.* **2001**, *276*, 44495–44501. [CrossRef] [PubMed]
22. Wu, J.J.; Roth, R.J.; Anderson, E.J.; Hong, E.G.; Lee, M.K.; Choi, C.S.; Neufer, P.D.; Shulman, G.I.; Kim, J.K.; Bennett, A.M. Mice lacking MAP kinase phosphatase-1 have enhanced MAP kinase activity and resistance to diet-induced obesity. *Cell Metab.* **2006**, *4*, 61–73. [CrossRef] [PubMed]
23. Xiong, Y.; Collins, Q.F.; An, J.; Lupo, E., Jr.; Liu, H.Y.; Liu, D.; Robidoux, J.; Liu, Z.; Cao, W. p38 mitogen-activated protein kinase plays an inhibitory role in hepatic lipogenesis. *J. Biol. Chem.* **2007**, *282*, 4975–4982. [CrossRef] [PubMed]
24. Zhou, Y.; Lee, J.; Reno, C.M.; Sun, C.; Park, S.W.; Chung, J.; Lee, J.; Fisher, S.J.; White, M.F.; Biddinger, S.B.; et al. Regulation of glucose homeostasis through a XBP-1-FoxO1 interaction. *Nat. Med.* **2011**, *17*, 356–365. [CrossRef] [PubMed]
25. Lee, J.; Sun, C.; Zhou, Y.; Lee, J.; Gokalp, D.; Herrema, H.; Park, S.W.; Davis, R.J.; Ozcan, U. p38 MAPK-mediated regulation of Xbp1s is crucial for glucose homeostasis. *Nat. Med.* **2011**, *17*, 1251–1260. [CrossRef]
26. Diabetes Prevention Program Research Group. Reduction in the incidence of type 2 diabetes with lifestyle intervention or metformin. *N. Engl. J. Med.* **2002**, *346*, 393–403. [CrossRef]
27. Borst, S.E. Interventions for sarcopenia and muscle weakness in older people. *Age Ageing* **2004**, *33*, 548–555. [CrossRef]
28. Egan, B.; Zierath, J.R. Exercise metabolism and the molecular regulation of skeletal muscle adaptation. *Cell Metab.* **2013**, *17*, 162–184. [CrossRef]
29. Sigal, R.J.; Kenny, G.P.; Boule, N.G.; Wells, G.A.; Prud'homme, D.; Fortier, M.; Reid, R.D.; Tulloch, H.; Coyle, D.; Phillips, P.; et al. Effects of aerobic training, resistance training, or both on glycemic control in type 2 diabetes: A randomized trial. *Ann. Intern. Med.* **2007**, *147*, 357–369. [CrossRef]
30. Keren, A.; Tamir, Y.; Bengal, E. The p38 MAPK signaling pathway: A major regulator of skeletal muscle development. *Mol. Cell. Endocrinol.* **2006**, *252*, 224–230. [CrossRef]
31. Boppart, M.D.; Hirshman, M.F.; Sakamoto, K.; Fielding, R.A.; Goodyear, L.J. Static stretch increases c-Jun NH2-terminal kinase activity and p38 phosphorylation in rat skeletal muscle. *Am. J. Physiol. Cell Physiol.* **2001**, *280*, C352–C358. [CrossRef] [PubMed]
32. Coffey, V.G.; Zhong, Z.; Shield, A.; Canny, B.J.; Chibalin, A.V.; Zierath, J.R.; Hawley, J.A. Early signaling responses to divergent exercise stimuli in skeletal muscle from well-trained humans. *FASEB J. Off. Publ. Fed. Am. Soc. Exp. Biol.* **2006**, *20*, 190–192. [CrossRef] [PubMed]
33. Yu, M.; Stepto, N.K.; Chibalin, A.V.; Fryer, L.G.; Carling, D.; Krook, A.; Hawley, J.A.; Zierath, J.R. Metabolic and mitogenic signal transduction in human skeletal muscle after intense cycling exercise. *J. Physiol.* **2003**, *546*, 327–335. [CrossRef]

34. Mounier, R.; Theret, M.; Lantier, L.; Foretz, M.; Viollet, B. Expanding roles for AMPK in skeletal muscle plasticity. *Trends Endocrinol. Metab. TEM* **2015**, *26*, 275–286. [CrossRef] [PubMed]
35. Chambers, M.A.; Moylan, J.S.; Smith, J.D.; Goodyear, L.J.; Reid, M.B. Stretch-stimulated glucose uptake in skeletal muscle is mediated by reactive oxygen species and p38 MAP-kinase. *J. Physiol.* **2009**, *587*, 3363–3373. [CrossRef] [PubMed]
36. Combes, A.; Dekerle, J.; Webborn, N.; Watt, P.; Bougault, V.; Daussin, F.N. Exercise-induced metabolic fluctuations influence AMPK, p38-MAPK and CaMKII phosphorylation in human skeletal muscle. *Physiol. Rep.* **2015**, *3*, e12462. [CrossRef]
37. Kim, N.; Lee, J.O.; Lee, H.J.; Lee, Y.W.; Kim, H.I.; Kim, S.J.; Park, S.H.; Lee, C.S.; Ryoo, S.W.; Hwang, G.S.; et al. AMPK, a metabolic sensor, is involved in isoeugenol-induced glucose uptake in muscle cells. *J. Endocrinol.* **2016**, *228*, 105–114. [CrossRef]
38. Lemieux, K.; Konrad, D.; Klip, A.; Marette, A. The AMP-activated protein kinase activator AICAR does not induce GLUT4 translocation to transverse tubules but stimulates glucose uptake and p38 mitogen-activated protein kinases alpha and beta in skeletal muscle. *FASEB J. Off. Publ. Fed. Am. Soc. Exp. Biol.* **2003**, *17*, 1658–1665. [CrossRef]
39. Zetser, A.; Gredinger, E.; Bengal, E. p38 mitogen-activated protein kinase pathway promotes skeletal muscle differentiation. Participation of the Mef2c transcription factor. *J. Biol. Chem.* **1999**, *274*, 5193–5200. [CrossRef]
40. Zhao, M.; New, L.; Kravchenko, V.V.; Kato, Y.; Gram, H.; di Padova, F.; Olson, E.N.; Ulevitch, R.J.; Han, J. Regulation of the MEF2 family of transcription factors by p38. *Mol. Cell Biol.* **1999**, *19*, 21–30. [CrossRef]
41. McGee, S.L.; Hargreaves, M. Exercise and myocyte enhancer factor 2 regulation in human skeletal muscle. *Diabetes* **2004**, *53*, 1208–1214. [CrossRef] [PubMed]
42. Somwar, R.; Perreault, M.; Kapur, S.; Taha, C.; Sweeney, G.; Ramlal, T.; Kim, D.Y.; Keen, J.; Cote, C.H.; Klip, A.; et al. Activation of p38 mitogen-activated protein kinase alpha and beta by insulin and contraction in rat skeletal muscle: Potential role in the stimulation of glucose transport. *Diabetes* **2000**, *49*, 1794–1800. [CrossRef] [PubMed]
43. Antonescu, C.N.; Huang, C.; Niu, W.; Liu, Z.; Eyers, P.A.; Heidenreich, K.A.; Bilan, P.J.; Klip, A. Reduction of insulin-stimulated glucose uptake in L6 myotubes by the protein kinase inhibitor SB203580 is independent of p38MAPK activity. *Endocrinology* **2005**, *146*, 3773–3781. [CrossRef]
44. Ribe, D.; Yang, J.; Patel, S.; Koumanov, F.; Cushman, S.W.; Holman, G.D. Endofacial competitive inhibition of glucose transporter-4 intrinsic activity by the mitogen-activated protein kinase inhibitor SB203580. *Endocrinology* **2005**, *146*, 1713–1717. [CrossRef] [PubMed]
45. Somwar, R.; Koterski, S.; Sweeney, G.; Sciotti, R.; Djuric, S.; Berg, C.; Trevillyan, J.; Scherer, P.E.; Rondinone, C.M.; Klip, A. A dominant-negative p38 MAPK mutant and novel selective inhibitors of p38 MAPK reduce insulin-stimulated glucose uptake in 3T3-L1 adipocytes without affecting GLUT4 translocation. *J. Biol. Chem.* **2002**, *277*, 50386–50395. [CrossRef] [PubMed]
46. Fujishiro, M.; Gotoh, Y.; Katagiri, H.; Sakoda, H.; Ogihara, T.; Anai, M.; Onishi, Y.; Ono, H.; Funaki, M.; Inukai, K.; et al. MKK6/3 and p38 MAPK pathway activation is not necessary for insulin-induced glucose uptake but regulates glucose transporter expression. *J. Biol. Chem.* **2001**, *276*, 19800–19806. [CrossRef]
47. Geiger, P.C.; Wright, D.C.; Han, D.H.; Holloszy, J.O. Activation of p38 MAP kinase enhances sensitivity of muscle glucose transport to insulin. *Am. J. Physiol. Endocrinol. Metab.* **2005**, *288*, E782–E788. [CrossRef]
48. Kim, J.H.; Lee, J.O.; Lee, S.K.; Moon, J.W.; You, G.Y.; Kim, S.J.; Park, S.H.; Park, J.M.; Lim, S.Y.; Suh, P.G.; et al. The glutamate agonist homocysteine sulfinic acid stimulates glucose uptake through the calcium-dependent AMPK-p38 MAPK-protein kinase C zeta pathway in skeletal muscle cells. *J. Biol. Chem.* **2011**, *286*, 7567–7576. [CrossRef]
49. Sylow, L.; Nielsen, I.L.; Kleinert, M.; Moller, L.L.; Ploug, T.; Schjerling, P.; Bilan, P.J.; Klip, A.; Jensen, T.E.; Richter, E.A. Rac1 governs exercise-stimulated glucose uptake in skeletal muscle through regulation of GLUT4 translocation in mice. *J. Physiol.* **2016**, *594*, 4997–5008. [CrossRef]
50. Henriquez-Olguin, C.; Knudsen, J.R.; Raun, S.H.; Li, Z.; Dalbram, E.; Treebak, J.T.; Sylow, L.; Holmdahl, R.; Richter, E.A.; Jaimovich, E.; et al. Cytosolic ROS production by NADPH oxidase 2 regulates muscle glucose uptake during exercise. *Nat. Commun.* **2019**, *10*, 4623. [CrossRef]
51. Ho, R.C.; Alcazar, O.; Fujii, N.; Hirshman, M.F.; Goodyear, L.J. p38gamma MAPK regulation of glucose transporter expression and glucose uptake in L6 myotubes and mouse skeletal muscle. *Am. J. Physiol. Regul. Integr. Comp. Physiol.* **2004**, *286*, R342–R349. [CrossRef] [PubMed]

52. Lin, J.; Handschin, C.; Spiegelman, B.M. Metabolic control through the PGC-1 family of transcription coactivators. *Cell Metab.* **2005**, *1*, 361–370. [CrossRef] [PubMed]
53. Baar, K.; Wende, A.R.; Jones, T.E.; Marison, M.; Nolte, L.A.; Chen, M.; Kelly, D.P.; Holloszy, J.O. Adaptations of skeletal muscle to exercise: Rapid increase in the transcriptional coactivator PGC-1. *FASEB J. Off. Publ. Fed. Am. Soc. Exp. Biol.* **2002**, *16*, 1879–1886. [CrossRef] [PubMed]
54. Calvo, J.A.; Daniels, T.G.; Wang, X.; Paul, A.; Lin, J.; Spiegelman, B.M.; Stevenson, S.C.; Rangwala, S.M. Muscle-specific expression of PPARgamma coactivator-1alpha improves exercise performance and increases peak oxygen uptake. *J. Appl. Physiol.* **2008**, *104*, 1304–1312. [CrossRef] [PubMed]
55. Ji, L.L.; Kang, C. Role of PGC-1alpha in sarcopenia: Etiology and potential intervention—A mini-review. *Gerontology* **2015**, *61*, 139–148. [CrossRef] [PubMed]
56. Akimoto, T.; Pohnert, S.C.; Li, P.; Zhang, M.; Gumbs, C.; Rosenberg, P.B.; Williams, R.S.; Yan, Z. Exercise stimulates Pgc-1alpha transcription in skeletal muscle through activation of the p38 MAPK pathway. *J. Biol. Chem.* **2005**, *280*, 19587–19593. [CrossRef]
57. Perez-Schindler, J.; Summermatter, S.; Santos, G.; Zorzato, F.; Handschin, C. The transcriptional coactivator PGC-1alpha is dispensable for chronic overload-induced skeletal muscle hypertrophy and metabolic remodeling. *Proc. Natl. Acad. Sci. USA* **2013**, *110*, 20314–20319. [CrossRef]
58. McKinsey, T.A.; Zhang, C.L.; Olson, E.N. Activation of the myocyte enhancer factor-2 transcription factor by calcium/calmodulin-dependent protein kinase-stimulated binding of 14-3-3 to histone deacetylase 5. *Proc. Natl. Acad. Sci. USA* **2000**, *97*, 14400–14405. [CrossRef]
59. Wu, H.; Rothermel, B.; Kanatous, S.; Rosenberg, P.; Naya, F.J.; Shelton, J.M.; Hutcheson, K.A.; DiMaio, J.M.; Olson, E.N.; Bassel-Duby, R.; et al. Activation of MEF2 by muscle activity is mediated through a calcineurin-dependent pathway. *EMBO J.* **2001**, *20*, 6414–6423. [CrossRef]
60. Puigserver, P.; Adelmant, G.; Wu, Z.; Fan, M.; Xu, J.; O'Malley, B.; Spiegelman, B.M. Activation of PPARgamma coactivator-1 through transcription factor docking. *Science* **1999**, *286*, 1368–1371. [CrossRef]
61. Knutti, D.; Kressler, D.; Kralli, A. Regulation of the transcriptional coactivator PGC-1 via MAPK-sensitive interaction with a repressor. *Proc. Natl. Acad. Sci. USA* **2001**, *98*, 9713–9718. [CrossRef]
62. Puigserver, P.; Rhee, J.; Lin, J.; Wu, Z.; Yoon, J.C.; Zhang, C.Y.; Krauss, S.; Mootha, V.K.; Lowell, B.B.; Spiegelman, B.M. Cytokine stimulation of energy expenditure through p38 MAP kinase activation of PPARgamma coactivator-1. *Mol. Cell* **2001**, *8*, 971–982. [CrossRef]
63. Lin, J.; Wu, H.; Tarr, P.T.; Zhang, C.Y.; Wu, Z.; Boss, O.; Michael, L.F.; Puigserver, P.; Isotani, E.; Olson, E.N.; et al. Transcriptional co-activator PGC-1 alpha drives the formation of slow-twitch muscle fibres. *Nature* **2002**, *418*, 797–801. [CrossRef]
64. Handschin, C.; Rhee, J.; Lin, J.; Tarr, P.T.; Spiegelman, B.M. An autoregulatory loop controls peroxisome proliferator-activated receptor gamma coactivator 1alpha expression in muscle. *Proc. Natl. Acad. Sci. USA* **2003**, *100*, 7111–7116. [CrossRef]
65. Pogozelski, A.R.; Geng, T.; Li, P.; Yin, X.; Lira, V.A.; Zhang, M.; Chi, J.T.; Yan, Z. p38gamma mitogen-activated protein kinase is a key regulator in skeletal muscle metabolic adaptation in mice. *PLoS ONE* **2009**, *4*, e7934. [CrossRef]
66. Foster, W.H.; Tidball, J.G.; Wang, Y. p38gamma activity is required for maintenance of slow skeletal muscle size. *Muscle Nerve* **2012**, *45*, 266–273. [CrossRef]
67. Meyer, C.; Dostou, J.M.; Welle, S.L.; Gerich, J.E. Role of human liver, kidney, and skeletal muscle in postprandial glucose homeostasis. *Am. J. Physiol. Endocrinol. Metab.* **2002**, *282*, E419–E427. [CrossRef]
68. Bonen, A.; Parolin, M.L.; Steinberg, G.R.; Calles-Escandon, J.; Tandon, N.N.; Glatz, J.F.; Luiken, J.J.; Heigenhauser, G.J.; Dyck, D.J. Triacylglycerol accumulation in human obesity and type 2 diabetes is associated with increased rates of skeletal muscle fatty acid transport and increased sarcolemmal FAT/CD36. *FASEB J. Off. Publ. Fed. Am. Soc. Exp. Biol.* **2004**, *18*, 1144–1146. [CrossRef]
69. Fisher-Wellman, K.H.; Neufer, P.D. Linking mitochondrial bioenergetics to insulin resistance via redox biology. *Trends Endocrinol. Metab. TEM* **2012**, *23*, 142–153. [CrossRef]
70. Yu, C.; Chen, Y.; Cline, G.W.; Zhang, D.; Zong, H.; Wang, Y.; Bergeron, R.; Kim, J.K.; Cushman, S.W.; Cooney, G.J.; et al. Mechanism by which fatty acids inhibit insulin activation of insulin receptor substrate-1 (IRS-1)-associated phosphatidylinositol 3-kinase activity in muscle. *J. Biol. Chem.* **2002**, *277*, 50230–50236. [CrossRef]

71. Henriksen, E.J.; Diamond-Stanic, M.K.; Marchionne, E.M. Oxidative stress and the etiology of insulin resistance and type 2 diabetes. *Free Radic. Biol. Med.* **2011**, *51*, 993–999. [CrossRef]
72. Gehart, H.; Kumpf, S.; Ittner, A.; Ricci, R. MAPK signalling in cellular metabolism: Stress or wellness? *EMBO Rep.* **2010**, *11*, 834–840. [CrossRef]
73. de Alvaro, C.; Teruel, T.; Hernandez, R.; Lorenzo, M. Tumor necrosis factor alpha produces insulin resistance in skeletal muscle by activation of inhibitor kappaB kinase in a p38 MAPK-dependent manner. *J. Biol. Chem.* **2004**, *279*, 17070–17078. [CrossRef]
74. Kim, S.J.; Roy, R.R.; Kim, J.A.; Zhong, H.; Haddad, F.; Baldwin, K.M.; Edgerton, V.R. Gene expression during inactivity-induced muscle atrophy: Effects of brief bouts of a forceful contraction countermeasure. *J. Appl. Physiol.* **2008**, *105*, 1246–1254. [CrossRef]
75. Dokken, B.B.; Saengsirisuwan, V.; Kim, J.S.; Teachey, M.K.; Henriksen, E.J. Oxidative stress-induced insulin resistance in rat skeletal muscle: Role of glycogen synthase kinase-3. *Am. J. Physiol. Endocrinol. Metab.* **2008**, *294*, E615–E621. [CrossRef]
76. Koistinen, H.A.; Chibalin, A.V.; Zierath, J.R. Aberrant p38 mitogen-activated protein kinase signalling in skeletal muscle from Type 2 diabetic patients. *Diabetologia* **2003**, *46*, 1324–1328. [CrossRef]
77. Archuleta, T.L.; Lemieux, A.M.; Saengsirisuwan, V.; Teachey, M.K.; Lindborg, K.A.; Kim, J.S.; Henriksen, E.J. Oxidant stress-induced loss of IRS-1 and IRS-2 proteins in rat skeletal muscle: Role of p38 MAPK. *Free Radic. Biol. Med.* **2009**, *47*, 1486–1493. [CrossRef]
78. Crunkhorn, S.; Dearie, F.; Mantzoros, C.; Gami, H.; da Silva, W.S.; Espinoza, D.; Faucette, R.; Barry, K.; Bianco, A.C.; Patti, M.E. Peroxisome proliferator activator receptor gamma coactivator-1 expression is reduced in obesity: Potential pathogenic role of saturated fatty acids and p38 mitogen-activated protein kinase activation. *J. Biol. Chem.* **2007**, *282*, 15439–15450. [CrossRef]
79. Kawamoto, E.; Koshinaka, K.; Yoshimura, T.; Masuda, H.; Kawanaka, K. Immobilization rapidly induces muscle insulin resistance together with the activation of MAPKs (JNK and p38) and impairment of AS160 phosphorylation. *Physiol. Rep.* **2016**, *4*, e12876. [CrossRef]
80. Brown, A.E.; Palsgaard, J.; Borup, R.; Avery, P.; Gunn, D.A.; De Meyts, P.; Yeaman, S.J.; Walker, M. p38 MAPK activation upregulates proinflammatory pathways in skeletal muscle cells from insulin-resistant type 2 diabetic patients. *Am. J. Physiol. Endocrinol. Metab.* **2015**, *308*, E63–E70. [CrossRef]
81. Talbot, N.A.; Wheeler-Jones, C.P.; Cleasby, M.E. Palmitoleic acid prevents palmitic acid-induced macrophage activation and consequent p38 MAPK-mediated skeletal muscle insulin resistance. *Mol. Cell. Endocrinol.* **2014**, *393*, 129–142. [CrossRef]
82. Daugaard, J.R.; Nielsen, J.N.; Kristiansen, S.; Andersen, J.L.; Hargreaves, M.; Richter, E.A. Fiber type-specific expression of GLUT4 in human skeletal muscle: Influence of exercise training. *Diabetes* **2000**, *49*, 1092–1095. [CrossRef]
83. Lawan, A.; Min, K.; Zhang, L.; Canfran-Duque, A.; Jurczak, M.J.; Camporez, J.P.G.; Nie, Y.; Gavin, T.P.; Shulman, G.I.; Fernandez-Hernando, C.; et al. Skeletal Muscle-Specific Deletion of MKP-1 Reveals a p38 MAPK/JNK/Akt Signaling Node That Regulates Obesity-Induced Insulin Resistance. *Diabetes* **2018**, *67*, 624–635. [CrossRef]
84. Roth, R.J.; Le, A.M.; Zhang, L.; Kahn, M.; Samuel, V.T.; Shulman, G.I.; Bennett, A.M. MAPK phosphatase-1 facilitates the loss of oxidative myofibers associated with obesity in mice. *J. Clin. Investig.* **2009**, *119*, 3817–3829. [CrossRef]
85. Bost, F.; Aouadi, M.; Caron, L.; Even, P.; Belmonte, N.; Prot, M.; Dani, C.; Hofman, P.; Pages, G.; Pouyssegur, J.; et al. The extracellular signal-regulated kinase isoform ERK1 is specifically required for in vitro and in vivo adipogenesis. *Diabetes* **2005**, *54*, 402–411. [CrossRef]
86. Yamauchi, T.; Kamon, J.; Minokoshi, Y.; Ito, Y.; Waki, H.; Uchida, S.; Yamashita, S.; Noda, M.; Kita, S.; Ueki, K.; et al. Adiponectin stimulates glucose utilization and fatty-acid oxidation by activating AMP-activated protein kinase. *Nat. Med.* **2002**, *8*, 1288–1295. [CrossRef]
87. Yamauchi, T.; Kamon, J.; Ito, Y.; Tsuchida, A.; Yokomizo, T.; Kita, S.; Sugiyama, T.; Miyagishi, M.; Hara, K.; Tsunoda, M.; et al. Cloning of adiponectin receptors that mediate antidiabetic metabolic effects. *Nature* **2003**, *423*, 762–769. [CrossRef]
88. Yoon, M.J.; Lee, G.Y.; Chung, J.J.; Ahn, Y.H.; Hong, S.H.; Kim, J.B. Adiponectin increases fatty acid oxidation in skeletal muscle cells by sequential activation of AMP-activated protein kinase, p38 mitogen-activated protein kinase, and peroxisome proliferator-activated receptor alpha. *Diabetes* **2006**, *55*, 2562–2570. [CrossRef]

89. Mao, X.; Kikani, C.K.; Riojas, R.A.; Langlais, P.; Wang, L.; Ramos, F.J.; Fang, Q.; Christ-Roberts, C.Y.; Hong, J.Y.; Kim, R.Y.; et al. APPL1 binds to adiponectin receptors and mediates adiponectin signalling and function. *Nat. Cell Biol.* **2006**, *8*, 516–523. [CrossRef]
90. Sayeed, M.; Gautam, S.; Verma, D.P.; Afshan, T.; Kumari, T.; Srivastava, A.K.; Ghosh, J.K. A collagen domain-derived short adiponectin peptide activates APPL1 and AMPK signaling pathways and improves glucose and fatty acid metabolisms. *J. Biol. Chem.* **2018**, *293*, 13509–13523. [CrossRef]
91. Parker, L.; Stepto, N.K.; Shaw, C.S.; Serpiello, F.R.; Anderson, M.; Hare, D.L.; Levinger, I. Acute High-Intensity Interval Exercise-Induced Redox Signaling Is Associated with Enhanced Insulin Sensitivity in Obese Middle-Aged Men. *Front. Physiol.* **2016**, *7*, 411. [CrossRef]
92. Somwar, R.; Kim, D.Y.; Sweeney, G.; Huang, C.; Niu, W.; Lador, C.; Ramlal, T.; Klip, A. GLUT4 translocation precedes the stimulation of glucose uptake by insulin in muscle cells: Potential activation of GLUT4 via p38 mitogen-activated protein kinase. *Biochem. J.* **2001**, *359*, 639–649. [CrossRef]
93. Thong, F.S.; Derave, W.; Urso, B.; Kiens, B.; Richter, E.A. Prior exercise increases basal and insulin-induced p38 mitogen-activated protein kinase phosphorylation in human skeletal muscle. *J. Appl. Physiol.* **2003**, *94*, 2337–2341. [CrossRef]
94. Ishiki, M.; Klip, A. Minireview: Recent developments in the regulation of glucose transporter-4 traffic: New signals, locations, and partners. *Endocrinology* **2005**, *146*, 5071–5078. [CrossRef]
95. Klip, A. The many ways to regulate glucose transporter 4. *Appl. Physiol. Nutr. Metab.* **2009**, *34*, 481–487. [CrossRef]
96. Foretz, M.; Guigas, B.; Bertrand, L.; Pollak, M.; Viollet, B. Metformin: From mechanisms of action to therapies. *Cell Metab.* **2014**, *20*, 953–966. [CrossRef]
97. Ding, H.; Zhang, G.; Sin, K.W.; Liu, Z.; Lin, R.K.; Li, M.; Li, Y.P. Activin A induces skeletal muscle catabolism via p38beta mitogen-activated protein kinase. *J. Cachexia Sarcopenia Muscle* **2017**, *8*, 202–212. [CrossRef]
98. Paul, P.K.; Gupta, S.K.; Bhatnagar, S.; Panguluri, S.K.; Darnay, B.G.; Choi, Y.; Kumar, A. Targeted ablation of TRAF6 inhibits skeletal muscle wasting in mice. *J. Cell Biol.* **2010**, *191*, 1395–1411. [CrossRef]
99. Qin, W.; Pan, J.; Wu, Y.; Bauman, W.A.; Cardozo, C. Protection against dexamethasone-induced muscle atrophy is related to modulation by testosterone of FOXO1 and PGC-1alpha. *Biochem. Biophys. Res. Commun.* **2010**, *403*, 473–478. [CrossRef]
100. Zhang, G.; Jin, B.; Li, Y.P. C/EBPbeta mediates tumour-induced ubiquitin ligase atrogin1/MAFbx upregulation and muscle wasting. *EMBO J.* **2011**, *30*, 4323–4335. [CrossRef]
101. Zhang, G.; Li, Y.P. p38beta MAPK upregulates atrogin1/MAFbx by specific phosphorylation of C/EBPbeta. *Skelet. Muscle* **2012**, *2*, 20. [CrossRef] [PubMed]
102. Odeh, M.; Tamir-Livne, Y.; Haas, T.; Bengal, E. P38alpha MAPK coordinates the activities of several metabolic pathways that together induce atrophy of denervated muscles. *FEBS J.* **2020**, *287*, 73–93. [CrossRef] [PubMed]
103. Papaconstantinou, J.; Wang, C.Z.; Zhang, M.; Yang, S.; Deford, J.; Bulavin, D.V.; Ansari, N.H. Attenuation of p38alpha MAPK stress response signaling delays the in vivo aging of skeletal muscle myofibers and progenitor cells. *Aging* **2015**, *7*, 718–733. [CrossRef] [PubMed]
104. Yuasa, K.; Okubo, K.; Yoda, M.; Otsu, K.; Ishii, Y.; Nakamura, M.; Itoh, Y.; Horiuchi, K. Targeted ablation of p38alpha MAPK suppresses denervation-induced muscle atrophy. *Sci. Rep.* **2018**, *8*, 9037. [CrossRef] [PubMed]
105. Feng, Y.J.; Li, Y.Y. The role of p38 mitogen-activated protein kinase in the pathogenesis of inflammatory bowel disease. *J. Dig. Dis.* **2011**, *12*, 327–332. [CrossRef] [PubMed]
106. Jing, Y.; Liu, W.; Cao, H.; Zhang, D.; Yao, X.; Zhang, S.; Xia, H.; Li, D.; Wang, Y.C.; Yan, J.; et al. Hepatic p38alpha regulates gluconeogenesis by suppressing AMPK. *J. Hepatol.* **2015**, *62*, 1319–1327. [CrossRef]
107. Hegarty, B.D.; Cooney, G.J.; Kraegen, E.W.; Furler, S.M. Increased efficiency of fatty acid uptake contributes to lipid accumulation in skeletal muscle of high fat-fed insulin-resistant rats. *Diabetes* **2002**, *51*, 1477–1484. [CrossRef]
108. James, A.M.; Collins, Y.; Logan, A.; Murphy, M.P. Mitochondrial oxidative stress and the metabolic syndrome. *Trends Endocrinol. Metab. TEM* **2012**, *23*, 429–434. [CrossRef]
109. Bonnard, C.; Durand, A.; Peyrol, S.; Chanseaume, E.; Chauvin, M.A.; Morio, B.; Vidal, H.; Rieusset, J. Mitochondrial dysfunction results from oxidative stress in the skeletal muscle of diet-induced insulin-resistant mice. *J. Clin. Investig.* **2008**, *118*, 789–800. [CrossRef]

110. Holloszy, J.O. "Deficiency" of mitochondria in muscle does not cause insulin resistance. *Diabetes* **2013**, *62*, 1036–1040. [CrossRef]
111. Koh, J.H.; Johnson, M.L.; Dasari, S.; LeBrasseur, N.K.; Vuckovic, I.; Henderson, G.C.; Cooper, S.A.; Manjunatha, S.; Ruegsegger, G.N.; Shulman, G.I.; et al. TFAM Enhances Fat Oxidation and Attenuates High-Fat Diet-Induced Insulin Resistance in Skeletal Muscle. *Diabetes* **2019**, *68*, 1552–1564. [CrossRef] [PubMed]
112. Sandri, M.; Lin, J.; Handschin, C.; Yang, W.; Arany, Z.P.; Lecker, S.H.; Goldberg, A.L.; Spiegelman, B.M. PGC-1alpha protects skeletal muscle from atrophy by suppressing FoxO3 action and atrophy-specific gene transcription. *Proc. Natl. Acad. Sci. USA* **2006**, *103*, 16260–16265. [CrossRef] [PubMed]
113. Handschin, C.; Choi, C.S.; Chin, S.; Kim, S.; Kawamori, D.; Kurpad, A.J.; Neubauer, N.; Hu, J.; Mootha, V.K.; Kim, Y.B.; et al. Abnormal glucose homeostasis in skeletal muscle-specific PGC-1alpha knockout mice reveals skeletal muscle-pancreatic beta cell crosstalk. *J. Clin. Investig.* **2007**, *117*, 3463–3474. [CrossRef]
114. Izumiya, Y.; Hopkins, T.; Morris, C.; Sato, K.; Zeng, L.; Viereck, J.; Hamilton, J.A.; Ouchi, N.; LeBrasseur, N.K.; Walsh, K. Fast/Glycolytic muscle fiber growth reduces fat mass and improves metabolic parameters in obese mice. *Cell Metab.* **2008**, *7*, 159–172. [CrossRef] [PubMed]
115. LeBrasseur, N.K.; Walsh, K.; Arany, Z. Metabolic benefits of resistance training and fast glycolytic skeletal muscle. *Am. J. Physiol. Endocrinol. Metab.* **2011**, *300*, E3–E10. [CrossRef] [PubMed]
116. Pospisilik, J.A.; Knauf, C.; Joza, N.; Benit, P.; Orthofer, M.; Cani, P.D.; Ebersberger, I.; Nakashima, T.; Sarao, R.; Neely, G.; et al. Targeted deletion of AIF decreases mitochondrial oxidative phosphorylation and protects from obesity and diabetes. *Cell* **2007**, *131*, 476–491. [CrossRef] [PubMed]
117. Brunmair, B.; Staniek, K.; Gras, F.; Scharf, N.; Althaym, A.; Clara, R.; Roden, M.; Gnaiger, E.; Nohl, H.; Waldhausl, W.; et al. Thiazolidinediones, like metformin, inhibit respiratory complex I: A common mechanism contributing to their antidiabetic actions? *Diabetes* **2004**, *53*, 1052–1059. [CrossRef]
118. Wessels, B.; Ciapaite, J.; van den Broek, N.M.; Nicolay, K.; Prompers, J.J. Metformin impairs mitochondrial function in skeletal muscle of both lean and diabetic rats in a dose-dependent manner. *PLoS ONE* **2014**, *9*, e100525. [CrossRef]

International Journal of
Molecular Sciences

Article

P38 Regulates Kainic Acid-Induced Seizure and Neuronal Firing via Kv4.2 Phosphorylation

Jia-hua Hu †, Cole Malloy † and Dax A. Hoffman *

Molecular Neurophysiology and Biophysics Section, The Eunice Kennedy Shriver National Institute of Child Health and Human Development, Bethesda, MD 20892, USA; jia-hua.hu@nih.gov (J.-h.H.); cole.malloy@nih.gov (C.M.)

* Correspondence: hoffmand@mail.nih.gov

† These authors contributed equally to this work.

Received: 30 July 2020; Accepted: 14 August 2020; Published: 18 August 2020

Abstract: The subthreshold, transient A-type K^+ current is a vital regulator of the excitability of neurons throughout the brain. In mammalian hippocampal pyramidal neurons, this current is carried primarily by ion channels comprising Kv4.2 α-subunits. These channels occupy the somatodendritic domains of these principle excitatory neurons and thus regulate membrane voltage relevant to the input–output efficacy of these cells. Owing to their robust control of membrane excitability and ubiquitous expression in the hippocampus, their dysfunction can alter network stability in a manner that manifests in recurrent seizures. Indeed, growing evidence implicates these channels in intractable epilepsies of the temporal lobe, which underscores the importance of determining the molecular mechanisms underlying their regulation and contribution to pathologies. Here, we describe the role of p38 kinase phosphorylation of a C-terminal motif in Kv4.2 in modulating hippocampal neuronal excitability and behavioral seizure strength. Using a combination of biochemical, single-cell electrophysiology, and in vivo seizure techniques, we show that kainic acid-induced seizure induces p38-mediated phosphorylation of Thr607 in Kv4.2 in a time-dependent manner. The pharmacological and genetic disruption of this process reduces neuronal excitability and dampens seizure intensity, illuminating a cellular cascade that may be targeted for therapeutic intervention to mitigate seizure intensity and progression.

Keywords: Kv4.2; seizure; p38 MAPK; temporal lobe epilepsy; hippocampus; neuronal firing and excitability

1. Introduction

Seizures are a common manifestation of a host of neurological disorders, including epilepsy. Affecting an estimated 1% of the US population [1], these events are characterized by the anomalous synchronization of electrical activity in the brain due to the hyperexcitability of individual neurons and neural networks. Although relatively indiscriminate in their anatomical localization, seizures often occur in the hippocampus and surrounding cortical areas of the temporal lobe. Recurrent seizures in this brain region underlie temporal lobe epilepsy (TLE), which is thought to be the most common epilepsy syndrome in adults [2]. In addition to being a prevalent form of epilepsy, TLE is frequently the most difficult to treat [3]. These seizures are often resistant to antiepileptic drugs, which makes the affected region susceptible to ongoing, intractable epilepsy [4–6]. For up to one-third of patients, surgical intervention is encouraged for mitigation [1,7,8]. As a result, an emphasis on the identification of additional mechanisms underlying TLE, which can serve as potential targets for therapy, is requisite.

While the etiology of seizures underlying TLE is multifaceted, a common feature is the dysfunction of voltage-gated ion channels in hippocampal pyramidal neurons. A host of channelopathies—both inherited and acquired—are associated with TLE in this principal cell type, including in hyperpolarization-activated cyclic nucleotide-gated (HCN) channels [9–13], voltage-gated Na^+ channels (Na_v) [14,15], and voltage-gated Ca^{2+} channels (Ca_v) [16,17]. Featured most prominently in TLE pathologies are abnormalities in the function of voltage-gate K^+ channels (K_v), which are critical regulators of the intrinsic excitability of neurons throughout the brain (reviewed in [18,19]). Although members of all classes of K^+ channels have been shown to be altered in various epilepsy syndromes [19], there has been a steady increase in findings linking A-type K^+ channels (Shal subfamily) to TLE. Chief among the members of this family is Kv4.2, which has been heavily implicated in TLE in both animal models [19] and humans [20–22]. Kv4.2 is the primary pore-forming K_v channel subunit underlying the rapidly activating and inactivating somatodendritic A-current (I_A) in CA1 pyramidal neurons of the hippocampus [23–25]. Operating at subthreshold voltages, Kv4.2 regulates action potential (AP) repolarization and repetitive firing, dampens AP backpropagation into dendrites, and shapes synaptic potentials, thus acting as a powerful modulator of the input–output efficacy of pyramidal neurons [23,26,27]. Although mutations in the Kv4.2 gene that impart defects intrinsic to channel function are associated with TLE [20–23], it is also evident that disruptions/modifications in Kv4.2 channel properties occur *in response* to seizures, suggesting that regulation of these channels likely contributes to the intractability of TLE. Indeed, TLE has been shown to decrease Kv4.2 availability [28,29], however, the molecular mechanisms underlying this activity-induced downregulation remain unclear.

Substantial evidence supports the notion that Kv4.2 channels function in macromolecular complexes with auxiliary subunits, including the K^+ channel-interacting proteins (KChIP1-4) and dipeptidyl peptidases 6 and 10 (DPP6 and DPP10) [30]. Both KChIPs and DPPs work together to exert the strong modification of Kv4.2 expression, membrane surface localization, and channel kinetics [31–34]. Evidence of increased seizure susceptibility is present in mice harboring mutations in these auxiliary subunits, including KChIP 2 [35], suggesting the maintenance of channel complexes is a key factor in moderating seizures. Likely modulators of Kv4.2 complex dynamics are protein kinases. The phosphorylation of Kv4.2 by protein kinase A (PKA), protein kinase C (PKC), and extracellular signal-regulated kinase/mitogen-activated protein kinase (ERK/MAPK) downregulates I_A [36–38]. In pyramidal neurons, this downregulation facilitates an increase in somatodendritic excitability, enhancing susceptibility to network hyperexcitability in the hippocampus [39–41]. We have recently identified a specific MAPK, p38α, as a potent regulator of the Kv4.2 complex [42]. The p38 phosphorylation of Kv4.2 C-terminal motifs triggers a molecular cascade that facilitates the dissociation of Kv4.2 from its auxiliary subunit DPP6 [42]. This cascade is particularly intriguing in the context of TLE, as it occurs in an activity-dependent fashion, representing a novel mechanism that may be integral in regulating seizure susceptibility [42].

In the present study, we expand on our previous findings and address how p38 kinase modulates seizure susceptibility and neuronal excitability. We use biochemical, electrophysiological, and in vivo seizure techniques in WT and a novel mouse model harboring a point mutation preventing p38 phosphorylation of Kv4.2 at C-terminal Thr607 (Kv4.2TA) to illuminate the role of p38 phosphorylation of Kv4.2 in regulating the intrinsic excitability of hippocampal pyramidal neurons and seizure intensity. We show that p38 phosphorylation of Kv4.2 at C-terminal Thr607 is integral in modulating seizure strength and may contribute to the progression of seizure intensity over time. Furthermore, we confirm previous findings that a molecular cascade triggered by p38 phosphorylation alters Kv4.2-mediated excitability of hippocampal pyramidal neurons, illuminating a novel molecular mechanism involved in network hyperexcitability in mice. The combined pharmacological and genetic manipulation of the cellular cascade described here offers insight into various avenues through which therapeutic intervention to curtail seizure progression can be pursued.

2. Results

2.1. p38 MAPK Contributes to Kainic Acid-Induced Seizure in WT but Not Kv4.2TA Mice

We have generated a mutant mouse, Kv4.2TA, with abolished dynamic Thr 607 phosphorylation of Kv4.2 and isomerization of Kv4.2 by Pin1 [42]. Kv4.2TA mice displayed increased I_A, decreased neuronal excitability, and improved cognitive flexibility [42]. Here, we examine if acute behavioral seizure is altered following the systemic injection of kainic acid (KA) in the Kv4.2TA mice. KA (25 mg/kg) was injected intraperitoneally into Kv4.2TA mice (n = 13) and littermate controls (n = 15) and behavioral seizure responses were scored using the modified Racine scale [43] for 60 min post injection. We observed a significant difference in behavior seizure scores, with Kv4.2TA mice showing significantly reduced seizure intensity over the full 1 h period following KA injection (Figure 1A,B). Our previous work showed that p38 can phosphorylate Kv4.2 at T607 in response to seizure induced by pentylenetetrazol (PTZ) or KA [42]. In addition, acute behavioral seizure in p38 knockout mice is significantly decreased compared to WT mice [44]. Therefore, we hypothesized that the effect of p38 on seizure intensity is dependent on the T607 site of Kv4.2. We injected p38 inhibitor SB 203580 (20 mg/kg, i.p.) 15 min ahead of KA injection. The result showed that control mice with SB 203580 injection (n = 14) exhibited significantly reduced behavioral seizure intensity compared to those with the control injection (Figure 1A,B). Interestingly, Kv4.2TA mice with SB 203580 injection (n = 13) did not display a significant reduction in behavioral seizure intensity compared to those injected with vehicle (Figure 1A,B). This decrease in sensitivity to KA-induced seizure was also reflected in the latency to stage 3 seizure (Figure 1C). Taken together, these data support the notion that p38 phosphorylation of Kv4.2 at T607 contributes to KA-induced seizure.

2.2. Seizure Induced by Kainic Acid Triggers Kv4.2 T607 Phosphorylation in a Time-Dependent Manner

It has been reported that Kv4.2 availability is altered in response to seizure, suggesting these events may trigger a molecular cascade leading to the functional downregulation of I_A [28,29]. Our collective analyses indicate that this cascade is initiated by p38 kinase. We reason that the prolonged activation of p38 and subsequent phosphorylation of Kv4.2 in response to continued seizures (kindling) may be an important factor underlying the intractability of seizures in the temporal lobe. Therefore, in order to study the timing of Kv4.2 phosphorylation in response to seizure, we examined Kv4.2 phosphorylation at various times following KA injection. Since p38 can phosphorylate both T602 and T607 of Kv4.2 [42], we examined both phosphorylation sites. KA induced Kv4.2 phosphorylation at T607 at 15 min after KA injection but not at 5 min (Figure 2A,B). T607 phosphorylation peaked 3 h after KA injection and the induction effect lasted, even at 5 days post injection (Figure 2A,B). Kv4.2 phosphorylation at T602 was also induced 3 h post KA injection but not at earlier time points (Figure 2A,C). These data show that seizure induced by KA triggers a long-lasting effect on Kv4.2 that may contribute to on-going seizure progression.

2.3. Kainic Acid-Induced Kv4.2 Phosphorylation at T607 Is Dependent on p38 MAPK

Next, we wanted to know if p38 is required for KA-induced Kv4.2 phosphorylation at T607. The p38 inhibitor SB 203580 (20 mg/kg, i.p.) was injected 15 min ahead of KA injection (25 mg/kg, i.p.). We found that the p38 inhibitor SB 203580 blocked the induction of Kv4.2 phosphorylation 15 min after KA injection in the mouse hippocampus (Figure 3A,B), suggesting p38 contributed to KA-induced Kv4.2 phosphorylation at T607.

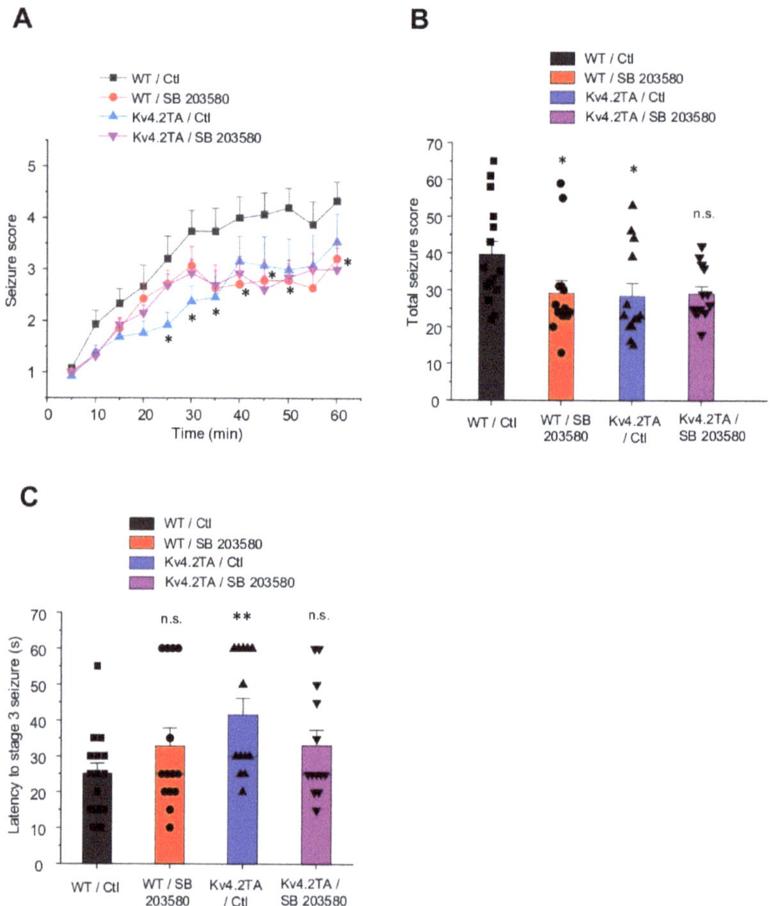

Figure 1. p38 mitogen-activated protein kinase (MAPK) contributed to kainic acid-induced seizure in WT mice but not Kv4.2TA mice. (**A**) Time course of mean behavioral seizure score following kainic acid injection. The mean behavioral seizure score was significantly reduced in Kv4.2TA mice compared to WT mice. Furthermore, p38 inhibitor SB 203580 significantly reduced behavioral seizure score following kainic acid injection in WT mice but not in Kv4.2TA mice, $n = 13$–15 for each group, two-way ANOVA, * $p < 0.05$. (**B**) Total behavioral seizure score for each group, $n = 13$–15 for each group, *t*-test, * $p < 0.05$. (**C**) Latency to stage 3 seizure for each group. $n = 13$–15 for each group, *t*-test, ** $p < 0.01$.

2.4. p38 MAPK Colocalizes with Kv4.2

Since p38 phosphorylates Kv4.2, we wanted to see if it colocalized with Kv4.2 in a heterologous system and in the mouse brain. First, HEK 293T cells were double stained after 2 days of transfection with p38 and Kv4.2. The result showed that p38 partially colocalized with Kv4.2 (Figure 4A). High magnification images and line scan confirmed this result (Figure 4B,C). In addition, we performed the double staining of phosphorylated p38 (pp38) and Kv4.2 on mouse brain sections. Phospho-p38 is mainly localized in the cell body of hippocampal pyramidal neurons but also localized in dendrites, while Kv4.2 is mainly localized in dendrites (Figure 4C). High magnification images showed pp38 partially colocalized with Kv4.2, as indicated by arrow heads (Figure 4D).

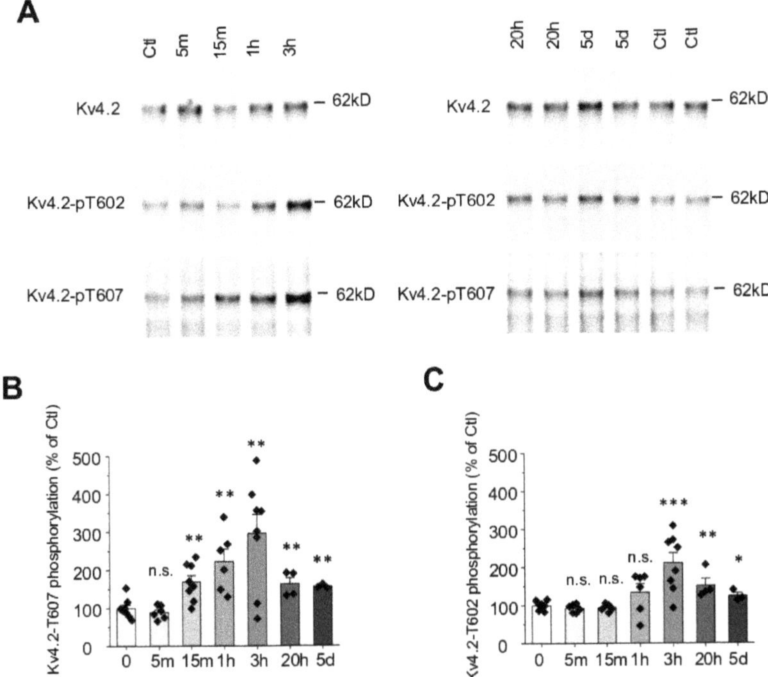

Figure 2. Seizure induced by kainic acid triggers Kv4.2 T607 phosphorylation in a time-dependent manner in mouse hippocampus. (**A**) Time course of Kv4.2 phosphorylation at Thr602 and Thr607 by kainic acid administration (25 mg/kg, i.p.) in mouse hippocampus. (**B**) Statistical analysis of kainic acid-induced phosphorylation of Kv4.2 at Thr607 in mouse hippocampus, $n = 3$–8 in each group, *t*-test, ** $p < 0.01$. (**C**) Statistical analysis of kainic acid-induced phosphorylation of Kv4.2 at Thr602 in mouse hippocampus, $n = 3$–8 in each group, *t*-test, * $p < 0.05$, ** $p < 0.01$, *** $p < 0.001$.

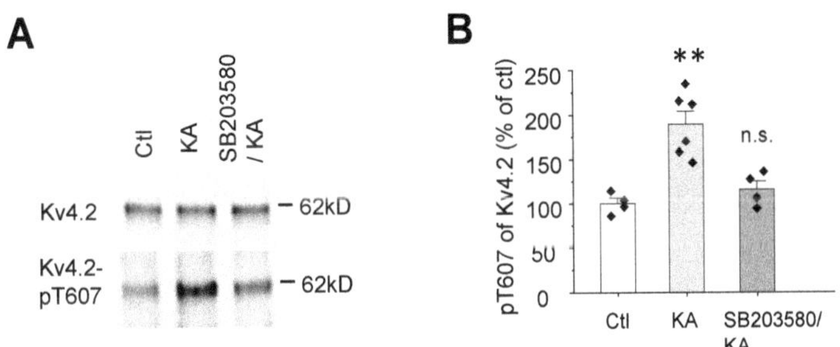

Figure 3. p38 MAPK contributes to kainic acid-induced Kv4.2 phosphorylation at T607. (**A**) SB 203580, a potent p38 inhibitor (20 mg/kg, i.p., 15 min), blocked kainic acid-induced phosphorylation of Kv4.2 T607 in mouse hippocampus. (**B**) Statistical analysis of the effect of SB 203580 on kainic acid-induced phosphorylation of Kv4.2 at Thr607 in mouse hippocampus, $n = 4$–6 in each group, *t*-test, ** $p < 0.01$.

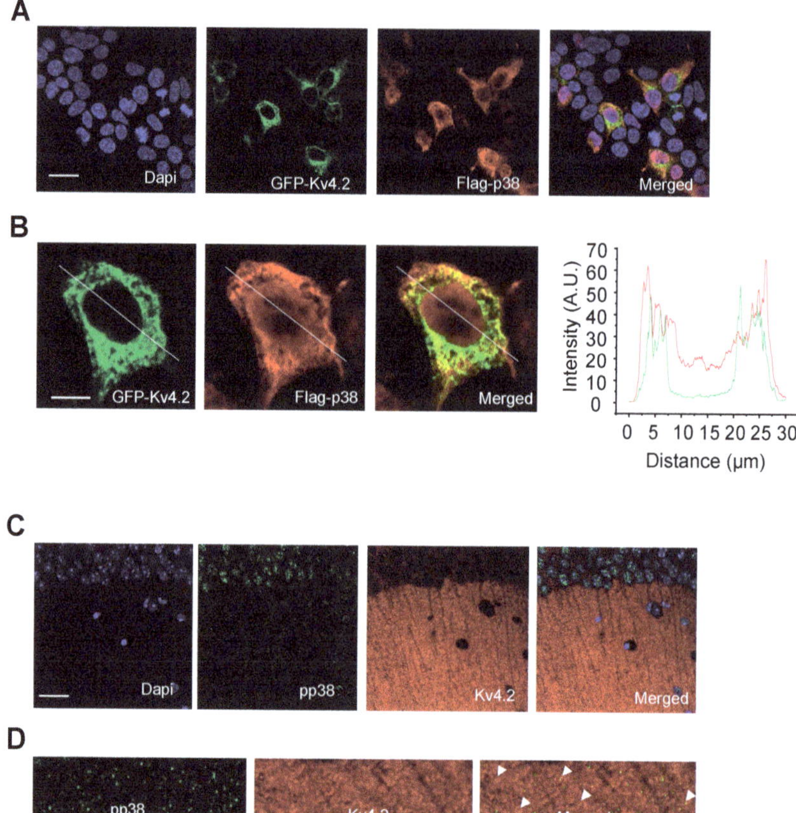

Figure 4. p38 MAPK colocalizes with Kv4.2. (**A**) HEK293T cells were transfected with GFP-Kv4.2 and Flag-p38. Cells were fixed and stained with GFP and Flag to show co-localization. Scale bar: 20 μm. (**B**) High magnification images and line scan analysis of colocalization. Scale bar: 5 μm. (**C**) Mouse brains were co-stained with Kv4.2 and pp38 antibody. Phosphorylated p38 is localized in the cell body and dendrites as well. Scale bar: 20 μm. (**D**) High magnification images showing Kv4.2 and pp38 colocalized in dendrites, as indicated with arrow heads. Scale bar: 5 μm.

2.5. Kainic Acid Activates p38 MAPK in both WT and Kv4.2TA Mice

We next assessed whether KA-induced seizure activated p38 in both WT mice and Kv4.2TA mice. Mouse brain sections were stained with pp38. The pp38 level is significantly increased after KA injection (1 h) in the cell body of hippocampal pyramidal neurons (Figure 5A). Furthermore, we examined pp38 by western blot. The pp38 level but not p38 level is significantly increased after KA injection (30 min) in WT mouse hippocampus (Figure 5B). A similar result was found in Kv4.2TA mice (Figure 5B). These data indicate that the initiation of p38 kinase activity by KA is similar in WT and Kv4.2TA mice, and the observed effects on seizure intensity can be ascribed to the inability of p38 to phosphorylate Thr607.

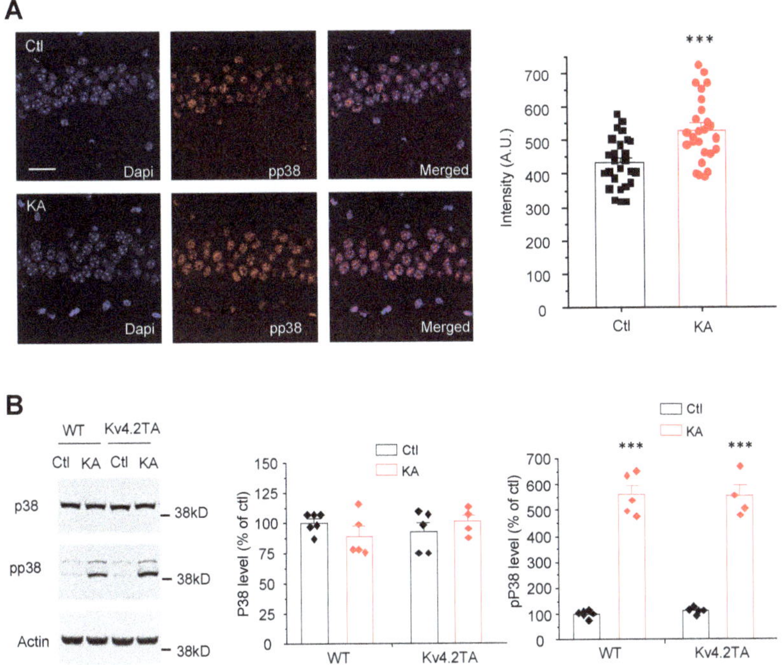

Figure 5. Kainic acid activates p38 MAPK in both WT and Kv4.2TA mice. (**A**) Immunostaining analysis showed p38 phosphorylation increased with kainic acid administration (25 mg/kg, i.p., 30 min) in mouse hippocampus, $n = 26$ cells in each group, t-test, *** $p < 0.001$. (**B**) Western blot analysis showed p38 phosphorylation increased with kainic acid administration (25 mg/kg, i.p., 30 min) in hippocampus in both WT and Kv4.2TA mice, $n = 4$–6 cells in each group, t-test, *** $p < 0.001$.

2.6. p38 MAPK Modulates Neuronal Excitability through Kv4.2

In our previous study, we reported that hippocampal pyramidal cells from acute hippocampal slices of Kv4.2TA mice exhibited a nearly two-fold reduction in AP firing frequency in response to somatic current injection relative to WT mice. We determined that this was due to an enhancement of I_A as a result of the T607A mutation blocking the dissociation of Kv4.2 from its auxiliary subunit DPP6 and subsequent functional downregulation [42]. Furthermore, we found the slicing process largely activated p38 kinase and induced Kv4.2 phosphorylation, revealing that this procedure acts similarly to kainic acid-induced seizure in altering the phospho-state of Kv4.2 in the hippocampus [42]. Therefore, we sought to investigate how a pharmacological blockade of p38 (SB 203580) altered the excitability of the principal hippocampal neurons in area CA1 of acute hippocampal slices. In light of the time-dependency of phosphorylation of Kv4.2 at T602 and T607 in response to seizure induction (peak ~3 h), we incubated slices in recovery solution with pharmacological treatment or vehicle (0.1% DMSO) for 2 h and continued their exposure during electrophysiological recordings (3–4 h total exposure). In response to stepped somatic current injections, we identified that treatment with 5 μM SB 203580 reduced AP firing frequency of pyramidal neurons at each magnitude above rheobase in WT slices (Figure 6A,B). At maximum current injection (+300 pA), the difference in firing frequency reached a statistically significant level. Specifically, a +300 pA injection induced a firing rate of 23.7 Hz in the presence of DMSO, representing a~33% increase relative to the rate recorded in the presence of 5 μM SB 203580 (16.9 Hz), shown in Figure 6B. This alteration in excitability was limited to suprathreshold properties, as subthreshold excitability was largely unaffected by pharmacological p38 blockade (Table 1). Parameters measured from ramp current injections (400 pA/s), including AP threshold

(Table 1), rheobase (Figure 6D,E), and latency to AP onset (Figure 6F), were similar in conditions with or without SB 203580, although a modest, but non-statistically significant, increase in AP threshold and rheobase was observed in slices from WT mice with pharmacological p38 blockade (Table 1 and Figure 6E).

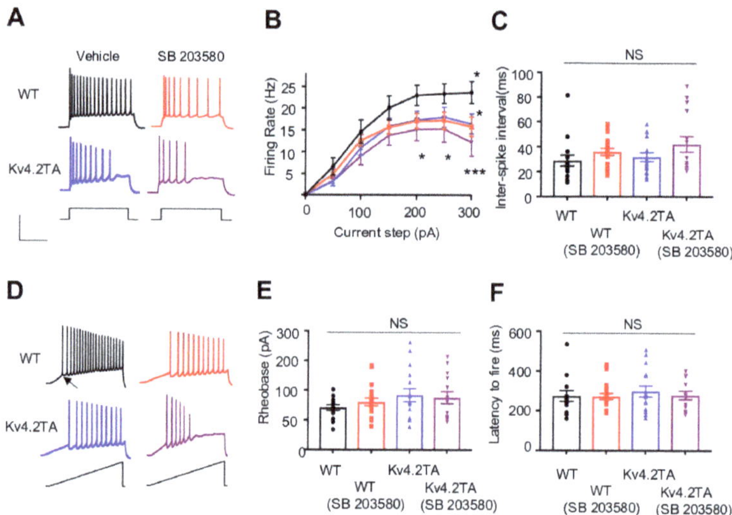

Figure 6. p38 impacts hippocampal pyramidal neuron excitability through Kv4.2. (**A**) Current step of +300 pA induces repetitive firing in pyramidal neurons recorded from WT and Kv4.2TA mice with or without SB 203580 treatment. Scale 40 mV/250 ms. Square current inset 300 pA. (**B**) Sequential somatic current injections increasing in magnitude reveal p38 kinase inhibition reduces AP firing frequency in WT hippocampal neurons at +300 pA relative to vehicle ($n = 15$ in vehicle, $n = 19$ in treatment; two-way ANOVA, * $p < 0.05$). Kv4.2TA neurons display reduced firing frequency at +300 pA relative to WT in vehicle, which is augmented in the presence of SB 203580 such that current magnitudes of +200 and +250 pA also exhibit significant differences ($n = 18$ in vehicle, $n = 14$ in SB 203580; two-way ANOVA, * $p < 0.05$; *** $p < 0.001$). (**C**) Inter-spike intervals measured between the first two spikes in a train evoked by 150 pA injection display no significant difference among groups. Kruskal–Wallis test, $p > 0.05$. (**D**) Ramp current injections evoke repetitive firing in all pyramidal neurons recorded in each condition. Arrow indicates point at which action potential (AP) threshold, rheobase, and latency to fire were measured. Ramp current inset 400 pA/s. (**E**) Minimum current to elicit AP firing at threshold (rheobase) is not significantly different among the populations. One-way ANOVA, $p > 0.05$. (**F**) Latency to fire in response to ramp injection is not significantly different among populations. Kruskal–Wallis test, $p > 0.05$.

Table 1. Passive membrane properties and single action potential (AP) parameters (mean ± SEM).

Parameter	WT	WT (SB 203580)	Kv4.2TA	Kv4.2TA (SB 203580)
RMP (mv)	−60.5 ± 0.86	−60.1 ± 0.72	−58.8 ± 0.62	−58.9 ± 0.60
Whole-cell capacitance (pF)	16.2 ± 1.0	22.4 ± 1.6	17.1 ± 0.77	17.8 ± 1.3
R_{input} (MΩ)	228.8 ± 18.2	273.4 ± 19.7	231.2 ± 13.4	228.9 ± 16.0
Time to AP Peak (ms)	0.84 ± 0.1	1.07 ± 0.1	0.9 ± 0.1	0.89 ± 0.1
AP amplitude (mV)	81.0 ± 3.3	79.6 ± 3.7	76.7 ± 2.8	75.1 ± 3.3
AP half-width (ms)	1.5 ± 0.1	1.7 ± 0.2	1.6 ± 0.1	1.6 ± 0.1
AP threshold (mV)	−40.1 ± 1.1	−35.9 ± 1.5	−34.9 ± 1.4 [a]	−37.7 ± 1.4

[a] $p < 0.05$.

We next tested the effect of SB 203580 on slices obtained from Kv4.2TA mice. Because p38-mediated phosphorylation of Kv4.2 is significantly reduced in these mice, we anticipated the mutation would occlude the impact of SB 203580 on neuronal firing if p38 mediates excitability primarily through its regulation of I_A. Generally consistent with previous observations, pyramidal neurons from Kv4.2TA slices exhibited reduced AP firing frequency relative to WT in multiple experimental conditions (Figure 6A,B). AP firing frequency was reduced at each current magnitude, with a statistically significant reduction exhibited at peak injection in the presence of DMSO (17.5 Hz vs. 23.7 Hz, Kv4.2TA vs. WT, respectively; $p < 0.05$, Figure 6B). The significant decrease in AP firing frequency corresponded with a significant increase in AP threshold in Kv4.2TA neurons relative to WT in this condition (Table 1). Additionally, AP firing mirrored that of WT neurons in the presence of 5 μM SB 203580 (Figure 6B). Importantly, SB 203580 treatment in Kv4.2TA slices did not significantly reduce suprathreshold excitability, contrary to its impact in WT pyramidal neurons (Figure 6B). This suggests that p38 modulation of AP firing can primarily be traced to its regulation of Kv4.2. Furthermore, as noted previously, Kv4.2TA neurons in this condition displayed significant pauses in repetitive firing, which correlated with slight, non-statistically significant, increases in fast after-hyperpolarization amplitudes and inter-spike intervals (Table 1 and Figure 6C, respectively). Therefore, taken together, pharmacological blockade of p38 kinase reduced the suprathreshold excitability of hippocampal pyramidal neurons. The T607A mutation occludes the action of SB 203580, suggesting its impact on AP firing frequency is mediated predominantly through its modulation of Kv4.2-mediated I_A.

3. Discussion

The present study describes a novel mechanism of KA-induced seizure which involves p38-dependent phosphorylation of Kv4.2 at T607. Both T602 and T607 are phosphorylated by KA but the induction timing is different (Figure 2A–C). KA induces T607 phosphorylation relatively promptly (about 15 min) while T602 phosphorylation is relatively delayed (about 3 h). Both T602 and T607 phosphorylation were sustained for at least 5 days, the longest data point measured (Figure 2A–C). Kindling is a commonly used model for the development of seizures and epilepsy in which the duration and behavioral involvement of induced seizures increases after seizures are induced repeatedly [45]. Repeated seizure could boost Kv4.2 phosphorylation levels at T602 and T607, leading to downregulation and enhanced excitability. Therefore, the long-term phosphorylation response to seizure could be a mechanism of kindling.

We have found that Pin1 binds to T607 of Kv4.2 and isomerizes the T607-P bond to modulate the function of Kv4.2 [42]. The dual phosphorylation of T602 and T607 increases the Pin1 binding ability and therefore improves Kv4.2 modulation. KA triggers Kv4.2 phosphorylation at both sites after 3 h and lasts for 5 days (Figure 2A–C), suggesting that the Pin1 effect could be long lasting as well. The excitability of CA1 pyramidal neuron dendrites was increased in TLE because of the decreased availability of A-type potassium ion channels [28]. Pin1's long-lasting effect could eventually lead to reduced availability of Kv4.2, which fits with the notion that seizure decreases I_A. Furthermore, the continued phosphorylation at these sites and persistent Pin1 activity in response to prolonged seizure may exacerbate their severity, manifesting as a positive feedback loop, promoting further downregulation of I_A. Thus, the activation of p38 and subsequent isomerization of Kv4.2 may serve as a mechanism underlying the intractability of seizure progression often associated with severe epilepsy in the temporal lobe [1–3].

Kv4.2 T602 and T607 phosphorylation occur via both p38 MAPK (Figure 2A–C) and ERK MAPK [38,46]. However, the p38 inhibitor SB 203,580 blocked Kv4.2 phosphorylation induced by KA (Figure 3A,B), suggesting the primary role of p38 in seizure. Since ERK MAPK can be activated by KA [47], ERK may also have a contributing effect on Kv4.2. In addition, KA activates p38 in Kv4.2TA mice at similar levels as in WT mice, suggesting the reduced seizure phenotype in Kv4.2TA mice (Figure 1) is not because of the differential induction of p38 but the deficiency of Kv4.2 phosphorylation at T607.

At the cellular level, Kv4.2 phosphorylation by p38 at T607 alters the excitability of hippocampal pyramidal neurons of area CA1. The results presented here indicate that the role of p38 in altering membrane excitability is primarily through a reduction in suprathreshold excitability. While Kv4.2 can impact subthreshold properties in pyramidal neurons, including latency to AP onset [48] and rheobase [42], its role in regulating the frequency of repetitive AP firing is well documented [48,49]. Its robust control of membrane potential fluctuations in response to depolarizing currents from resting potential permits a significant functional interaction with voltage-gated Na^+ channels [48]. Indeed, we identified a significant increase in AP threshold in Kv4.2TA pyramidal neurons. Moreover, a slight increase in after-hyperpolarization amplitude in these cells is likely contributory in delaying the recruitment of Na^+ channel activation to stepped current injection, which was evidenced in this study and our previous analysis [42]. It is likely that the enhancement of I_A amplitude that results with p38-Pin1 blockade reduces the precision of repetitive spiking through the modulation of voltage-gated channels, driving spiking in hippocampal pyramidal neurons.

It is clear that the T607A mutation induces a reduction of pyramidal cell excitability. The trend in reduced excitability of Kv4.2TA neurons was persistent in this analysis, strengthening our previous findings [42]. While slight variations in neuronal responsiveness to current injections of various magnitudes in Kv4.2TA mice relative to WT were observed in this study relative to [42], the overall trend toward a reduction in repetitive firing remained. Slight alterations in the slice recovery and recording conditions are likely to underlie these differences. Furthermore, by complementing the T607A mutation with pharmacological blockade of p38, we uncovered multiple means by which to lower hippocampal pyramidal cell excitability through Kv4.2. The application of SB 203580 altered the firing mode of WT neurons in a manner that mirrored that of Kv4.2TA neurons in its absence. This suggests that p38 regulation of membrane excitability is primarily mediated through Kv4.2. We did, however, note that, although p38 blockade did not significantly alter AP firing properties in Kv4.2TA slices, a reduction in excitability was generally augmented in its presence. This may imply that the inhibition of phosphorylation of both T602 and T607 sites may, together, produce an additive effect in Kv4.2 regulation of membrane properties. As noted, the dual phosphorylation of these motifs provides an environment particularly conducive to binding multiple domains of Pin1 upon p38 phosphorylation, which could facilitate dissociation of Kv4.2 from DPP6 and downregulation [42]. While this likely contributes to the observed phenotype, we cannot, however, rule out the involvement of additional ion channels that may also be impacted by broad pharmacological p38 blockade.

Furthermore, the breadth of this work and that of our previous study focuses on somatic Kv4.2 channel activity. While somatic Kv4 channels are capable of impacting neuronal firing modes, their privileged distribution in dendrites suggests their control of membrane potentials may be more impactful in these domains [23,50–52]. It is possible that p38 phosphorylation of Kv4.2 may contribute to the alteration of dendritic excitability by modulating the coupling of synaptic inputs and AP output. Indeed, in the context of the regulation of DPP6-Kv4.2 dynamics, p38 activity in the apical dendrites may be of particular significance. DPP6 knockout mice exhibit a non-uniformity in their manifestation of alteration in membrane excitability, with dendritic excitability being predominantly impacted [53]. Whether p38 phosphorylation of dendritic Kv4.2 channels may impact dendritic function, and further contribute to increased network hyperexcitability and seizure severity in the hippocampus, is a topic of future investigation.

4. Materials and Methods

4.1. Animals

C57/BL6J wild-type mice were used in this study. Kv4.2TA mice were generated as described before [42]. Mice were group housed in plastic mouse cages with free access to standard rodent chow and water. The colony room was maintained at 22 ± 2 °C with a 12 h: 12 h light: dark cycle. Kv4.2TA mice were backcrossed at least three generations onto C57/Bl6J mice. Age-matched male adult WT and

Kv4.2TA were used. All animal procedures were performed in accordance with guidelines approved by the *Eunice Kennedy Shriver* National Institute of Child Health and Human Development Animal Care and Use Committee and in accordance with NIH guidelines (20-042, 3 April 2020).

4.2. Expression Constructs

The human Myc-DDK-Kv4.2 construct was purchased from Origene (Rockville, MD, USA, RC215266). The p38 construct was from Addgene (Watertown, MA, USA, 20351).

4.3. Chemicals

All chemicals were purchased: KA (Sigma, St. Louis, MO, USA, K0250) and SB 203580 (Tocris, Minneapolis, MN, USA, 1202). For injections, KA was dissolved in saline; SB 203580 was dissolved in DMSO and 10% Tween 80.

4.4. Antibodies

Mouse anti-Kv4.2 (NeuroMab, Davis, CA, USA, 75-016) was used at 1:2000 for western blot, 1:200 for immunostaining, rabbit anti-Kv4.2 (Sigma, St. Louis, MO, USA, P0233) was used at 1:2000 for western blot, pT602 (Santa Cruz, Dallas, TX, USA, SC-16983-R) was used at 1:1000 for western blot, pT607 (Santa Cruz, Dallas, TX, USA, SC-22254-R) was used at 1:500 for western blot, and pp38 (Cell Signaling, Danvers, MA, USA, 4511s) at 1:1000 for western blot, 1:100 for immunostaining. Flag (Sigma, St. Louis, MO, USA, F3165) was used at 1:300 for immunostaining, actin (Sigma, St. Louis, MO, USA, A-1978) was used at 1:10,000 for western blot; Alexa Fluor 488 goat anti-mouse (Invitrogen, Carlsbad, CA, USA, A-11029) was used at 1:500; Alexa Fluor 488 goat anti-rabbit (Invitrogen, Carlsbad, CA, USA, A-11034) was used at 1:500; Alexa Fluor 555 goat anti-mouse (Invitrogen, Carlsbad, CA, USA, A-21424) was used at 1:500; Alexa Fluor 555 goat anti-rabbit (Invitrogen, Carlsbad, CA, USA, A-21429) was used at 1:500; Alexa Fluor 680 goat anti-mouse (Invitrogen, Carlsbad, CA, USA, A-21057) was used at 1:10,000; Alexa Fluor 680 goat anti-rabbit (Invitrogen, Carlsbad, CA, USA, A-21076) was used at 1:10,000; IRDye 800CW goat anti-mouse (Licor, Licoln, NE, USA, 926-32210) was used at 1:10,000, IRDye 800CW goat anti-rabbit (Licor, Licoln, NE, USA, 926-32211) was used at 1:10,000.

4.5. Cell Culture and Transfection

HEK-293T cells used in biochemistry experiments were obtained from Dr. Paul Worley's lab. HEK-293T cells were cultured in DMEM medium containing 10% FBS. Transfections were performed with X-tremeGENE 9 (Sigma, St. Louis, MO, USA, XTG9-RO) according to the manufacturer's specifications. Cells were harvested about 40 h after transfection.

4.6. Western Blot and Quantification

Protein samples were mixed with 4x LDS sample buffer (Invitrogen, Carlsbad, CA, USA, NP0007) and 10x sample reducing agent (Invitrogen, Carlsbad, CA, USA, NP0007) to a final concentration of 1x. Samples were loaded on 4%–12% Bis-Tris gradient gel (Invitrogen, Carlsbad, CA, USA, 12-well, NP0322; 15-well, NP0323). The proteins were transferred to an Immobilon-FL PVDF membrane (EMD Millipore, Burlington, MA, USA, IPFL00010). The membrane was blocked with Odyssey blocking buffer (Li-COR, Licoln, NE, USA, 927-40000) for 1 h at room temperature, followed by incubation with primary antibody in PBS overnight at 4 °C. The membrane was then washed with PBST (PBS, pH 7.4 and 0.1% Tween-20) three times and incubated with secondary antibody in PBS for another hour. After three washes with PBS, the membrane was scanned using an Odyssey imaging system (LI-COR, Licoln, NE, USA) according to the manufacturer's protocol. Quantification of western blots was carried out using the gel analysis function in ImageJ within the linear range of detection, which is determined by using serial dilutions of a representative sample.

4.7. Immunostaining

Mice were fixed with 4% PFA and brain sections were cut into 24-well plates. They were then blocked with 10% horse serum at RT for 1 h and then incubated with primary antibodies at 4 °C overnight. After washing, sections were incubated with anti-mouse-555 and anti-rabbit-488 secondary antibodies at RT for 2 h. After washing, cells were then mounted on slides with anti-fade mounting medium containing 4′,6-diamidino-2-phenylindole (DAPI, Invitrogen, Carlsbad, CA, USA, P36962) and imaged using a Zeiss (Oberkochen, Germany) 710 laser scanning confocal microscope equipped with a 63 × objective.

4.8. Acute Hippocampal Slice Preparation

Adult male and female (5–6 weeks old) mice were used for all acute slice electrophysiological recordings. Mice were anesthetized with isoflurane and decapitated. Brains were rapidly removed and washed with ice-cold sucrose cutting solution. The sucrose solution was made up of the following (in mM): 60 NaCl, 3 KCl, 28 $NaHCO_3$, 1.25 NaH_2PO_4, 7.5 Glucose, 0.5 $CaCl_2$, 4.5 $MgCl_2$. Brain hemispheres were dissected and mounted following a 45° cut of the dorsal cerebral hemisphere(s). Modified transverse slices (300 μm) were made by a Leica (Wetzlar, Germany, VT1200S) vibrating microtome in ice-cold sucrose that was continuously bubbled with carbogen (95% O_2/5% CO_2). Slices were recovered at 32 °C in sucrose solution for 15 min, at which time the solution temperature was slowly lowered to room temperature where it remained for the remainder of the recording day. Slices were exposed to pharmacological treatments (treatment or vehicle) during the slicing procedure and in recovery.

4.9. Whole-Cell Current Clamp Recordings

Following a 2-h recovery in sucrose cutting solution with or without pharmacological treatment (SB 203580 5 μM and 0.1% DMSO vehicle, respectively), hippocampal slices were transferred to a recording chamber submerged in artificial cerebral spinal fluid (ACSF) with the temperature maintained at 33 °C (±1 °C). The ACSF consisted of the following (in mM): 125 NaCl, 2.5 KCl, 25 $NaHCO_3$, 1.25 NaH_2PO_4, 25 glucose, 2 $CaCl_2$, (pH 7.4). In select recordings, 5 μM SB 203580 was added to the bath solution. The recording chamber was continuously perfused with carbogen-bubbled ACSF at a rate of 3 mL/min. Somatic whole-cell patch clamp recordings were performed on identified somata of hippocampal CA1 pyramidal neurons, which were viewed using infrared differential interference contrast (DIC) on an upright Zeiss (Oberkochen, Germany) Examiner. Cells were patched with 4–5 MΩ borosilicate glass pipettes pulled from a Narishige (Amityville, NY, USA) vertical puller and filled with K^+ gluconate-based intracellular solution consisting of the following (in mM): 20 KCl, 130 K-gluconate, 2 MgCl, 0.1 EGTA, 2 Na_2ATP, 0.3 NaGTP, 10 HEPES, 10 Phosphocreatine with pH adjusted with KOH, and HCl to a final value of 7.25–7.30 and an osmolarity of 290–300 mOsm.

AP firing properties were measured from whole-cell recordings in current clamp mode in the conditions described above. All data were recorded with a Multiclamp 700B amplifier (Molecular Devices, San Jose, CA, USA) and a Digidata 1440A digitizer. Signals were low-pass filtered at 5 kHz and digitized at 10 kHz using Clampex 10.7 software and were acquired in bridge balance mode to compensate for series resistance. Liquid junction potential was not corrected for. Passive membrane properties were measured after initial break-in in order to avoid dialysis as a result of solution exchange. Whole-cell capacitance and series resistance were measured from a Multiclamp 700B commander (Molecular Devices, San Jose, CA, USA). A voltage step of −10 mV was used, and the decay tau of the whole-cell capacitive transient current was used to calculate these parameters. Recordings where series resistance exceeded 25 MΩ or resting membrane potential was more depolarized than −55 mV were discarded. Input resistance was calculated as the slope of the current-voltage (I-V) curve in response to current steps from −50 to 50 pA in 50 pA steps (three steps in total). To evoke action potentials in patched CA1 pyramidal neurons, square 500 ms current pulses were elicited in 50 pA steps with current injections ranging from −200 pA to +300 pA. Two sweeps at each magnitude were elicited and the average response of these sweeps was used for each cell. All measures of action potential waveform were taken from the first spike in response to a 150 pA injection and inter-spike measurements, including inter-spike interval, were recorded between the first two spikes in a train elicited by a 150 pA square current. This current magnitude was used as it was the minimum current that provoked AP firing in 100% of cells patched. Additionally, second-long ramp current injections were elicited at 400 pA/sec and the latency to fire was calculated as the time from the initiation of the current injection to AP threshold (first AP). AP threshold and rheobase were measured from ramp injections, with rheobase recorded as the current magnitude required to produce the first AP (at threshold voltage) in the ramp.

4.10. Seizure Behavioral Assays

Kainic acid (Sigma, St. Louis, MO, USA, K0250) was administered i.p. at a dose of 25 mg/kg. Animals were monitored for 60 min after the injection. Behavioral responses were recorded using a video camera and scored using the following: stage 0, normal behavior; stage 1, immobility and rigidity; stage 2, head bobbing; stage 3, forelimb clonus and rearing; stage 4, continuous rearing and falling; stage 5, clonic–tonic seizure; stage 6, death (Racine, 1972). Total seizure scores were calculated by summing up every five-minute score. The behavioral assessments described above were performed in a blind manner.

4.11. Statistical Analysis

Biochemistry and behavior data were analyzed by Origin 2018b (Northampton, MA, USA) by two-tailed Student's *t*-test and two-way ANOVA, respectively. Electrophysiological data were analyzed by GraphPad Prism 7 (San Diego, CA, USA, 7.0 d). For all electrophysiological experiments, the experimenter was blinded to the genotypes. For analysis of the pharmacological impact on single AP parameters and in response to ramp current injections in WT and Kv4.2TA slices, a one-way ANOVA (ordinary), or one-way ANOVA on ranks (Kruskal–Wallis test) was used and was corrected for multiple comparisons with Dunnett's test (ordinary) or Dunn's test (ranks). The use of parametric or non-parametric analysis was determined after testing for normal distribution of the data using the D'Agostino and Pearson normality test (alpha level = 0.05). Non-parametric statistics were used if the data failed normality testing. For all analysis of firing frequency in response to sequential current steps, a two-way ANOVA with Tukey's post hoc test was used. All the data are presented as mean ± SEM.

Author Contributions: J.-h.H., C.M., and D.A.H. conceived and planned the experiments; J.-h.H. and C.M. carried out the experiments; J.-h.H. and C.M. analyzed and interpreted the data; J.-h.H., C.M., and D.A.H. drafted the paper; All authors provided critical feedback and helped shape the research, analysis, and paper. All authors have read and agreed to the published version of the manuscript.

Funding: This research received no external funding.

Acknowledgments: We thank Vincent Schram at the NICHD imaging core facility for helping with imaging and members of the Hoffman lab for advice and suggestions. This work was supported by the *Eunice Kennedy Shriver* NICHD Intramural Research Program.

Conflicts of Interest: The authors state no competing interests.

References

1. Kwan, P.; Brodie, M.J. Early identification of refractory epilepsy. *N. Engl. J. Med.* **2000**, *342*, 314–319. [CrossRef]
2. Bien, C.G.; Kurther, M.; Baron, K.; Lux, S.; Helmstaedter, C.; Schramm, J.; Elger, C.E. Long-term seizure outcome and antiepileptic drug treatment in surgically treated temporal lobe epilepsy patients: A controlled study. *Epilepsia* **2001**, *42*, 1416–1421. [CrossRef] [PubMed]
3. Engel, J.J. Mesial temporal lobe epilepsy: What have we learned? *Neuroscientist* **2001**, *7*, 340–352. [CrossRef] [PubMed]

4. Helmstaedter, C.; Kockelmann, E. Cognitive outcomes in patients with chronic temporal lobe epilepsy. *Epilepsia* **2006**, *47*, 96–98. [CrossRef] [PubMed]
5. French, J.A. Refractory epilepsy: Clinical overview. *Epilepsia* **2007**, *48*, 3–7. [CrossRef] [PubMed]
6. Coan, A.C.; Cendes, F. Understanding the spectrum of temporal lobe epilepsy: Contributions for the development of individual therapies. *Expert Rev. Neurother.* **2013**, *13*, 1383–1394. [CrossRef]
7. Brodie, M.J.; Dichter, M.D. Antiepileptic drugs. *N. Engl. J. Med.* **1996**, *334*, 168–175. [CrossRef]
8. Ryvlin, P.; Rheims, S. Epilepsy surgery: Eligibility criteria and presurgical evaluation. *Dialogues Clin. Neurosci.* **2008**, *10*, 91–103.
9. Jung, S.; Jones, T.D.; Lugo, J.N.J.; Sheerin, A.H.; Miller, J.W.; D'Ambrosio, R.; Anderson, A.E.; Poolos, N.P. Progressive dendritic HCN channelopathy during epileptogenesis in the rat pilocarpine model of epilepsy. *J. Neurosci.* **2007**, *27*, 13012–13021. [CrossRef]
10. Powell, K.L.; Ng, C.; O'Brien, T.J.; Xu, S.H.; Williams, D.A.; Foote, S.J.; Reid, C.A. Decreases in HCN mRNA expression in the hippocampus after kindling and status epilepticus in adult rats. *Epilepsia* **2008**, *49*, 1686–1695. [CrossRef]
11. Marcelin, B.; Chauviere, L.; Becker, A.; Migliore, M.; Esclapez, M.; Bernard, C. H channel-dependent deficit of theta-oscillation resonance and phase shift in temporal lobe epilepsy. *Neurobiol. Dis.* **2009**, *33*, 436–447. [CrossRef] [PubMed]
12. Arnold, E.C.; McMurray, C.; Gray, R.; Johnston, D. Epilepsy-induced reduction in HCN channel expression contributes to an increased excitability in dorsal, but not ventral, hippocampal CA1 neurons. *eNeuro* **2019**, *6*, 6. [CrossRef] [PubMed]
13. Foote, K.M.; Lyman, K.A.; Han, Y.; Michailidis, I.E.; Heuermann, R.J.; Mandikian, D.; Trimmer, J.S.; Swanson, G.T.; Chetkovich, D.M. Phosphorylation of the HCN channel auxiliary subunit TRIP8b is altered in an animal model of temporal lobe epilepsy and modulates channel function. *J. Biol. Chem.* **2019**, *294*, 15743–15758. [CrossRef] [PubMed]
14. Ketelaars, S.O.; Gorter, J.A.; van Vliet, E.A.; Lopes da Silva, F.H.; Wadman, W.J. Sodium currents in isolated rat CA1 pyramidal and dentate granule neurones in the post-status epilepticus model of spilepsy. *Neuroscience* **2001**, *105*, 109–120. [CrossRef]
15. Kile, K.B.; Tian, N.; Durand, D.M. Scn2a channel mutation results in hyperexcitability in the hippocampus in vitro. *Epilepsia* **2008**, *49*, 488–499. [CrossRef]
16. Djamshidian, A.; Grassi, R.; Seltenhammer, M.; Czech, T.; Baumgartner, C.; Schmidbauer, M.; Ulrich, W.; Zimprich, F. Altered expression of voltage-gated calcium channel alpha(1) subunits in temporal lobe epilepsy with Ammon's horn sclerosis. *Neuroscience* **2002**, *111*, 57–69. [CrossRef]
17. Su, H.; Sochivko, D.; Becker, A.; Chen, J.; Jiang, Y.; Yaari, Y.; Beck, H. Upregulation of a T-type Ca2+ channel causes long-lasting modification of neuronal firing mode after status epilepticus. *J. Neurosci.* **2002**, *22*, 3645–3655. [CrossRef]
18. D'Adamo, M.C.; Catacuzzeno, L.; Di Giovanni, G.; Franciolini, F.; Pessia, M. K+ channelepsy: Progress in the neurobiology of potassium channels and epilepsy. *Front. Cell Neurosci.* **2013**, *7*, 134. [CrossRef]
19. Kohling, R.; Wolfart, J. Potassium channels in Epilepsy. *Cold Spring Harb. Perspect. Med.* **2016**, *6*, a022871. [CrossRef]
20. Singh, B.; Ogiwara, I.; Kaneda, M.; Tokonami, N.; Mazaki, E.; Baba, K.; Matsuda, K.; Inoue, Y.; Yamakawa, K. A Kv4.2 truncation mutation in a patient with temporal lobe epilepsy. *Neurobiol. Dis.* **2006**, *24*, 245–253. [CrossRef]
21. Lee, H.; Lin, M.C.; Kornblum, H.I.; Papazian, D.M.; Nelson, S.F. Exome sequencing identifies de novo gain of function missense mutation in KCND2 in identical twins with autism and seizures that slows potassium channel inactivation. *Hum. Mol. Genet.* **2014**, *23*, 3481–3489. [CrossRef] [PubMed]
22. Lin, M.A.; Cannon, S.C.; Papazian, D.M. Kv4.2 autism and epilepsy mutation enhances inactivation of closed channels but impairs access to inactivated state after opening. *Proc. Natl. Acad. Sci. USA* **2018**, *115*, E3559–E3568. [CrossRef] [PubMed]
23. Hoffman, D.A.; Magee, J.C.; Colbert, C.M.; Johnston, D. K+ channel regulation of signal propagation in dendrites of hippocampal pyramidal neurons. *Nature* **1997**, *387*, 869–875. [CrossRef] [PubMed]
24. Menegola, M.; Trimmer, J.S. Unanticipated region-and cell-specific downregulation of individual KChIP auxiliary subunit isotypes in Kv4.2 knock-out mouse brain. *J. Neurosci.* **2006**, *26*, 12137–12142. [CrossRef] [PubMed]

25. Chen, X.; Yuan, L.L.; Zhao, C.; Birnbaum, S.G.; Frick, A.; Jung, W.E.; Schwarz, T.L.; Sweatt, J.D.; Johnston, D. Deletion of Kv4.2 gene eliminates dendritic A-type K+ current and enhances induction of long-term potentiation in hippocampal CA1 pyramidal neurons. *J. Neurosci.* **2006**, *26*, 12143–12151. [CrossRef]
26. Magee, J.C.; Carruth, M. Dendritic voltage-gated ion channels regulate the action potential firing mode of hippocampal CA1 pyramidal neurons. *J. Neurophysiol.* **1999**, *82*, 1895–1901. [CrossRef]
27. Johnston, D.; Hoffman, D.A.; Magee, J.C.; Poolos, N.P.; Watanabe, S.; Colbert, C.M.; Migliore, M. Dendritic potassium channels in hippocampal pyramidal neurons. *J. Physiol.* **2000**, *525*, 75–81. [CrossRef]
28. Bernard, C.; Anderson, A.; Becker, A.; Poolos, N.P.; Beck, H.; Johnston, D. Acquired dendritic channelopathy in temporal lobe epilepsy. *Science* **2004**, *305*, 532–535. [CrossRef]
29. Hall, A.M.; Throesch, B.T.; Buckingham, S.C.; Markwardt, S.J.; Peng, Y.; Wang, Q.; Hoffman, D.A.; Roberson, E.D. Tau-Dependent depletion and dendritic hyperexcitability in a mouse model of Alzheimer's Disease. *J. Neurosci.* **2015**, *35*, 6221–6230. [CrossRef]
30. Pongs, O.; Schwarz, J.R. Ancillary subunits associated with voltage-gated K+ channels. *Physiol. Rev.* **2010**, *90*, 755–796. [CrossRef]
31. Nadal, M.S.; Ozaita, A.; Amarillo, Y.; Vega-Saenz de Miera, E.; Ma, Y.; Mo, W.; Goldberg, E.M.; Misumi, Y.; Ikehara, Y.; Neubert, T.; et al. The CD26-related dipeptidyl aminopeptidase-like protein DPPX is a critical component of neuronal A-type K+ channels. *Neuron* **2003**, *37*, 449–461. [CrossRef]
32. Rhodes, K.J.; Carroll, K.I.; Sung, M.A.; Doliveira, L.C.; Monaghan, M.M.; Burke, S.L.; Strassle, B.W.; Buchwalder, L.; Mengola, M.; Cao, J.; et al. KChIPs anf Kv4 α subunits as integral components of A-type potassium channels in mammalian brain. *J. Neurosci.* **2004**, *24*, 7903–7915. [CrossRef] [PubMed]
33. Jerng, H.H.; Pfaffinger, P.J.; Covarrubias, M. Molecular physiology and modulation of somatodendritic A-type potassium channels. *Mol. Cell Neurosci.* **2004**, *27*, 343–369. [CrossRef]
34. Jerng, H.H.; Kunjilwar, K.; Pfaffinger, P.J. Multiprotein assembly of Kv4.2, KChIP3, and DPP10 produces ternary channel complexes with ISA-like properties. *J. Physiol.* **2005**, *568*, 767–788. [CrossRef] [PubMed]
35. Wang, H.-G.; He, X.P.; Li, Q.; Madison, R.D.; Moore, S.D.; McNamara, J.O.; Pitt, G.S. The auxiliary subunit KChIP2 is an essential regulator of homeostatic excitability. *J. Biol. Chem.* **2013**, *288*, 13258–13268. [CrossRef]
36. Schrader, L.A.; Birnbaum, S.G.; Nadin, B.; Bui, D.; Anderson, A.E.; Sweatt, J.D. ERK/MAPK regulates the Kv4.2 potassium channel by direct phosphorylation of the pore-forming subunit. *Am. J. Physiol.* **2006**, *290*, 852–861. [CrossRef] [PubMed]
37. Hammond, R.S.; Lin, L.; Sidorov, M.S.; Wikenheiser, A.M.; Hoffman, D.A. Protein Kinase A mediates activity-dependent Kv4.2 channel trafficking. *J. Neurosci.* **2008**, *28*, 7513–7519. [CrossRef]
38. Schrader, L.A.; Ren, Y.-J.; Cheng, F.; Bui, D.; Sweatt, J.D.; Anderson, A.E. Kv4.2 is a locus for PKC and ERK/MAPK cross-talk. *Biochem. J.* **2009**, *41*, 705–715. [CrossRef]
39. Kim, J.; Jung, S.-C.; Clemens, A.M.; Petralia, R.S.; Hoffman, D.A. Regulation of dendritic excitability by activity-dependent trafficking of the A-type K+ channel subunit Kv4.2 in hippocampal neurons. *Neuron* **2007**, *54*, 933–947. [CrossRef]
40. Jung, S.-C.; Hoffman, D.A. Biphasic somatic A-type K channel downregulation mediates intrinsic plasticity in hippocampal CA1 pyramidal neurons. *PLoS ONE* **2009**, *4*, e6549. [CrossRef]
41. Rosenkranz, J.A.; Frick, A.; Johnston, D. Kinase-dependent modification of dendritic excitability after long-term potentiation. *J. Physiol.* **2009**, *587*, 115–125. [CrossRef]
42. Hu, J.H.; Malloy, C.; Tabor, G.T.; Gutzmann, J.J.; Liu, Y.; Abebe, D.; Karlsson, R.M.; Durrell, S.; Cameron, H.A.; Hoffman, D.A. Activity-dependent isomerization of Kv4.2 by Pin1 regulates cognitive flexibility. *Nat. Commun.* **2020**, *11*, 1567. [CrossRef]
43. Racine, R.J. Modification of seizure activity by electrical stimulation: II. Motor seizure. *Electroencephalogr. Clin. Neurophysiol.* **1972**, *32*, 281–294. [CrossRef]
44. Namiki, K.; Nakamura, A.; Furuya, M.; Mizuhashi, S.; Matsuo, Y.; Tokuhara, N.; Sudo, T.; Hama, H.; Kuwaki, T.; Yano, S.; et al. Involvement of p38alpha in kainite-induced seizure and neuronal cell damage. *J. Recept. Signal Tranduct.* **2007**, *27*, 99–111. [CrossRef] [PubMed]
45. Bertram, E. The relevance of kindling for human epilepsy. *Epilepsia* **2007**, *48* (Suppl. 2), 65–74. [CrossRef]
46. Adams, J.P.; Sweatt, J.D. Molecular psychology: Roles for ERK MAP kinase cascade in memory. *Annu. Rev. Pharmacol. Toxicol.* **2002**, *42*, 135–163. [CrossRef] [PubMed]
47. Jeon, S.H.; Kim, Y.S.; Bae, C.D.; Park, J.B. Activation of JNK and p38 in rat hippocampus after kainic acid induced seizure. *Exp. Mol. Med.* **2000**, *32*, 227–230. [CrossRef] [PubMed]

48. Kim, J.; Wei, D.-S.; Hoffman, D.A. Kv4 potassium channel subunits control action potential repolarization and frequency-dependent broadening in rat hippocampal CA1 pyramidal neurons. *J. Physiol.* **2005**, *569 Pt 1*, 41–57. [CrossRef] [PubMed]
49. Carrasquillo, Y.; Burkhalter, A.; Nerbonne, J.M. A-type K+ channels encoded by Kv4.2, Kv4.3 and Kv1.4 differentially regulate intrinsic excitability of cortical pyramidal neurons. *J. Physiol.* **2012**, *590*, 3877–3890. [CrossRef]
50. Menegola, M.; Misonou, H.; Vacher, H.; Trimmer, J.S. Dendritic A-type potassium channel subunit expression in CA1 hippocampal interneurons. *Neuroscience* **2008**, *154*, 953–964. [CrossRef]
51. Trimmer, J.S. Subcellular localization of K+ channels in mammalian brain neurons: Remarkable precision in the midst of extraordinary complexity. *Neuron* **2015**, *85*, 238–256. [CrossRef] [PubMed]
52. Alfaro-Ruíz, R.; Aguado, C.; Martín-Belmonte, A.; Moreno-Martínez, A.E.; Luján, R. Expression, cellular and subcellular localization of Kv4.2 and Kv4.3 channels in the rodent hippocampus. *Int. J. Mol. Sci.* **2019**, *20*, 246. [CrossRef] [PubMed]
53. Sun, W.; Maffie, J.K.; Lin, L.; Petralia, R.S.; Rudy, B.; Hoffman, D.A. DPP6 establishes the A-type K+ current gradient critical for the regulation of dendritic excitability in CA1 hippocampal neurons. *Neuron* **2011**, *71*, 1102–1115. [CrossRef] [PubMed]

International Journal of
Molecular Sciences

Review

P38α MAPK Signaling—A Robust Therapeutic Target for Rab5-Mediated Neurodegenerative Disease

Ursula A. Germann and John J. Alam *

EIP Pharma, Inc., Boston, MA 02116, USA; ugermann@eippharma.com
* Correspondence: jalam@eippharma.com

Received: 2 July 2020; Accepted: 30 July 2020; Published: 31 July 2020

Abstract: Multifactorial pathologies, involving one or more aggregated protein(s) and neuroinflammation are common in major neurodegenerative diseases, such as Alzheimer's disease and dementia with Lewy bodies. This complexity of multiple pathogenic drivers is one potential explanation for the lack of success or, at best, the partial therapeutic effects, respectively, with approaches that have targeted one specific driver, e.g., amyloid-beta, in Alzheimer's disease. Since the endosome-associated protein Rab5 appears to be a convergence point for many, if not all the most prominent pathogenic drivers, it has emerged as a major therapeutic target for neurodegenerative disease. Further, since the alpha isoform of p38 mitogen-activated protein kinase (p38α) is a major regulator of Rab5 activity and its effectors, a biology that is distinct from the classical nuclear targets of p38 signaling, brain-penetrant selective p38α kinase inhibitors provide the opportunity for significant therapeutic advances in neurogenerative disease through normalizing dysregulated Rab5 activity. In this review, we provide a brief summary of the role of Rab5 in the cell and its association with neurodegenerative disease pathogenesis. We then discuss the connection between Rab5 and p38α and summarize the evidence that through modulating Rab5 activity there are therapeutic opportunities in neurodegenerative diseases for p38α kinase inhibitors.

Keywords: p38 MAPK; p38α; Rab5; endosome; Alzheimer's; Lewy Bodies; amyloid-β; tau; α-synuclein

---

1. Introduction

Due to the scarcity of effective treatments for neurodegenerative diseases, urgent searches for candidate cellular and molecular mechanisms to develop therapeutic interventions are underway [1–8]. Neurodegenerative diseases, including Alzheimer's disease (AD), Parkinson's disease (PD), dementia with Lewy bodies (DLB), frontotemporal dementia (FTD), amyotrophic lateral sclerosis (ALS), and Huntington's disease (HD) are typically defined by aberrant accumulations of one or more specific protein(s) and by loss of certain neuronal populations resulting in anatomic vulnerability. However, it is becoming increasingly clear that different neurodegenerative diseases exhibit common, central processes associated with progressive neuronal dysfunction and death, revealing multifactorial pathologies, including proteotoxic stress, neuroinflammation, and other abnormalities [3,5,6,9,10]. Even though the species of accumulating proteins are distinct in different neurodegenerative diseases, increasing evidence indicates that defects in the protein clearance system play a central role in the gradual accumulation of protein aggregates. Emerging genetic and biological evidence suggests that the endo-lysosomal protein degradation machinery, which is part of a unified pathway together with the autophagosomal machinery, is dysfunctional across a broad spectrum of neurodegenerative diseases, including AD, PD, ALS, HD, and others [1,11–13].

In this review we focus the discussion on the abnormal activity of the Ras-related protein Rab5, the master regulatory guanosine triphosphatase (GTPase) in early endosomes and highlight its role as a mediator of AD and other neurodegenerative diseases. We also discuss the relevance of Rab5 as a target that is affected by the p38α isoform of p38 mitogen-activated protein kinase (MAPK), hence, can be modulated by specific p38α inhibitors. The main objectives of this review are as follows: (1) we briefly review the role of p38α in the cell, including the neuron; (2) we review the role of Rab5 in the cell and discuss the association of dysregulated Rab5 activity with neurodegenerative disease pathogenesis; (3) we discuss the connection between Rab5 and p38α; (4) we provide evidence that through modulating Rab5 activity there are therapeutic opportunities in neurodegenerative diseases for brain-penetrant, selective p38α kinase inhibitors; and (5) we offer ideas for further investigations to increase the understanding of the mechanism of action of p38α kinase inhibitors on Rab5 in neurodegenerative disease.

2. Overview of the p38α Isoform as a Member of the p38 MAPK Family

The p38 MAPK family consists of four members that are encoded by separate genes and are known as p38α/MAPK14, p38β/MAPK11, p38γ/MAPK12/ERK-6/SAPK3, and p38δ/MAPK13/SAPK4 [14]. These four major isoforms differ in their organ, tissue, or cellular expression patterns, and it is becoming increasingly clear that they exert distinct biological functions [4,15–23]. Among the p38 MAPK family members, p38α was discovered first as a stress-activated protein kinase that plays a central role in inflammation [24,25]. P38α is also the best characterized isoform to date as a central nervous system drug discovery target [19,26,27].

In the adult mouse, p38α is highly expressed in different brain areas, including the cerebral cortex, hippocampus, cerebellum, and a few nuclei of the brainstem [28]. Neuronal cells are the predominant cell type expressing p38α [28]. At the subcellular level, p38α is distributed in dendrites and in cytoplasmic and nuclear regions of the cell body of neurons [28].

Many studies that have characterized p38α isoform function (frequently together with p38β function) have shown that it is an intracellular protein kinase involved in transducing intracellular (e.g., DNA damage) and extracellular (e.g., osmotic stress, infection) signals into a cellular response (e.g., inflammation or activation of other cellular stress responses) [17,29]. The major signal transduction pathway for p38α has been extensively studied and involves upstream activators (e.g., the mitogen-activated protein kinase kinases MKK6, MKK3) and downstream targets (e.g., mitogen-activated protein kinase activated kinase 2, also known as MAPKAPK2 or MK2) [17,29]. In the classic pathway for activating the proinflammatory response, some studies showed that inactive p38α (as well as p38β) is sequestered in the cytoplasm through its binding to MK2; upon activation, p38α phosphorylates MK2, leading to its dissociation [17,29]. Once dissociated, p38α translocates to the nucleus where it phosphorylates transcriptional machinery targets (e.g., histones, mitogen- and stress-activated kinases MSK1/2) in the proinflammatory context around nuclear factor NF-κB-associated targets [17,29]. In terms of therapeutics development, this understanding has led to a range of efforts to develop p38α kinase inhibitors (many of them primarily inhibiting p38α and p38β activity) as anti-inflammatory agents for chronic inflammatory disorders, including rheumatoid arthritis (RA), inflammatory bowel disease (IBD), and chronic obstructive pulmonary disease (COPD) [30,31].

However, p38α signaling has many targets aside from this classical pathway [32] and biologic effects other than regulation of proinflammatory cytokine production. With regard to neurodegenerative diseases, a large variety of biological roles have been attributed to p38α in brain pathology which depend on the type and stage of central nervous system (CNS) disease, brain region, cell type [4,18,19,26]. These roles include modulation of proinflammatory cytokine, e.g., interleukin-1beta (IL-1β) and tumor necrosis factor alpha (TNFα) production and signaling (e.g., in glia, microglia, astrocytes, neurons), as well as orchestration of neurotoxicity, neuroinflammation, and/or synaptic dysfunction, among others [4,18,19,26].

While the first study to characterize the role of p38α in regulating stress-induced endocytosis and early endosomal biology via modulating the activity of the endosomal protein Rab5 was published nearly two decades ago [33], this effect has not been a focus of intense follow-up research. Nevertheless, this biology, specifically within the neuron, has come to the forefront as potentially the most relevant for AD and other neurodegenerative disorders as will be discussed in Section 5.

During the last two decades, deregulated p38α has emerged as a leading therapeutic target for AD and has also been associated with the pathology of other neurodegenerative disorders, including PD, DLB, or ALS [4,19,34–42]. These therapeutic opportunities for treatment with selective, brain-penetrant p38α inhibitors will be discussed in more details in Section 6.

3. The Endosome-Associated Protein Rab5

Endocytosis represents the process to internalize diverse cargos (e.g., extracellular macromolecules, viruses, bacteria, membrane proteins) into cells (including neurons) through vesicles that bud off from the plasma membrane [43]. After their internalization into the cytosol, the endocytic vesicles are rapidly targeted to and fused with the early endosome [44–46]. This functions as the primary sorting organelle from which endocytosed cargo (e.g., select receptors) is either recycled back to the plasma membrane, or delivered to the lysosome/vacuole for degradation after maturation of the early endosome into a late endosome [44–46]. Important roles of early endosomes include nutrient uptake, degradation of metabolic by-products, transport of materials to specific compartments in the cell, and regulating the cell-surface expression of receptors and transporters [44–46].

3.1. Overview of Rab5 Roles

The Ras-related protein Rab5 is a small GTPase that is a major regulator of early steps of endocytosis, and subsequent endosomal membrane trafficking, sorting and endosomal fusion [47–49]. Further, through interacting with effector proteins Rab5 has a critical role in regulating the docking and fusion of endosomal membranes, endosomal mobility and intracellular signal transduction [48,50]. Rab5 effector protein include EEA1 (early endosomal autoantigen 1), APPL (adaptor protein, phosphotyrosine interacting with pleckstrin homology (PH) domain and leucine zipper 1), PI3K (phosphatidylinositol-3-kinase), Rabenosyn-5/hVPS45 (human Sec1p-like vacuolar protein sorting), or Rabaptin-5/Rabex-5, among others. Additionally, a role of Rab5 in regulating the internalization and trafficking of membrane receptors by regulating vesicle fusion and receptor sorting in the early endosomes is emerging [49]. Rab5, which actually comprises three different isoforms, is among the best characterized endosomal markers, in part because of its abundant expression and ubiquitous tissue distribution, including neurons [51,52].

3.2. Rab5 Importance for Neuronal Function

It is clear that proper Rab GTPase function is critical for normal (wild-type) neuronal function, including trafficking for pre- and post-synaptic function as well as dendritic trafficking [52,53]. Studies in *Drosophila* have demonstrated that Rab5 is required for synaptic endosomal integrity, synaptic vesicle exo-/endocytosis rates, and neurotransmitter probability [54]. Furthermore, an essential function is that Rab5-dependent endosomal sorting may regulate the uniformity of synaptic vesicle size [55].

The neuron may be particularly sensitive to dysregulation of Rab5 activity for at least two main reasons: (1) Endocytosis and subsequent recycling (or not) regulate the concentration of neurotransmitter receptor density on the cell surface, determining signal strength [53,56,57]. For example, α-amino-3-hydroxy-5-methyl-4-isoxazolepropionic acid receptor (AMPAR) endocytosis in hippocampal neurons leads to long-term depression (LTD), and Rab5 is essential in this process [56,58,59]; and (2) neurotrophin signaling from synapses is dependent on endocytosis, retrograde transport of endosomes along axons, and endosomal signaling [1,48,52,60].

3.3. Rab5 Therapeutic Targeting Strategies

The activity of Rab5 is coordinately regulated and, therefore, can be therapeutically targeted at several levels through modulation of Rab5 regulatory proteins. Firstly, Rab5 is shuttled between membranes by the general Rab regulator GDP dissociation inhibitor (GDI) [61]. This serves to release

Rab5 that is bound to GDP, Rab5(GDP), from membranes to maintain Rab5 in the cytoplasm, and to recycle it back to donor membranes [61]. Thus, factors that increase formation of the Rab5-GDI complex also increase delivery of Rab5 to the plasma membrane where it can act [61]. Secondly, at the membrane, the activity of Rab5 is regulated by guanine nucleotide exchange factors (GEFs) and GTPase-activating proteins (GAPs) that determine the proportion of Rab5 bound to either GDP (Rab5(GDP); inactive state) or GTP (Rab5(GTP); active state) [62]. Thirdly, Rab5 activity is modulated by other factors that impact the effectors; for example, the phosphorylation of, and activity of, PI3K or EEA1 [63–66]. Additionally, the druggability of membrane-bound Rab5 itself, the selective inhibition of Rab5 GTPase activity, or blocking membrane recruitment through inhibition of Rab5 prenylation, or targeting Rab5-associated signaling pathways can be explored [67–69].

4. Role of Dysregulated Rab5 in the Pathogenesis of Neurodegenerative Disease

Dysregulated Rab5 activity has been defined as a major pathogenic driver in AD [1,48,70]. Moreover, a pathogenic role of aberrant Rab5 is emerging in many of the same other neurodegenerative diseases that are being targeted by p38α inhibitor programs, including PD, DLB, ALS, and HD [71–73]. Rab5 is a member of a large family of Rab proteins involved with neuronal function [53,71] and a number of other Rab proteins have been connected to neurodegenerative disease. However, as will be discussed in Section 5, Rab5 activity has been robustly connected to p38 MAPK signaling, while no such connection has been established for the other Rab proteins. Therefore, this review is focused on Rab5, and the reader is referred to a number of other excellent recent reviews on the broader family of Rab proteins and their relation to the pathogenesis of neurodegenerative disease [71–73].

4.1. Dysregulated Rab5 as Therapeutic Target in AD

Neuronal endocytic pathway activation is a specific and very early response in AD that precedes amyloid-beta (Aβ) deposition in sporadic AD, hence, the role of dysregulated Rab5 in AD has been extensively studied and reviewed elsewhere [1,48,70,71]. It will be discussed briefly here.

In a large series of experiments during more than two decades, Nixon and colleagues have documented specific impairments of the endosomal-lysosomal system at the earliest stage of AD and linked the genetic drivers that cause AD directly to functions within endocytic and autophagic pathways of the lysosomal system. They demonstrated that abnormal Rab5-positive endosome enlargement is the earliest pathologic event in sporadic AD patients [74,75]. They also showed that abnormal Rab5-positive endosome enlargement is the earliest pathologic event in Down syndrome (DS) patients [74,75]. DS patients are individuals with trisomy for all or part of third copy of chromosome 21 (which carries the β-Amyloid Precursor Protein (APP) gene among others), who nearly uniformly develop progressive AD after age 40 [74,75]. Importantly, Nixon and colleagues also defined the mechanistic basis of the endosome enlargement induced by APP to be Rab5 hyperactivation [70]. They also linked functional neuronal deficits and, where evident, subsequent neuronal loss in animal models of AD and DS to Rab5 hyperactivation [70].

Among other lines of evidence, Nixon and colleagues showed that the β-cleaved carboxy-terminal fragment of APP, termed β-CTF, recruits APPL1 to Rab5 endosomes [76]. There APPL1 stabilizes active Rab5(GTP), leading to pathologically accelerated endocytosis, endosome swelling and selectively impaired axonal transport of Rab5 endosomes [76]. Importantly, in DS fibroblasts an APPL1 knockdown corrected these endosomal anomalies [76]. β-CTF levels were also shown to be elevated in AD brain, which was accompanied by abnormally high recruitment of APPL1 to Rab5 endosomes, as was observed in DS fibroblasts [76]. Moreover, in a separate report, Nixon and colleagues [77] showed that partial reduction of β-APP cleaving enzyme 1 (BACE1) through genetic means in a transgenic mouse model (Ts2) of DS normalized both APP-β-CTF levels and Rab5 activation. This prevented age-related development of Rab5-positive endosomal enlargement (which is usually evident at approximately four months of age in the Ts2 mice) and subsequent loss of cholinergic neurons in the basal forebrain (which otherwise follows the Rab5-positive endosomal enlargement within approximately one to two months in the Ts2 mice) [77].

In complementary work, Xu et al. [78] demonstrated that full-length wild-type APP and β-CTF both, in vitro in three different relevant cell model systems, induced early endosomes enlargement and disrupted nerve growth factor (NGF) signaling and axonal trafficking. Moreover, β-CTF alone induced atrophy of cultured rat basal forebrain cholinergic neurons that was rescued by a dominant-negative Rab5 mutant [78]. Finally, expression of a dominant negative Rab5 construct markedly reduced APP-induced axonal blockage in *Drosophila* [78].

This earlier work indicated that Rab5 was necessary for APP-induced endosomal enlargement and cholinergic neuronal loss. Recently, Nixon and colleagues demonstrated that abnormal Rab5 activation is sufficient to induce endosomal enlargement and a neurodegenerative phenotype which mimics that seen with APP overexpression [79]. Specifically, modest neuron-specific transgenic Rab5 (PA-Rab5) expression in mice [79] induced increased Rab5 expression and abnormal activation of Rab5 comparable to that in AD brain [80,81]. PA-Rab5 reproduced AD-like Rab5-endosomal enlargement and mis-trafficking without impacting APP metabolism (i.e., no increase in Aβ levels) [79]. The PA-Rab5 mice also exhibited hippocampal synaptic plasticity deficits via accelerated AMPAR endocytosis and dendritic spine loss [79]. Moreover, they showed tau hyperphosphorylation [79]. Importantly, with further aging the PA-Rab5 mice developed progressive cholinergic neurodegeneration and impaired hippocampal-dependent memory subsequent to the observed Rab5-mediated endosomal dysfunction [79].

That Rab5 hyperactivity and endosome enlargement, rather than APP per se, are the critical factors in inducing degenerative AD-related changes is further supported by the following findings. Age-dependent Rab5-positive early endosome enlargement and endo-lysosomal dysfunction were observed in an AD-vulnerable brain region of targeted replacement mice expressing the human Apolipoprotein E4 (ApoE4) gene, the dominant genetic factor for the development of late-onset Alzheimer's disease, under the control of the endogenous murine promoter [82]. Similarly, depletion of another late onset AD risk gene sortilin-related receptor 1 (SORL1) [83] in human induced pluripotent stem cell (iPSC)-derived neurons leads to enlargement of early endosomes (i.e., endosomes staining positive for Rab5 and EEA1). Hence, genetic influences that increase AD risk, such as ApoE4 and SORL1, may do so by dysregulating the Rab5 impact on endosome dynamics and cell signaling.

A recent study that analyzed a comprehensive panel of iPSC-derived neuronal lines relevant to familial AD also demonstrated translatability of Rab5-mediated neurodegeneration to human AD [84]. The only consistent, intra-neuronal physiologic defect identified was enlargement of Rab5-positive early endosomes, mediated by APP-β-CTFs, not Aβ, and associated with endosomal/endocytic dysfunction [84].

In a very simplified view of many AD research data taken together, the small GTPase Rab5 can be depicted as a convergence point (Figure 1) for multiple established pathogenic drivers of neurodegeneration in AD (e.g., downstream of APP, APP-β-CTF, β-CTF, Aβ, ApoE4, and others). Dysregulation of the endo-lysosomal system represents the important early cellular phenotype of pathogenesis for AD that leads to disrupted AMPAR trafficking, tau pathology, synaptic dysfunction, and neurodegeneration. Elevated Rab5 activity (hyperactivation or overexpression) plays the key role in mediating these processes, hence, may promise high potential as a therapeutic target. Since endosome dysfunction occurs very early in the AD pathogenic process, therapies that turn Rab5 activity back to normal (e.g., via Rab5 therapeutic targeting strategies already discussed in Section 3.3.) may help slow or halt AD development before irreversible damage occurs. Additionally, the potential of p38α inhibition as a therapeutic lever to reduce Rab5 activity will be discussed in Section 6.

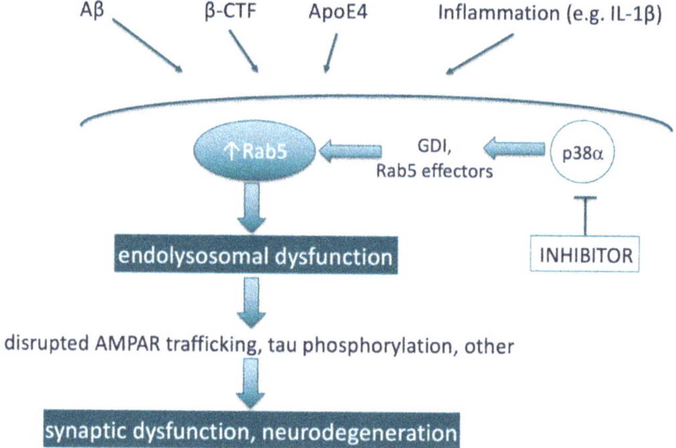

Figure 1. Simplified scheme representing Rab5 as a therapeutic target through being a convergence point for multiple pathogenic drivers of neurodegeneration in AD and the potential of p38α inhibition as a therapeutic lever to reduce Rab5 activity. Note: Aβ and β-CTF are both derived from proteolytic processing of APP.

4.2. Dysregulated Rab5 Associated with Abnormal α-Synuclein in PD, DLB, and AD

The enlargement of Rab5-positive early endosomes that is seen in AD is not observed during the development of PD and DLB. Nevertheless, the toxicity of the key pathologic protein associated with these two neurodegenerative diseases, abnormal α-synuclein (a pre-synaptic protein), has also been linked to Rab5 [71,85,86]. Specifically, the neurotoxicity of α-synuclein was shown to be dependent on Rab5-mediated entry into the cell via endocytosis [87,88]. Expression of a GTPase-deficient Rab5a protein led to a decrease in the cytotoxicity of α-synuclein through impairing its endocytosis [87]. Rab5 also appears to play a role in intracellular trafficking of α-synuclein [89,90]. Additionally, studying embryonic cortical neurons from a mouse model of Parkinson's disease, transgenic overexpression of α-synuclein was observed to increase the levels of activated Rab5 and Rab7 [91]. This impaired retrograde axonal transport of brain-derived neurotrophic factor (BDNF) and led to neuronal atrophy [91]. Therefore, the authors suggested that α-synuclein-induced neuronal dysfunction is a result of impaired endocytosis and endosomal dysfunction associated with aberrant activation of the two Rab proteins [91].

It is interesting to note that accumulating evidence suggests that the α-synuclein might also play a role as driver of pathophysiology in AD [92]. Intriguingly, α-synuclein and APP appear to be interconnected in terms of their activation of Rab5 and neurotoxicity, since genetic reduction of endogenous α-synuclein in an APP transgenic mouse model normalized Rab5 (and Rab3) activity and prevented cholinergic neuronal loss [93].

4.3. Dysregulated Rab5 in ALS

Evidence for the pathological role of Rab proteins has also been provided in ALS as another example of a neurodegenerative disease involving endosomal-lysosomal trafficking and signaling defects [71].

In the context of ALS, defects in endosomal trafficking have been consistently seen in transgenic mouse models based on identified human genetic defects [94]. In particular, Alsin, deficiency of which is associated with an autosomal recessive juvenile form of ALS called ALS 2, is a Rab5 exchange factor [95–97]. The primary biological effects of Alsin deficiency have been linked to aberrant activation of Rab5-mediated endosomal trafficking [98]. Rab5 interaction with Alsin has also been suggested to modulate the signaling of neurotrophic factors [96]. The analysis of Alsin-null mice, an animal model of ALS2, revealed that Rab5-dependent endosome fusion activity and endosomal transport of insulin-like growth factor 1 (IGF1) and BDNF receptors were affected [99]. It was suggested that these alterations in trophic receptor trafficking in the neurons of the Alsin-null mice may lead to the observed reduced size of the cortical neurons as well as animal hypoactivity, and that this may translate to the pathogenesis of ALS2 [99].

Moreover, the protein product of hexanucleotide GGGGCC repeat in the chromosome 9 open reading frame 72 (C9ORF72), which represents a major genetic cause of familial ALS (33% of familial cases) and FTD, has been co-localized with Rab5 in endosomes [100,101]. It was described to possess Rab GEF activity and function as a regulator of endosomal trafficking [100].

4.4. Dysregulated Rab5 in HD

Finally, Rab proteins also have a key role in HD [52,71]. It was reported that the upregulated Huntingtin (Htt)-associated protein 40 (HAP40) is an effector of Rab5 that mediates the recruitment of Htt to early endosomes and is affecting early endosomal motility [102]. As Rab5-positive endosomes are involved in retrograde transport of activated neurotrophin/receptor complexes and due to indication of altered axonal transport in HD [103,104], it is possible that impaired Rab5-mediated trafficking of neurotrophins affects neurotrophin signaling and might also contribute to HD pathogenesis [105,106]. Moreover, Rab5 overexpression reduces toxicity of the Htt mutant protein, while inhibition of Rab5 increases toxicity via macroautophagy regulation [107].

Taken together, overactivated Rab5 and subsequent endo-lysosomal dysfunction have emerged as a major driving force of degenerative and cognitive deficits during the development of AD [1,48,71,79] and alterations in Rab5 also seem to play an important role in other types of neurodegenerative diseases [52,71].

5. p38α Is a Major Regulator of Rab5 Activity

It is well-established in the scientific literature that p38α regulates Rab5 activity. This includes the research in the context of neuronal synaptic plasticity. Most of the findings were published in the early and mid-2000s. First, p38α was shown to be a regulator of endocytosis through phosphorylating GDI and stimulating the formation of cytosolic Rab5-GDI complex, thereby increasing the concentration of Rab5 in the plasma membrane (Figure 1) [33,108]. Moreover, in a genome-wide screen of human kinase-mediated regulation of endocytosis, ablation of a number of kinases increased endocytosis in association with increasing phosphorylation of p38 MAPK (i.e., activated p38 MAPK) and recruiting phospho-p38α to the endosome [109]. These authors also showed by confocal microscopy that the MAPK14 gene product was observed on endosomal structures [109]. Prior to this, p38 had been co-localized via a sucrose gradient with the Rab5- and NGF-containing early endosome fraction prepared from rat dorsal root ganglion (DRG) neurons, and was shown to be part of early endosome signaling pathways for conveying NGF signals from the target of nociceptive neurons to their cell bodies [110]. In addition, expression of an activated Rab5 mutant increased μ opioid receptor endocytosis in wild-type cells but not in p38α -/- cells [111]. In the same report, p38α was also shown to phosphorylate the Rab5 effectors EEA1 (on Thr-1392) and Rabenosyn-5 (on Ser-215), which led to increased recruitment of these proteins to membranes; providing a mechanism other than modulating GDI by which p38α increases Rab5 action. Both in the human kinase screen [109] and the μ opioid receptor endocytosis studies [111] it is noted that the effects of p38α on endocytosis are evident under basal (physiologic) conditions, and not just under conditions of cellular stress, whereas the role of p38α in relation to endocytosis has been suggested to be related to its role in responding to oxidative stress [112]. Collectively, the studies indicate that p38α regulates levels of both the basal and induced Rab5 activity, irrespective of other inputs to Rab5 activation state. As such, p38α inhibition provides an approach to reduce Rab5 activity in a diverse range of disease states that may have different drivers of Rab5 activation (Figure 1).

In the context of neuronal function, a critical component of synaptic plasticity is the maintenance and/or recycling of AMPAR from the surface of synapses, [59] a process in which Rab proteins, including and particularly Rab5, through increasing endocytosis play a prominent role (Figure 1) [58]. In particular for the aspect of synaptic plasticity termed LTD, p38 MAPK activation facilitating AMPAR removal through increasing endocytosis via the Rab5-GDI complex has been demonstrated [56].

In other studies, in which Rab5 activation leading to AMPAR removal from the surface was thought be a critical player in the process of NMDA-triggered LTD induction in the hippocampus, there was associated phosphorylation of p38 MAPK, i.e., p38 MAPK activation; though there was a temporal lag, which may reflect different kinetics of p38 MAPK activation at the plasma membrane versus the cell as a whole [58]. Serotonin-induced LTD has also been shown to be dependent on both p38 MAPK and Rab5, activation of which together led to enhanced AMPAR internalization via endocytosis during the process of LTD [113]. In a subsequent report [114] the same group demonstrated low dose serotonin and norepinephrine reuptake inhibitors (SNRIs), by acting on 5-HT1A and 2-adrenergic receptors, synergistically reduced AMPAR-mediated excitatory postsynaptic currents and AMPAR surface expression in prefrontal cortex pyramidal neurons via a mechanism involving Rab5/dynamin-mediated endocytosis of AMPAR. As this effect of SNRIs was dependent on p38 kinase activity, and their prior work, they hypothesized that SNRI activation of p38 MAPK accelerates AMPAR endocytosis by stimulating the formation of Rab5-GDI complex. However, they did not directly demonstrate this.

6. Therapeutic Potential of Dampening Rab5 Activity through Inhibiting p38α Signaling

In parallel with Rab5 emerging as a therapeutic target for neurodegenerative disease, p38α has also emerged as a promising therapeutic target for AD and other neurodegenerative disorders [4,19,34–42].

6.1. Therapeutic Potential in AD

From a mechanistic perspective, expression of p38α in the neuron is associated with formation of pathological Aβ-, inflammation- (e.g., IL-1β) and tau-induced impaired synaptic plasticity (Figure 1), as well dendritic spine loss [115–119]. Furthermore, studies in several distinct animal models driven by Aβ, inflammation, or tau showed that spatial learning and working memory deficits are reversed with small molecule inhibitors of p38α kinase activity [120–122], providing direct evidence that inhibition of p38α activity has therapeutic potential in AD. Specifically, the compound MW150 was active in APP-transgenic and tau-transgenic mice [121,123], the compounds MW181 and SB2399063 in aged tauopathy mice [122], and neflamapimod/VX-745 in aged rats [120]. In addition, a very recent publication demonstrated that oral administration of a selective p38 α/β inhibitor, NJK14047, to 9-month old 5XFAD (APP) transgenic mice reduced levels of amyloid-beta deposits, reduced spatial memory loss and reduced the number of degenerating neurons labeled with Fluoro-Jade B [41]. Moreover, genetic reduction of neuronal p38α in APP overexpressing transgenic mice improved synaptic transmission, decreased memory loss and reduced amyloid pathology [124,125]. P38 MAPK has also been identified as a therapeutic target for PD and DLB, i.e., α-synuclein mediated neurodegenerative diseases [27,39].

Since none of the published studies assessed Rab5 activity and/or endosomal pathology at present, the literature does not definitively establish that the aformentioned effects of p38α in animal models of neurodegenerative disease are via modulating Rab5 activity. However, several arguments suggest that a major component of the therapeutic effects of p38α is through targeting Rab5. First, in the AD context AMPAR removal is necessary and sufficient for both impaired synaptic plasticity and dendritic spine loss [126], the critical first steps in the neurodegenerative process associated with AD; and as discussed previously, p38α and Rab5 are intimately linked in the process of AMPAR endocytosis and removal from the cell surface. Second, across the variety of biological effects of modulating either Rab5 or p38α activity there is a similarity of effects (including directionality) that, given the known connection between the two, is unlikely to be due to chance. For example, decreasing p38α activity in neurons reduces Aβ production [125], while Rab5 activation increases Aβ production [127]. That is, aberrant activation of either p38 MAPK [115,128] or Rab5 [79] are associated with increased Aβ production.

Further, aberrant activation of Rab5 resulting in a block in endosomal maturation is considered to underlie impaired autophagy in AD [1], while inhibition of p38α has been identified as an approach to reversing impaired autophagy in AD [11]. Third, in the Rab5-overexpressing mouse a downstream biological marker in the neuron of Rab5 activation is tau phosphorylation [79], while p38α inhibitors in aged tauopathy mice improved working memory and, at same time, reduced tau phosphorylation [122]. As p38α is not a major tau kinase [129], we believe that those results provide indirect evidence that p38α inhibition reduces Rab5 activity in parallel with improving memory.

Recently [42], the effects of neuronal deficiency of p38α in neurodegenerative disease models were further evaluated by mating human APP transgenic mice and human P301S Tau-transgenic mice with *mapk14*-(gene for p38α)-floxed and neuron-specific Cre-knock-in mice. Deletion of p38α in neurons through this approach led to improvement of cognition in both the APP transgenic mice and the P301S Tau transgenic mice, associated with decreased Aβ and phosphorylated tau in the brain of the respective models. As along with normal Rab5 adequate calcium influx is essential to, and intimately associated with AMPAR endocytosis [130,131], it is particularly intriguing that neuronal deficiency of p38α in these models regulated the transcription of calcium homeostasis genes and deletion of p38α inhibited NMDA-triggered calcium influx in vitro [42].

More direct evidence on the contribution of Rab5 inhibition towards therapeutic effects of p38α inhibition have been presented at scientific meetings but are not available, yet, as primary research publications. In those studies [132,133] we and our collaborators showed that blocking Rab5 over-activation with a selective p38α inhibitor [37,134] rescued Rab5-positive-endosomal enlargement and cholinergic neurodegeneration in a mouse model of DS (Ts2) as effectively as reversing elevated APP-β-CTF levels [77]. The results directly support the role of p38α in regulating Rab5, as the compound utilized had previously been shown by an independent academic research group to have ~25-fold selectivity for p38α (Kd = 2.8 nM) versus p38β (Kd = 74 nM), as well as its >300-fold selectivity versus 445 other kinases (Kd ≥ 1100 nM) [135]. In addition, the compound had been recommended by yet another research group as the small molecule compound to utilize in experimental studies that have the objective of understanding the biologic effects of inhibiting p38α kinase activity [136].

6.2. Therapeutic Potential in ALS

In the context of ALS, it should also be noted that overexpression of the Rab5 GEF Alsin suppresses superoxide dismutase 1 (SOD1) neurotoxicity [137]. The link to p38α was recently established in studies that showed that p38α kinase inhibitors rescued the axonal transport defects in the SOD1^{G93A} mouse model of ALS [40]. In those studies, p38 MAPKs were found to enhance axonal transport of signaling endosomes in a pharmacological screen of a library of small molecule kinase inhibitors that was designed to identify molecules that would enhance that activity. Moreover, in vitro knockdown revealed that the alpha isoform (i.e., p38α) was the sole isoform responsible for the SOD1^{G93A}-induced transport deficits and acute treatment with p38α inhibitors restored the physiological rate of axonal retrograde transport in vivo in early symptomatic SOD1^{G93A} mice [10].

7. Future Directions for Research and Prospects

From a therapeutics development perspective, the most relevant preclinical mechanistic study to assess the potential of p38α inhibition as a treatment approach for Rab5-mediated neurodegenerative disease would be to evaluate the effects of a p38α inhibitor in the Rab5-overexpressing mouse. Such studies have recently been initiated and the results are anticipated to be published within the next 12 months [138]. From a mechanistic standpoint, a number of open lines of inquiry around the connection between p38α and Rab5 could be explored. For one, in the AD context, whether APPL1 and p38α may act sequentially or in parallel to increase Rab5 activation has not been defined (Figure 2). On one hand, APPL1 as a scaffolding protein could stabilize active Rab5 on the endosome, while p38α activation leading to increased levels of Rab5-GDI complex would deliver Rab5 to the endosome to associate with APPL1. In this "parallel" construct, APPL1 and p38α could have different upstream drivers; for example, p38α being activated upstream by a well-known activator IL-1β, rather than by β-CTF. However, APPL1 has also been shown to act as a scaffold to the p38MAPK signaling pathway [139] and so may act also upstream of p38α, i.e., sequentially, rather than in parallel to activate Rab5. From a therapeutics development model, the two models could impact the context, e.g., the disease in which p38α inhibitors would be most active.

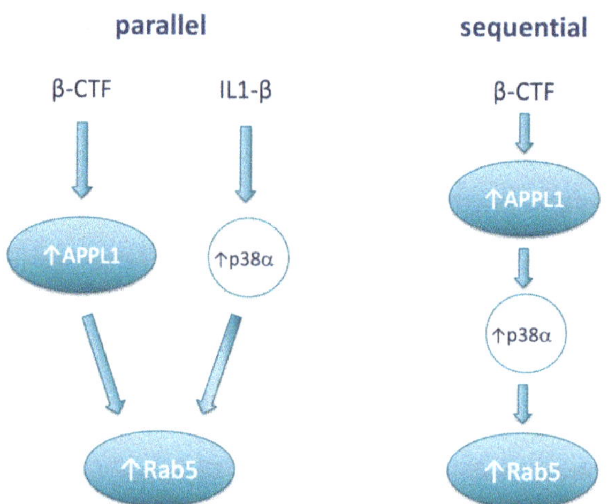

Figure 2. Potential models of relationships between APPL1, Rab5, and p38α.

Another open question is the specific mechanism by which Rab5 activation leads to defects in endosomal signaling and trafficking, an effect that is, to a certain extent, paradoxical as increasing endocytosis and endosome formation would be expected to increase endosomal signaling and potentially increase the number of endosomes delivered from the synapse back to the nucleus via axonal transport. While some specific mechanisms have been proposed [1], the more general hypothesis is that axonal transport and endosomal degradation via lysosomal pathways are rate-limiting and have to be well-matched to the rate of endocytosis. As a result, aberrantly-increased endocytosis overwhelms the rate-limiting disposal pathways, leading to a block in trafficking/degradation and endosomal enlargement. While compelling, this hypothesis has not been definitely established as the reason for the reduction in axonal transport of endosomes that is seen in AD. Further understanding of these mechanisms might identify additional therapeutic targets.

More generally, with respect to p38 MAPK signaling, the roles that regulation of endocytosis and endosomal biology play in the stress response that is otherwise mediated by p38α, or other p38 MAPK isoforms, remain to be fully defined. With respect to the proinflammatory activity of p38α, along with increasing cytokine production, activation of p38α increases cytokine signaling. Classically receptor endocytosis is thought to shut off the signal from the receptor. However, there are increasing examples, including in the context of cytokine signaling, that receptor endocytosis can increase signaling [140–142]. Further, in the context of the neuron endosomal signaling after axonal retrograde signaling, both the signaling pathways distinct from neurotrophins [60] and the cross-talk on the endosome between kinase pathways [143] are underexplored in terms of understanding their roles in modulating p38α (or p38 MAPK) signaling.

The ultimate proof of the therapeutic value of targeting Rab5 with p38α will be in the clinic. Towards that end, results were presented recently [144] from a 24-week 161 patient double-blind, placebo controlled clinical trial of a p38α inhibitor neflamapimod in early-stage AD (https://clinicaltrials.gov/ct2/show/NCT03402659). This study demonstrated the effectiveness of p38α inhibition relative to a placebo in significantly reducing cerebrospinal fluid (CSF) levels of p-tau and tau. Given that as discussed previously tau phosphorylation is a downstream marker of Rab5 hyperactivation, the results provide indirect evidence that p38α inhibition impacts Rab5 activity in humans. The study also showed plasma concentration-dependent effects on episodic memory function, though the dose level utilized led to sub-therapeutic levels in the majority of the patients. A higher dose regimen that achieves the identified therapeutic plasma drug concentration range in ~75% if patients is being utilized in a 16-week randomized, double-blind, placebo-controlled, clinical study of the same p38α inhibitor in patients exhibiting dementia with Lewy bodies (https://clinicaltrials.gov/ct2/show/NCT04001517). The results of this study should further inform on the potential of p38α inhibition as an approach to treat neurodegenerative disease.

Author Contributions: U.A.G. and J.J.A. wrote, reviewed, and edited the manuscript. All authors have read and agreed to the published version of the manuscript.

Funding: This research received no external funding.

Acknowledgments: Sylvie Grégoire for critical reading of the manuscript and input.

Conflicts of Interest: John Alam is the scientific founder and CEO of EIP Pharma, Inc., a private company based in Boston, Massachusetts, USA that is developing neflamapimod, a p38α kinase inhibitor, as a treatment for Alzheimer's disease and related dementias. Ursula Germann is a scientific advisor contracted by EIP Pharma, Inc.

Abbreviations

Aβ	Amyloid-β
AD	Alzheimer's disease
ALS	Amyotrophic lateral sclerosis
AMPAR	α-Amino-3-hydroxy-5-methyl-4-isoxazolepropionic acid receptor
ApoE	Apolipoprotein E
APP	β-Amyloid precursor protein
APPL	Adaptor protein, phosphotyrosine interacting with PH domain and leucine zipper 1
BACE-1	β-APP-cleaving enzyme 1
BDNF	Brain-derived neurotrophic factor
β-CTF	Carboxy-terminal APP fragment generated by BACE-1
C9ORF72	Chromosome 9 open reading frame 72
CK	Casein kinase
CNS	Central nervous system
COPD	Chronic obstructive pulmonary disease
CSF	Cerebrospinal fluid
DLB	Dementia with Lewy bodies
DRG	Dorsal root ganglion
DS	Down syndrome
EEA	Early endosomal autoantigen
ERK	Extracellular signal-regulated kinase
FTD	Frontotemporal dementia
GAP	GTPase activating protein
GDI	GDP dissociation inhibitor
GEF	Guanine nucleotide exchange factor
GTPase	Guanosine triphosphatase
HAP40	Htt-associated protein 40
HD	Huntington's disease
Htt	Huntingtin
hVPS45	Human Sec1p-like vacuolar protein sorting
IBD	Inflammatory bowel disease
IGF1	Insulin-like growth factor 1
IL-1β	Interleukin-1β
iPSC	Induced pluripotent stem cell

LTD	Long-term depression
LTP	Long-term potentiation
MAPK	Mitogen-activated protein kinase
MAPKAPK2	MAPK-activated protein kinase 2
MK2	MAPK-activated protein kinase 2
MKK	Mitogen-activated protein kinase kinase
MSK	Mitogen and stress-activated kinase
NGF	Nerve growth factor
NMDA	N-methyl-d-aspartate
PD	Parkinson's disease
PH	Pleckstrin homology
PI3K	Phosphatidylinositol-3-kinase
RA	Rheumatoid arthritis
Rab5	Ras-related protein Rab5
SAPK	Stress-activated protein kinase
SNRI	Serotonin and norepinephrine reuptake inhibitor
SOD1	Superoxide dismutase 1
SORL1	Sortilin-related receptor 1
TNFα	Tumor necrosis factor α

References

1. Nixon, R.A. Amyloid Precursor Protein and Endosomal-Lysosomal Dysfunction in Alzheimer's Disease: Inseparable Partners in a Multifactorial Disease. *FASEB J.* **2017**, *31*, 2729–2743. [CrossRef] [PubMed]
2. Balducci, C.; Forloni, G. Novel Targets in Alzheimer's Disease: A Special Focus on Microglia. *Pharmacol. Res.* **2018**, *130*, 402–413. [CrossRef]
3. Chung, C.G.; Lee, H.; Lee, S.B. Mechanisms of Protein Toxicity in Neurodegenerative Diseases. *Cell. Mol. Life Sci.* **2018**, *75*, 3159–3180. [CrossRef] [PubMed]
4. Lee, J.K.; Kim, N.J. Recent Advances in the Inhibition of p38 MAPK as a Potential Strategy for the Treatment of Alzheimer's Disease. *Molecules* **2017**, *22*, 1287. [CrossRef] [PubMed]
5. Guzman-Martinez, L.; Maccioni, R.B.; Andrade, V.; Navarrete, L.P.; Pastor, M.G.; Ramos-Escobar, N. Neuroinflammation as a Common Feature of Neurodegenerative Disorders. *Front. Pharmacol.* **2019**, *10*, 1008. [CrossRef]
6. Chen, X.Q.; Mobley, W.C. Exploring the Pathogenesis of Alzheimer Disease in Basal Forebrain Cholinergic Neurons: Converging Insights from Alternative Hypotheses. *Front. Neurosci.* **2019**, *13*, 446. [CrossRef] [PubMed]
7. Huang, L.K.; Chao, S.P.; Hu, C.J. Clinical Trials of New Drugs for Alzheimer Disease. *J. Biomed. Sci.* **2020**, *27*, 18. [CrossRef] [PubMed]
8. Lee, G.; Cummings, J.; Decourt, B.; Leverenz, J.B.; Sabbagh, M.N. Clinical Drug Development for Dementia with LEWY Bodies: Past and Present. *Expert Opin. Investig. Drugs* **2019**, *28*, 951–965. [CrossRef]
9. Dugger, B.N.; Dickson, D.W. Pathology of Neurodegenerative Diseases. *Cold Spring Harb. Perspect. Biol.* **2017**, *9*, a028035. [CrossRef]
10. Gan, L.; Cookson, M.R.; Petrucelli, L.; La Spada, A.R. Converging pathways in neurodegeneration, from genetics to mechanisms. *Nat. Neurosci.* **2018**, *21*, 1300–1309. [CrossRef]
11. Alam, J.; Scheper, W. Targeting Neuronal MAPK14/p38alpha Activity to Modulate Autophagy in the Alzheimer Disease Brain. *Autophagy* **2016**, *12*, 2516–2520. [CrossRef] [PubMed]
12. Malik, B.R.; Maddison, D.C.; Smith, G.A.; Peters, O.M. Autophagic and Endo-Lysosomal Dysfunction in Neurodegenerative Disease. *Mol. Brain* **2019**, *12*, 100. [CrossRef] [PubMed]
13. Koh, J.Y.; Kim, H.N.; Hwang, J.J.; Kim, Y.H.; Park, S.E. Lysosomal Dysfunction in Proteinopathic Neurodegenerative Disorders: Possible Therapeutic Roles of cAMP and Zinc. *Mol. Brain* **2019**, *12*, 18. [CrossRef] [PubMed]
14. Papaconstantinou, J.; Hsieh, C.-C.; DeFord, J.H. p38 MAPK Family. In *Encyclopedia of Signaling Molecules*; Choi, S., Ed.; Springer International Publishing: Cham, Switzerland, 2018; pp. 3728–3739. [CrossRef]

15. Pramanik, R.; Qi, X.; Borowicz, S.; Choubey, D.; Schultz, R.M.; Han, J.; Chen, G. p38 Isoforms Have Opposite Effects on AP-1-Dependent Transcription through Regulation of c-Jun. The Determinant Roles of the Isoforms in the p38 MAPK Signal Specificity. *J. Biol. Chem.* **2003**, *278*, 4831–4839. [CrossRef] [PubMed]
16. Korb, A.; Tohidast-Akrad, M.; Cetin, E.; Axmann, R.; Smolen, J.; Schett, G. Differential Tissue Expression and Activation of p38 MAPK Alpha, Beta, Gamma, and Delta Isoforms in Rheumatoid Arthritis. *Arthr. Rheum* **2006**, *54*, 2745–2756. [CrossRef]
17. Cuenda, A.; Rousseau, S. p38 MAP-Kinases Pathway Regulation, Function and Role in Human Diseases. *Biochim. Biophys. Acta* **2007**, *1773*, 1358–1375. [CrossRef]
18. Bachstetter, A.D.; Van Eldik, L.J. The p38 MAP Kinase Family as Regulators of Proinflammatory Cytokine Production in Degenerative Diseases of the CNS. *Aging Dis.* **2010**, *1*, 199–211.
19. Correa, S.A.; Eales, K.L. The Role of p38 MAPK and Its Substrates in Neuronal Plasticity and Neurodegenerative Disease. *J. Signal Transduct.* **2012**, *2012*, 649079. [CrossRef]
20. O'Callaghan, C.; Fanning, L.J.; Barry, O.P. p38delta MAPK: Emerging Roles of a Neglected Isoform. *Int. J. Cell Biol.* **2014**, *2014*, 272689. [CrossRef]
21. Escos, A.; Risco, A.; Alsina-Beauchamp, D.; Cuenda, A. p38gamma and p38delta Mitogen Activated Protein Kinases (MAPKs), New Stars in the MAPK Galaxy. *Front. Cell Dev. Biol.* **2016**, *4*, 31. [CrossRef]
22. Yang, C.; Cao, P.; Gao, Y.; Wu, M.; Lin, Y.; Tian, Y.; Yuan, W. Differential Expression of p38 MAPK Alpha, Beta, Gamma, Delta Isoforms in Nucleus Pulposus Modulates Macrophage Polarization in Intervertebral Disc Degeneration. *Sci. Rep.* **2016**, *6*, 22182. [CrossRef] [PubMed]
23. Cuenda, A.; Sanz-Ezquerro, J.J. p38gamma and p38delta: From Spectators to Key Physiological Players. *Trends Biochem. Sci.* **2017**, *42*, 431–442. [CrossRef] [PubMed]
24. Han, J.; Lee, J.D.; Bibbs, L.; Ulevitch, R.J. A MAP Kinase Targeted by Endotoxin and Hyperosmolarity in Mammalian Cells. *Science* **1994**, *265*, 808–811. [CrossRef] [PubMed]
25. Lee, J.C.; Laydon, J.T.; McDonnell, P.C.; Gallagher, T.F.; Kumar, S.; Green, D.; McNulty, D.; Blumenthal, M.J.; Heys, J.R.; Landvatter, S.W.; et al. A Protein Kinase Involved in the Regulation of Inflammatory Cytokine Biosynthesis. *Nature* **1994**, *372*, 739–746. [CrossRef]
26. Borders, A.S.; de Almeida, L.; Van Eldik, L.J.; Watterson, D.M. The p38alpha Mitogen-Activated Protein Kinase as a Central Nervous System Drug Discovery Target. *BMC Neurosci.* **2008**, *9* (Suppl. 2), S12. [CrossRef]
27. He, J.; Zhong, W.; Zhang, M.; Zhang, R.; Hu, W. P38 Mitogen-Activated Protein Kinase and Parkinson's Disease. *Transl. Neurosci.* **2018**, *9*, 147–153. [CrossRef]
28. Lee, S.H.; Park, J.; Che, Y.; Han, P.L.; Lee, J.K. Constitutive Activity and Differential Localization of p38alpha and p38beta MAPKs in Adult Mouse Brain. *J. Neurosci. Res.* **2000**, *60*, 623–631. [CrossRef]
29. Cuadrado, A.; Nebreda, A.R. Mechanisms and Functions of p38 MAPK Signalling. *Biochem. J.* **2010**, *429*, 403–417. [CrossRef]
30. Goldstein, D.M.; Kuglstatter, A.; Lou, Y.; Soth, M.J. Selective p38alpha Inhibitors Clinically Evaluated for the Treatment of Chronic Inflammatory Disorders. *J. Med. Chem.* **2010**, *53*, 2345–2353. [CrossRef]
31. Singh, D. P38 Inhibition in COPD; Cautious Optimism. *Thorax* **2013**, *68*, 705–706. [CrossRef]
32. Prikas, E.; Poljak, A.; Ittner, A. Mapping p38alpha Mitogen-Activated Protein Kinase Signaling by Proximity-Dependent Labeling. *Protein Sci.* **2020**, *29*, 1196–1210. [CrossRef] [PubMed]
33. Cavalli, V.; Vilbois, F.; Corti, M.; Marcote, M.J.; Tamura, K.; Karin, M.; Arkinstall, S.; Gruenberg, J. The Stress-Induced MAP Kinase p38 Regulates Endocytic Trafficking via the GDI: Rab5 complex. *Mol. Cell* **2001**, *7*, 421–432. [CrossRef]
34. Bendotti, C.; Bao Cutrona, M.; Cheroni, C.; Grignaschi, G.; Lo Coco, D.; Peviani, M.; Tortarolo, M.; Veglianese, P.; Zennaro, E. Inter- and Intracellular Signaling in Amyotrophic Lateral Sclerosis: Role of p38 Mitogen-Activated Protein Kinase. *Neurodegener. Dis.* **2005**, *2*, 128–134. [CrossRef] [PubMed]
35. Yasuda, S.; Sugiura, H.; Tanaka, H.; Takigami, S.; Yamagata, K. p38 MAP Kinase Inhibitors as Potential Therapeutic Drugs for Neural Diseases. *Cent. Nerv. Syst. Agents Med. Chem.* **2011**, *11*, 45–59. [CrossRef] [PubMed]
36. Prieto, G.A.; Snigdha, S.; Baglietto-Vargas, D.; Smith, E.D.; Berchtold, N.C.; Tong, L.; Ajami, D.; LaFerla, F.M.; Rebek, J.J.; Cotman, C.W. Synapse-Specific IL-1 Receptor Subunit Reconfiguration Augments Vulnerability to IL-1beta in the Aged Hippocampus. *Proc. Natl. Acad. Sci. USA* **2015**, *112*, E5078–E5087. [CrossRef] [PubMed]

37. Alam, J.; Blackburn, K.; Patrick, D. Neflamapimod: Clinical Phase 2b-Ready Oral Small Molecule Inhibitor of p38alpha to Reverse Synaptic Dysfunction in Early Alzheimer's Disease. *J. Prev. Alzheimers Dis.* **2017**, *4*, 273–278. [CrossRef]
38. Kheiri, G.; Dolatshahi, M.; Rahmani, F.; Rezaei, N. Role of p38/MAPKs in Alzheimer's Disease: Implications for Amyloid Beta Toxicity Targeted Therapy. *Rev. Neurosci.* **2018**, *30*, 9–30. [CrossRef]
39. Obergasteiger, J.; Frapporti, G.; Pramstaller, P.P.; Hicks, A.A.; Volta, M. A New Hypothesis for Parkinson's Disease Pathogenesis: GTPase-p38 MAPK Signaling and Autophagy as Convergence Points of Etiology and Genomics. *Mol. Neurodegener.* **2018**, *13*, 40. [CrossRef]
40. Gibbs, K.L.; Kalmar, B.; Rhymes, E.R.; Fellows, A.D.; Ahmed, M.; Whiting, P.; Davies, C.H.; Greensmith, L.; Schiavo, G. Inhibiting p38 MAPK Alpha Rescues Axonal Retrograde Transport Defects in a Mouse Model of ALS. *Cell Death Dis.* **2018**, *9*, 596. [CrossRef]
41. Gee, M.S.; Son, S.H.; Jeon, S.H.; Do, J.; Kim, N.; Ju, Y.J.; Lee, S.J.; Chung, E.K.; Inn, K.S.; Kim, N.J.; et al. A Selective p38alpha/beta MAPK Inhibitor Alleviates Neuropathology and Cognitive Impairment, and Modulates Microglia Function in 5XFAD Mouse. *Alzheimers Res. Ther.* **2020**, *12*, 45. [CrossRef]
42. Schnoder, L.; Gasparoni, G.; Nordstrom, K.; Schottek, A.; Tomic, I.; Christmann, A.; Schafer, K.H.; Menger, M.D.; Walter, J.; Fassbender, K.; et al. Neuronal Deficiency of p38alpha-MAPK Ameliorates Symptoms and Pathology of APP or Tau-Transgenic Alzheimer's Mouse Models. *FASEB J.* **2020**, *34*, 9628–9649. [CrossRef] [PubMed]
43. Hinze, C.; Boucrot, E. Endocytosis in Proliferating, Quiescent and Terminally Differentiated Cells. *J. Cell Sci.* **2018**, *131*, jcs216804. [CrossRef]
44. Scott, C.C.; Vacca, F.; Gruenberg, J. Endosome Maturation, Transport and Functions. *Semin. Cell Dev. Biol.* **2014**, *31*, 2–10. [CrossRef] [PubMed]
45. Kaur, G.; Lakkaraju, A. Early Endosome Morphology in Health and Disease. *Adv. Exp. Med. Biol.* **2018**, *1074*, 335–343. [CrossRef] [PubMed]
46. Naslavsky, N.; Caplan, S. The Enigmatic Endosome—Sorting the Ins and Outs of Endocytic Trafficking. *J. Cell Sci.* **2018**, *131*, jcs216499. [CrossRef] [PubMed]
47. Li, G.; Marlin, M.C. Rab Family of GTPases. *Methods Mol. Biol.* **2015**, *1298*, 1–15. [CrossRef]
48. Xu, W.; Fang, F.; Ding, J.; Wu, C. Dysregulation of Rab5-Mediated Endocytic Pathways in Alzheimer's Disease. *Traffic* **2018**, *19*, 253–262. [CrossRef]
49. Yuan, W.; Song, C. The Emerging Role of Rab5 in Membrane Receptor Trafficking and Signaling Pathways. *Biochem. Res. Int.* **2020**, *2020*, 4186308. [CrossRef]
50. Olchowik, M.; Miaczynska, M. Effectors of GTPase Rab5 in Endocytosis and Signal Transduction. *Postep. Biochem.* **2009**, *55*, 171–180.
51. Bucci, C.; Lutcke, A.; Steele-Mortimer, O.; Olkkonen, V.M.; Dupree, P.; Chiariello, M.; Bruni, C.B.; Simons, K.; Zerial, M. Co-Operative Regulation of Endocytosis by Three Rab5 Isoforms. *FEBS Lett.* **1995**, *366*, 65–71. [CrossRef]
52. Bucci, C.; Alifano, P.; Cogli, L. The Role of Rab Proteins in Neuronal Cells and in the Trafficking of Neurotrophin Receptors. *Membranes (Basel)* **2014**, *4*, 642–677. [CrossRef] [PubMed]
53. Mignogna, M.L.; D'Adamo, P. Critical Importance of RAB Proteins for Synaptic Function. *Small GTPases* **2018**, *9*, 145–157. [CrossRef] [PubMed]
54. Wucherpfennig, T.; Wilsch-Brauninger, M.; Gonzalez-Gaitan, M. Role of Drosophila Rab5 during Endosomal Trafficking at the Synapse and Evoked Neurotransmitter Release. *J. Cell Biol.* **2003**, *161*, 609–624. [CrossRef] [PubMed]
55. Shimizu, H.; Kawamura, S.; Ozaki, K. An Essential Role of Rab5 in Uniformity of Synaptic Vesicle Size. *J. Cell Sci.* **2003**, *116*, 3583–3590. [CrossRef]
56. Huang, C.C.; You, J.L.; Wu, M.Y.; Hsu, K.S. Rap1-Induced p38 Mitogen-Activated Protein Kinase Activation Facilitates AMPA Receptor Trafficking via the GDI.Rab5 Complex. Potential Role in (S)-3,5-Dihydroxyphenylglycene-Induced Long Term Depression. *J. Biol. Chem.* **2004**, *279*, 12286–12292. [CrossRef]
57. Parkinson, G.T.; Hanley, J.G. Mechanisms of AMPA Receptor Endosomal Sorting. *Front. Mol. Neurosci.* **2018**, *11*, 440. [CrossRef]

58. Brown, T.C.; Tran, I.C.; Backos, D.S.; Esteban, J.A. NMDA Receptor-Dependent Activation of the Small GTPase Rab5 Drives the Removal of Synaptic AMPA Receptors during Hippocampal LTD. *Neuron* **2005**, *45*, 81–94. [CrossRef]
59. Hausser, A.; Schlett, K. Coordination of AMPA Receptor Trafficking by Rab GTPases. *Small GTPases* **2019**, *10*, 419–432. [CrossRef]
60. Goto-Silva, L.; McShane, M.P.; Salinas, S.; Kalaidzidis, Y.; Schiavo, G.; Zerial, M. Retrograde Transport of Akt by a Neuronal Rab5-APPL1 Endosome. *Sci. Rep.* **2019**, *9*, 2433. [CrossRef]
61. Edler, E.; Stein, M. Recognition and Stabilization of Geranylgeranylated Human Rab5 by the GDP Dissociation Inhibitor (GDI). *Small GTPases* **2019**, *10*, 227–242. [CrossRef]
62. Rana, M.; Lachmann, J.; Ungermann, C. Identification of a Rab GTPase-Activating Protein Cascade that Controls Recycling of the Rab5 GTPase Vps21 from the Vacuole. *Mol. Biol. Cell* **2015**, *26*, 2535–2549. [CrossRef] [PubMed]
63. Simonsen, A.; Lippe, R.; Christoforidis, S.; Gaullier, J.M.; Brech, A.; Callaghan, J.; Toh, B.H.; Murphy, C.; Zerial, M.; Stenmark, H. EEA1 Links PI(3)K Function to Rab5 Regulation of Endosome Fusion. *Nature* **1998**, *394*, 494–498. [CrossRef] [PubMed]
64. Lawe, D.C.; Chawla, A.; Merithew, E.; Dumas, J.; Carrington, W.; Fogarty, K.; Lifshitz, L.; Tuft, R.; Lambright, D.; Corvera, S. Sequential Roles for Phosphatidylinositol 3-Phosphate and Rab5 in Tethering and Fusion of Early Endosomes via Their Interaction with EEA1. *J. Biol. Chem.* **2002**, *277*, 8611–8617. [CrossRef]
65. McKnight, N.C.; Zhong, Y.; Wold, M.S.; Gong, S.; Phillips, G.R.; Dou, Z.; Zhao, Y.; Heintz, N.; Zong, W.X.; Yue, Z. Beclin 1 Is Required for Neuron Viability and Regulates Endosome Pathways via the UVRAG-VPS34 Complex. *PLoS Genet.* **2014**, *10*, e1004626. [CrossRef]
66. Bresnick, A.R.; Backer, J.M. PI3Kbeta-A Versatile Transducer for GPCR, RTK, and Small GTPase Signaling. *Endocrinology* **2019**, *160*, 536–555. [CrossRef]
67. Agola, J.O.; Jim, P.A.; Ward, H.H.; Basuray, S.; Wandinger-Ness, A. Rab GTPases as Regulators of Endocytosis, Targets of Disease and Therapeutic Opportunities. *Clin. Genet.* **2011**, *80*, 305–318. [CrossRef]
68. Hong, L.; Sklar, L.A. Targeting GTPases in Parkinson's Disease: Comparison to the Historic Path of Kinase Drug Discovery and Perspectives. *Front. Mol. Neurosci.* **2014**, *7*, 52. [CrossRef] [PubMed]
69. Edler, E.; Stein, M. Probing the Druggability of Membrane-Bound Rab5 by Molecular Dynamics Simulations. *J. Enzym. Inhib. Med. Chem.* **2017**, *32*, 434–443. [CrossRef]
70. Colacurcio, D.J.; Pensalfini, A.; Jiang, Y.; Nixon, R.A. Dysfunction of Autophagy and Endosomal-Lysosomal Pathways: Roles in Pathogenesis of Down Syndrome and Alzheimer's Disease. *Free. Radic. Biol. Med.* **2018**, *114*, 40–51. [CrossRef]
71. Kiral, F.R.; Kohrs, F.E.; Jin, E.J.; Hiesinger, P.R. Rab GTPases and Membrane Trafficking in Neurodegeneration. *Curr. Biol.* **2018**, *28*, R471–R486. [CrossRef]
72. Guadagno, N.A.; Progida, C. Rab GTPases: Switching to Human Diseases. *Cells* **2019**, *8*, 909. [CrossRef] [PubMed]
73. Veleri, S.; Punnakkal, P.; Dunbar, G.L.; Maiti, P. Molecular Insights into the Roles of Rab Proteins in Intracellular Dynamics and Neurodegenerative Diseases. *Neuromol. Med.* **2018**, *20*, 18–36. [CrossRef] [PubMed]
74. Cataldo, A.M.; Peterhoff, C.M.; Troncoso, J.C.; Gomez-Isla, T.; Hyman, B.T.; Nixon, R.A. Endocytic Pathway Abnormalities Precede Amyloid Beta Deposition in Sporadic Alzheimer's Disease and Down Syndrome: Differential Effects of APOE Genotype and Presenilin Mutations. *Am. J. Pathol.* **2000**, *157*, 277–286. [CrossRef]
75. Lott, I.T.; Dierssen, M. Cognitive Deficits and Associated Neurological Complications in Individuals with Down's Syndrome. *Lancet Neurol.* **2010**, *9*, 623–633. [CrossRef]
76. Kim, S.; Sato, Y.; Mohan, P.S.; Peterhoff, C.; Pensalfini, A.; Rigoglioso, A.; Jiang, Y.; Nixon, R.A. Evidence That the Rab5 Effector APPL1 Mediates APP-betaCTF-Induced Dysfunction of Endosomes in Down Syndrome and Alzheimer's Disease. *Mol. Psychiatry* **2016**, *21*, 707–716. [CrossRef]
77. Jiang, Y.; Rigoglioso, A.; Peterhoff, C.M.; Pawlik, M.; Sato, Y.; Bleiwas, C.; Stavrides, P.; Smiley, J.F.; Ginsberg, S.D.; Mathews, P.M.; et al. Partial BACE1 Reduction in a Down Syndrome Mouse Model Blocks Alzheimer-Related Endosomal Anomalies and Cholinergic Neurodegeneration: Role of APP-CTF. *Neurobiol. Aging* **2016**, *39*, 90–98. [CrossRef]

78. Xu, W.; Weissmiller, A.M.; White, J.A., 2nd; Fang, F.; Wang, X.; Wu, Y.; Pearn, M.L.; Zhao, X.; Sawa, M.; Chen, S.; et al. Amyloid Precursor Protein-Mediated Endocytic Pathway Disruption Induces Axonal Dysfunction and Neurodegeneration. *J. Clin. Investig.* **2016**, *126*, 1815–1833. [CrossRef]
79. Pensalfini, A.; Kim, S.; Subbanna, S.; Bleiwas, C.; Goulbourne, C.N.; Stavrides, P.H.; Jiang, Y.; Lee, J.-H.; Darji, S.; Pawlik, M.; et al. Endosomal Dysfunction Induced by Directly Over-Activating Rab5 Recapitulates Prodromal and Neurodegenerative Features of Alzheimer's Disease (31 December, 2019). *CELL-REPORTS-D-19-05136*, 2019; Publication under review. [CrossRef]
80. Ginsberg, S.D.; Alldred, M.J.; Counts, S.E.; Cataldo, A.M.; Neve, R.L.; Jiang, Y.; Wuu, J.; Chao, M.V.; Mufson, E.J.; Nixon, R.A.; et al. Microarray Analysis of Hippocampal CA1 Neurons Implicates Early Endosomal Dysfunction during Alzheimer's Disease Progression. *Biol. Psychiatry* **2010**, *68*, 885–893. [CrossRef]
81. Ginsberg, S.D.; Mufson, E.J.; Alldred, M.J.; Counts, S.E.; Wuu, J.; Nixon, R.A.; Che, S. Upregulation of Select Rab GTPases in Cholinergic Basal Forebrain Neurons in Mild Cognitive Impairment and Alzheimer's Disease. *J. Chem. Neuroanat.* **2011**, *42*, 102–110. [CrossRef]
82. Nuriel, T.; Peng, K.Y.; Ashok, A.; Dillman, A.A.; Figueroa, H.Y.; Apuzzo, J.; Ambat, J.; Levy, E.; Cookson, M.R.; Mathews, P.M.; et al. The Endosomal-Lysosomal Pathway Is Dysregulated by APOE4 Expression In Vivo. *Front. Neurosci.* **2017**, *11*, 702. [CrossRef]
83. Knupp, A.; Mishra, S.; Martinez, R.; Braggin, J.E.; Szabo, M.; Kinoshita, C.; Hailey, D.W.; Small, S.A.; Jayadev, S.; Young, J.E. Depletion of the AD Risk Gene SORL1 Selectively Impairs Neuronal Endosomal Traffic Independent of Amyloidogenic APP Processing. *Cell Rep.* **2020**, *31*, 107719. [CrossRef] [PubMed]
84. Kwart, D.; Gregg, A.; Scheckel, C.; Murphy, E.A.; Paquet, D.; Duffield, M.; Fak, J.; Olsen, O.; Darnell, R.B.; Tessier-Lavigne, M. A Large Panel of Isogenic APP and PSEN1 Mutant Human iPSC Neurons Reveals Shared Endosomal Abnormalities Mediated by APP beta-CTFs, Not Abeta. *Neuron* **2019**, *104*, 256–270. [CrossRef]
85. Shi, M.M.; Shi, C.H.; Xu, Y.M. Rab GTPases: The Key Players in the Molecular Pathway of Parkinson's Disease. *Front. Cell Neurosci.* **2017**, *11*, 81. [CrossRef] [PubMed]
86. Bridi, J.C.; Hirth, F. Mechanisms of alpha-Synuclein Induced Synaptopathy in Parkinson's Disease. *Front. Neurosci.* **2018**, *12*, 80. [CrossRef] [PubMed]
87. Sung, J.Y.; Kim, J.; Paik, S.R.; Park, J.H.; Ahn, Y.S.; Chung, K.C. Induction of Neuronal Cell Death by Rab5A-Dependent Endocytosis of Alpha-Synuclein. *J. Biol. Chem.* **2001**, *276*, 27441–27448. [CrossRef] [PubMed]
88. Ngolab, J.; Trinh, I.; Rockenstein, E.; Mante, M.; Florio, J.; Trejo, M.; Masliah, D.; Adame, A.; Masliah, E.; Rissman, R.A. Brain-Derived Exosomes from Dementia with Lewy Bodies Propagate Alpha-Synuclein Pathology. *Acta Neuropathol. Commun.* **2017**, *5*, 46. [CrossRef]
89. Masaracchia, C.; Hnida, M.; Gerhardt, E.; Lopes da Fonseca, T.; Villar-Pique, A.; Branco, T.; Stahlberg, M.A.; Dean, C.; Fernandez, C.O.; Milosevic, I.; et al. Membrane Binding, Internalization, and Sorting of Alpha-Synuclein in the Cell. *Acta Neuropathol. Commun.* **2018**, *6*, 79. [CrossRef]
90. Eisbach, S.E.; Outeiro, T.F. Alpha-Synuclein and Intracellular Trafficking: Impact on the Spreading of Parkinson's Disease Pathology. *J. Mol. Med.* **2013**, *91*, 693–703. [CrossRef]
91. Fang, F.; Yang, W.; Florio, J.B.; Rockenstein, E.; Spencer, B.; Orain, X.M.; Dong, S.X.; Li, H.; Chen, X.; Sung, K.; et al. Synuclein Impairs Trafficking and Signaling of BDNF in a Mouse Model of Parkinson's Disease. *Sci. Rep.* **2017**, *7*, 3868. [CrossRef]
92. Twohig, D.; Nielsen, H.M. Alpha-Synuclein in the Pathophysiology of Alzheimer's Disease. *Mol. Neurodegener.* **2019**, *14*, 23. [CrossRef]
93. Spencer, B.; Desplats, P.A.; Overk, C.R.; Valera-Martin, E.; Rissman, R.A.; Wu, C.; Mante, M.; Adame, A.; Florio, J.; Rockenstein, E.; et al. Reducing Endogenous Alpha-Synuclein Mitigates the Degeneration of Selective Neuronal Populations in an Alzheimer's Disease Transgenic Mouse Model. *J. Neurosci.* **2016**, *36*, 7971–7984. [CrossRef] [PubMed]
94. Burk, K.; Pasterkamp, R.J. Disrupted Neuronal Trafficking in Amyotrophic Lateral Sclerosis. *Acta Neuropathol.* **2019**, *137*, 859–877. [CrossRef] [PubMed]
95. Yang, Y.; Hentati, A.; Deng, H.X.; Dabbagh, O.; Sasaki, T.; Hirano, M.; Hung, W.Y.; Ouahchi, K.; Yan, J.; Azim, A.C.; et al. The Gene Encoding Alsin, a Protein with Three Guanine-Nucleotide Exchange Factor Domains, Is Mutated in a Form of Recessive Amyotrophic Lateral Sclerosis. *Nat. Genet.* **2001**, *29*, 160–165. [CrossRef] [PubMed]

96. Topp, J.D.; Gray, N.W.; Gerard, R.D.; Horazdovsky, B.F. Alsin Is a Rab5 and Rac1 Guanine Nucleotide Exchange Factor. *J. Biol. Chem.* **2004**, *279*, 24612–24623. [CrossRef] [PubMed]
97. Otomo, A.; Hadano, S.; Okada, T.; Mizumura, H.; Kunita, R.; Nishijima, H.; Showguchi-Miyata, J.; Yanagisawa, Y.; Kohiki, E.; Suga, E.; et al. ALS2, a Novel Guanine Nucleotide Exchange Factor for the Small GTPase Rab5, Is Implicated in Endosomal Dynamics. *Hum. Mol. Genet.* **2003**, *12*, 1671–1687. [CrossRef]
98. Lai, C.; Xie, C.; Shim, H.; Chandran, J.; Howell, B.W.; Cai, H. Regulation of Endosomal Motility and Degradation by Amyotrophic Lateral Sclerosis 2/alsin. *Mol. Brain* **2009**, *2*, 23. [CrossRef]
99. Devon, R.S.; Orban, P.C.; Gerrow, K.; Barbieri, M.A.; Schwab, C.; Cao, L.P.; Helm, J.R.; Bissada, N.; Cruz-Aguado, R.; Davidson, T.L.; et al. Als2-Deficient Mice Exhibit Disturbances in Endosome Trafficking Associated with Motor Behavioral Abnormalities. *Proc. Natl. Acad. Sci. USA* **2006**, *103*, 9595–9600. [CrossRef]
100. Farg, M.A.; Sundaramoorthy, V.; Sultana, J.M.; Yang, S.; Atkinson, R.A.; Levina, V.; Halloran, M.A.; Gleeson, P.A.; Blair, I.P.; Soo, K.Y.; et al. C9ORF72, Implicated in Amytrophic Lateral Sclerosis and Frontotemporal Dementia, Regulates Endosomal Trafficking. *Hum. Mol. Genet.* **2014**, *23*, 3579–3595. [CrossRef]
101. Tang, B.L. C9orf72's Interaction with Rab GTPases-Modulation of Membrane Traffic and Autophagy. *Front. Cell Neurosci.* **2016**, *10*, 228. [CrossRef]
102. Pal, A.; Severin, F.; Lommer, B.; Shevchenko, A.; Zerial, M. Huntingtin-HAP40 Complex is a Novel Rab5 Effector That Regulates Early Endosome Motility and Is Up-Regulated in Huntington's Disease. *J. Cell Biol.* **2006**, *172*, 605–618. [CrossRef]
103. Her, L.S.; Goldstein, L.S. Enhanced Sensitivity of Striatal Neurons to Axonal Transport Defects Induced by Mutant Huntingtin. *J. Neurosci.* **2008**, *28*, 13662–13672. [CrossRef] [PubMed]
104. McGuire, J.R.; Rong, J.; Li, S.H.; Li, X.J. Interaction of Huntingtin-Associated Protein-1 with Kinesin Light Chain: Implications in Intracellular Trafficking in Neurons. *J. Biol. Chem.* **2006**, *281*, 3552–3559. [CrossRef]
105. Liot, G.; Zala, D.; Pla, P.; Mottet, G.; Piel, M.; Saudou, F. Mutant Huntingtin Alters Retrograde Transport of TrkB Receptors in Striatal Dendrites. *J. Neurosci.* **2013**, *33*, 6298–6309. [CrossRef] [PubMed]
106. Pla, P.; Orvoen, S.; Saudou, F.; David, D.J.; Humbert, S. Mood disorders in Huntington's Disease: From Behavior to Cellular and Molecular Mechanisms. *Front. Behav. Neurosci.* **2014**, *8*, 135. [CrossRef] [PubMed]
107. Ravikumar, B.; Imarisio, S.; Sarkar, S.; O'Kane, C.J.; Rubinsztein, D.C. Rab5 Modulates Aggregation and Toxicity of Mutant Huntingtin through Macroautophagy in Cell and Fly Models of Huntington Disease. *J. Cell Sci.* **2008**, *121*, 1649–1660. [CrossRef] [PubMed]
108. Felberbaum-Corti, M.; Cavalli, V.; Gruenberg, J. Capture of the small GTPase Rab5 by GDI: Regulation by p38 MAP Kinase. *Methods Enzymol.* **2005**, *403*, 367–381. [CrossRef]
109. Pelkmans, L.; Fava, E.; Grabner, H.; Hannus, M.; Habermann, B.; Krausz, E.; Zerial, M. Genome-Wide Analysis of Human Kinases in Clathrin- and Caveolae/Raft-Mediated Endocytosis. *Nature* **2005**, *436*, 78–86. [CrossRef]
110. Delcroix, J.D.; Valletta, J.S.; Wu, C.; Hunt, S.J.; Kowal, A.S.; Mobley, W.C. NGF Signaling in Sensory Neurons: Evidence that Early Endosomes Carry NGF Retrograde Signals. *Neuron* **2003**, *39*, 69–84. [CrossRef]
111. Mace, G.; Miaczynska, M.; Zerial, M.; Nebreda, A.R. Phosphorylation of EEA1 by p38 MAP Kinase Regulates Mu Opioid Receptor Endocytosis. *EMBO J.* **2005**, *24*, 3235–3246. [CrossRef]
112. Felberbaum-Corti, M.; Morel, E.; Cavalli, V.; Vilbois, F.; Gruenberg, J. The Redox Sensor TXNL1 Plays a Regulatory Role in Fluid Phase Endocytosis. *PLoS ONE* **2007**, *2*, e1144. [CrossRef]
113. Zhong, P.; Liu, W.; Gu, Z.; Yan, Z. Serotonin Facilitates Long-Term Depression Induction in Prefrontal Cortex via p38 MAPK/Rab5-Mediated Enhancement of AMPA Receptor Internalization. *J. Physiol.* **2008**, *586*, 4465–4479. [CrossRef]
114. Yuen, E.Y.; Qin, L.; Wei, J.; Liu, W.; Liu, A.; Yan, Z. Synergistic Regulation of Glutamatergic Transmission by Serotonin and Norepinephrine Reuptake Inhibitors in Prefrontal Cortical Neurons. *J. Biol. Chem.* **2014**, *289*, 25177–25185. [CrossRef] [PubMed]
115. Lynch, M.A. Age-Related Neuroinflammatory Changes Negatively Impact on Neuronal Function. *Front. Aging Neurosci.* **2010**, *1*, 6. [CrossRef] [PubMed]
116. Li, S.; Jin, M.; Koeglsperger, T.; Shepardson, N.E.; Shankar, G.M.; Selkoe, D.J. Soluble Abeta Oligomers Inhibit Long-Term Potentiation through a Mechanism Involving Excessive Activation of Extrasynaptic NR2B-Containing NMDA Receptors. *J. Neurosci.* **2011**, *31*, 6627–6638. [CrossRef]

117. Birnbaum, J.H.; Bali, J.; Rajendran, L.; Nitsch, R.M.; Tackenberg, C. Calcium Flux-Independent NMDA Receptor Activity is Required for Abeta Oligomer-Induced Synaptic Loss. *Cell Death Dis.* **2015**, *6*, e1791. [CrossRef] [PubMed]
118. Koppensteiner, P.; Trinchese, F.; Fa, M.; Puzzo, D.; Gulisano, W.; Yan, S.; Poussin, A.; Liu, S.; Orozco, I.; Dale, E.; et al. Time-Dependent Reversal of Synaptic Plasticity Induced by Physiological Concentrations of OLigomeric Abeta42: An Early Index of Alzheimer's Disease. *Sci. Rep.* **2016**, *6*, 32553. [CrossRef]
119. Bhaskar, K.; Konerth, M.; Kokiko-Cochran, O.N.; Cardona, A.; Ransohoff, R.M.; Lamb, B.T. Regulation of Tau Pathology by the Microglial fRactalkine Receptor. *Neuron* **2010**, *68*, 19–31. [CrossRef]
120. Alam, J.J. Selective Brain-Targeted Antagonism of p38 MAPKalpha Reduces Hippocampal IL-1beta Levels and Improves Morris Water Maze Performance in Aged Rats. *J. Alzheimer's Dis.* **2015**, *48*, 219–227. [CrossRef]
121. Roy, S.M.; Grum-Tokars, V.L.; Schavocky, J.P.; Saeed, F.; Staniszewski, A.; Teich, A.F.; Arancio, O.; Bachstetter, A.D.; Webster, S.J.; Van Eldik, L.J.; et al. Targeting Human Central Nervous System Protein Kinases: An Isoform Selective p38alphaMAPK Inhibitor That Attenuates Disease Progression in Alzheimer's Disease Mouse Models. *ACS Chem. Neurosci.* **2015**, *6*, 666–680. [CrossRef]
122. Maphis, N.; Jiang, S.; Xu, G.; Kokiko-Cochran, O.N.; Roy, S.M.; Van Eldik, L.J.; Watterson, D.M.; Lamb, B.T.; Bhaskar, K. Selective Suppression of the Alpha Isoform of p38 MAPK Rescues Late-Stage Tau Pathology. *Alzheimer's Res. Ther.* **2016**, *8*, 54. [CrossRef]
123. Roy, S.M.; Minasov, G.; Arancio, O.; Chico, L.W.; Van Eldik, L.J.; Anderson, W.F.; Pelletier, J.C.; Watterson, D.M. A Selective and Brain Penetrant p38alphaMAPK Inhibitor Candidate for Neurologic and Neuropsychiatric Disorders That Attenuates Neuroinflammation and Cognitive Dysfunction. *J. Med. Chem.* **2019**, *62*, 5298–5311. [CrossRef] [PubMed]
124. Colie, S.; Sarroca, S.; Palenzuela, R.; Garcia, I.; Matheu, A.; Corpas, R.; Dotti, C.G.; Esteban, J.A.; Sanfeliu, C.; Nebreda, A.R. Neuronal p38alpha Mediates Synaptic and Cognitive Dysfunction in an Alzheimer's Mouse Model by Controlling Beta-Amyloid Production. *Sci. Rep.* **2017**, *7*, 45306. [CrossRef] [PubMed]
125. Schnoder, L.; Hao, W.; Qin, Y.; Liu, S.; Tomic, I.; Liu, X.; Fassbender, K.; Liu, Y. Deficiency of Neuronal p38alpha MAPK Attenuates Amyloid Pathology in Alzheimer Disease Mouse and Cell Models through Facilitating Lysosomal Degradation of BACE1. *J. Biol. Chem.* **2016**, *291*, 2067–2079. [CrossRef]
126. Hsieh, H.; Boehm, J.; Sato, C.; Iwatsubo, T.; Tomita, T.; Sisodia, S.; Malinow, R. AMPAR Removal Underlies Abeta-Induced Synaptic Depression and Dendritic Spine Loss. *Neuron* **2006**, *52*, 831–843. [CrossRef]
127. Grbovic, O.M.; Mathews, P.M.; Jiang, Y.; Schmidt, S.D.; Dinakar, R.; Summers-Terio, N.B.; Ceresa, B.P.; Nixon, R.A.; Cataldo, A.M. Rab5-Stimulated Up-Regulation of the Endocytic Pathway Increases Intracellular Beta-Cleaved Amyloid Precursor Protein Carboxyl-Terminal Fragment Levels and Abeta Production. *J. Biol. Chem.* **2003**, *278*, 31261–31268. [CrossRef]
128. Tong, L.; Prieto, G.A.; Kramar, E.A.; Smith, E.D.; Cribbs, D.H.; Lynch, G.; Cotman, C.W. Brain-Derived Neurotrophic Factor-Dependent Synaptic Plasticity Is Suppressed by Interleukin-1beta via p38 Mitogen-Activated Protein Kinase. *J. Neurosci.* **2012**, *32*, 17714–17724. [CrossRef]
129. Dolan, P.J.; Johnson, G.V. The Role of Tau Kinases in Alzheimer's Disease. *Curr. Opin. Drug Discov. Dev.* **2010**, *13*, 595–603.
130. Beattie, E.C.; Carroll, R.C.; Yu, X.; Morishita, W.; Yasuda, H.; von Zastrow, M.; Malenka, R.C. Regulation of AMPA Receptor Endocytosis by a Signaling Mechanism Shared with LTD. *Nat. Neurosci.* **2000**, *3*, 1291–1300. [CrossRef]
131. Palmer, C.L.; Lim, W.; Hastie, P.G.; Toward, M.; Korolchuk, V.I.; Burbidge, S.A.; Banting, G.; Collingridge, G.L.; Isaac, J.T.; Henley, J.M. Hippocalcin Functions as a Calcium Sensor in Hippocampal LTD. *Neuron* **2005**, *47*, 487–494. [CrossRef]
132. Alam, J.; Jiang, Y.; Nixon, R.A. Antagonism of p38 MAPK Alpha (p38α) Reverses APP-Induced Endosomal Abnormalities and Improves Lysosomal Function in Down Syndrome Fibroblasts. *Alzheimer's Dement.* **2017**, *13*, P1496–P1497. [CrossRef]
133. Jiang, Y.; Stavrides, P.; Darji, S.; Yang, D.-S.; Bleiwas, C.; Smiley, J.F.; Germann, U.A.; Alam, J.; Nixon, R.A. Effects of p38α MAP Kinase Inhibition on the Neurodegenerative Phenotype of the Ts2 Down Syndrome Mouse Model. *Alzheimer's Dement.* **2019**, *15*, P1597. [CrossRef]
134. Duffy, J.P.; Harrington, E.M.; Salituro, F.G.; Cochran, J.E.; Green, J.; Gao, H.; Bemis, G.W.; Evindar, G.; Galullo, V.P.; Ford, P.J.; et al. The Discovery of VX-745: A Novel and Selective p38alpha Kinase Inhibitor. *ACS Med. Chem. Lett.* **2011**, *2*, 758–763. [CrossRef] [PubMed]

135. Davis, M.I.; Hunt, J.P.; Herrgard, S.; Ciceri, P.; Wodicka, L.M.; Pallares, G.; Hocker, M.; Treiber, D.K.; Zarrinkar, P.P. Comprehensive Analysis of Kinase Inhibitor Selectivity. *Nat. Biotechnol.* **2011**, *29*, 1046–1051. [CrossRef] [PubMed]
136. Uitdehaag, J.C.; Verkaar, F.; Alwan, H.; de Man, J.; Buijsman, R.C.; Zaman, G.J. A Guide to Picking the Most Selective Kinase Inhibitor Tool Compounds for Pharmacological Validation of Drug Targets. *Br. J. Pharmacol.* **2012**, *166*, 858–876. [CrossRef] [PubMed]
137. Kanekura, K.; Hashimoto, Y.; Niikura, T.; Aiso, S.; Matsuoka, M.; Nishimoto, I. Alsin, the Product of ALS2 Gene, Suppresses SOD1 Mutant Neurotoxicity through RhoGEF Domain by Interacting with SOD1 Mutants. *J. Biol. Chem.* **2004**, *279*, 19247–19256. [CrossRef] [PubMed]
138. Nixon, R.A. (The Nathan S. Kline Institute for Psychiatric Research, Orangeburg, NY, USA; New York University, New York, NY, USA). Personal Communication, 2020.
139. Xin, X.; Zhou, L.; Reyes, C.M.; Liu, F.; Dong, L.Q. APPL1 Mediates Adiponectin-Stimulated p38 MAPK Activation by Scaffolding the TAK1-MKK3-p38 MAPK Pathway. *Am. J. Physiol. Endocrinol. Metab.* **2011**, *300*, E103–E110. [CrossRef] [PubMed]
140. Cendrowski, J.; Maminska, A.; Miaczynska, M. Endocytic Regulation of Cytokine Receptor Signaling. *Cytokine Growth Factor Rev.* **2016**, *32*, 63–73. [CrossRef]
141. Kurgonaite, K.; Gandhi, H.; Kurth, T.; Pautot, S.; Schwille, P.; Weidemann, T.; Bokel, C. Essential Role of Endocytosis for Interleukin-4-Receptor-Mediated JAK/STAT Signalling. *J. Cell Sci.* **2015**, *128*, 3781–3795. [CrossRef]
142. Villasenor, R.; Kalaidzidis, Y.; Zerial, M. Signal Processing by the Endosomal System. *Curr. Opin. Cell Biol.* **2016**, *39*, 53–60. [CrossRef]
143. Palfy, M.; Remenyi, A.; Korcsmaros, T. Endosomal Crosstalk: Meeting Points for Signaling Pathways. *Trends Cell Biol.* **2012**, *22*, 447–456. [CrossRef]
144. Scheltens, P.; Alam, J.; Harrison, J.; Blackburn, K.; Prins, N. Efficacy and Safety Results of REVERSE-SD, Phase-2b Clinical Study of the Selective p38α Kinase Inhibitor Neflamapimod in Early-Stage Alzheimer's Disease (AD). In Proceedings of the Clinical Trials on Alzheimer's Disease, San Diego, CA, USA, 4–7 December 2019.

International Journal of
Molecular Sciences

Review

Involvement of p38 MAPK in Synaptic Function and Dysfunction

Chiara Falcicchia [1], Francesca Tozzi [2], Ottavio Arancio [3], Daniel Martin Watterson [4] and Nicola Origlia [1,*]

1 Institute of Neuroscience, Italian National Research Council, 56124 Pisa, Italy; chiara.falcicchia@in.cnr.it
2 Bio@SNS laboratory, Scuola Normale Superiore, 56124 Pisa, Italy; francesca.tozzi@sns.it
3 Taub Institute for Research on Alzheimer's Disease and the Aging Brain, Columbia University, New York, NY 10032, USA; oa1@cumc.columbia.edu
4 Department of Pharmacology, Northwestern University, Chicago, IL 60611, USA; d.m.watterson@gmail.com
* Correspondence: origlia@in.cnr.it; Tel.: +39-050-3153193

Received: 6 July 2020; Accepted: 5 August 2020; Published: 6 August 2020

Abstract: Many studies have revealed a central role of p38 MAPK in neuronal plasticity and the regulation of long-term changes in synaptic efficacy, such as long-term potentiation (LTP) and long-term depression (LTD). However, p38 MAPK is classically known as a responsive element to stress stimuli, including neuroinflammation. Specific to the pathophysiology of Alzheimer's disease (AD), several studies have shown that the p38 MAPK cascade is activated either in response to the Aβ peptide or in the presence of tauopathies. Here, we describe the role of p38 MAPK in the regulation of synaptic plasticity and its implication in an animal model of neurodegeneration. In particular, recent evidence suggests the p38 MAPK α isoform as a potential neurotherapeutic target, and specific inhibitors have been developed and have proven to be effective in ameliorating synaptic and memory deficits in AD mouse models.

Keywords: p38-MAPK α inhibitor; Alzheimer's disease; synaptic plasticity; neuroinflammation; β-amyloid; Tau

---

1. P38 Mitogen-Activated Protein Kinases (p38-MAPK)

The mitogen activated protein kinases (MAPKs) are serine and threonine protein kinases expressed in neuronal and non-neuronal cells in a mature central nervous system (CNS) during a dynamic state in response to various external stimuli, such as growth factors, glutamate and hormones, cellular stress, and pathogens [1]; they mediate proliferation, differentiation, and cell survival [2]. Depending on the context in which MAPKs are activated, they perform specific biological functions that can be therapeutically exploited. The basic module of MAPK cascades consists of three kinases that act in a sequential manner, namely, MAP kinase kinase kinase (MAPKKK) → MAP kinase kinase (MAPKK) → MAP kinase (MAPK) [3,4]. There are more than a dozen MAPK enzymes, but the best known are the extracellular signal-regulated kinases 1 and 2 (ERK1/2), ERK5, c-Jun amino-terminal kinases 1 to 3 (JNK1 to −3), and p38 (α, β, γ, and δ) families [5]. The latter two are also known as the stress-related protein kinases, because they are strongly activated in several pathologic processes, including β-amyloid neurodegeneration associated with Alzheimer's disease [6–9]. In particular, mammalian cells are known to express four different genes encoding p38 MAPK isoforms (p38α, p38β, p38γ, and p38δ), which retain a high sequence homology between each other; p38α is 75% identical to p38β and shares 62% and 61% identical protein sequences with p38γ and p38δ, respectively. In addition, p38γ shares around 70% identical sequence with the p38δ isoform. Among them, p38α and p38β are ubiquitously expressed and are mainly involved in inflammatory disorders, whereas p38γ and p38δ are expressed in a tissue-specific manner [10]. They all differ in their expression patterns, substrate specificities, and sensitivities to chemical inhibitors [11]. Each isoform of the p38 MAPK enzyme is activated by dual phosphorylation of the threonine and tyrosine residues. Dual phosphorylation, by either MAP kinase kinase 3 (MKK3) or MAP kinase kinase 6 (MKK6), induces global conformational reorganizations that allow for the binding of ATP and the desired substrate [2]. Many p38 MAPK targets have been described, including protein kinases (MAPK-activated protein kinases, MAPK- interacting kinase, and mitogen- and stress-activated kinase), which in turn phosphorylate transcription factors (p53, ATF-2, NFAT, and STAT1), cytoskeletal proteins (e.g., the microtubule-associated protein Tau), and other proteins with enzymatic activity, such as the glycogen synthase and cytosolic phospholipase A2 [1]. The lack of specific inhibitors for p38γ and p38δ have made the elucidation of the biological roles played by these two p38 isoforms compared to p38α and p38β more difficult. However, the use of knockout mouse models has allowed for demonstrating, for example, that p38γ can bind to the PDZ domain of a variety of proteins, such as PSD95, and modulate their phosphorylation state [12–14], while p38δ can phosphorylate Tau and seems to play a role in cytoskeletal remodeling [15]. Immunohistochemistry techniques have been used to study the localization of the main p38 MAPK isoforms in adult mice brains, which demonstrated the presence of p38α and p38β in different regions, including the cerebral cortex and the hippocampus [16]. Their different distribution among cell types was further characterized, showing a predominant neuronal expression for p38α, while p38β is also highly expressed in glial cells [16]. Regarding their subcellular localization in CA1 hippocampal neurons, p38α was found to be widely distributed in the different neuronal compartments, including dendrites, cytoplasm, and nucleus, while p38β was mostly localized at a nuclear level [2,17]. p38α plays a critical role in cellular response to infection related stressors (e.g., lipopolysaccharide (LPS)) [18] and became a drug development target in order to block cytokines production [19]. Moreover, the identification of roles independent of infections led to the extension of what has been called "sterile inflammation" (e.g., injury, illness, or aging). In particular, the activity of p38α has been associated with (a) the progression of the expression of protein markers of the aging phenotype [20–22]; (b) the development of inflammation and oxidative stress [10,23] associated with neurodegeneration, including Alzheimer's [24–26], lipopolysaccharide (LPS) [27,28], and Parkinson's [29,30] diseases; cardiovascular [31] and musculoskeletal diseases; diabetes [32]; rheumatoid arthritis [33]; and toxin-induced preterm birth [34]. Importantly, small molecule inhibitors of the p38 MAPK family have been developed, and show efficacy in blocking the production of proinflammatory cytokines, such as interleukin (IL)-1β and tumor necrosis factor alpha (TNF-α) [35]. Moreover, translational studies identified p38 α as one of several pathophysiology biomarkers in acute brain injury, progressive neurodegenerative disease, psychiatric disorders, and therapy induced drug-resistance [36]. In the present review, we provide an overview of the involvement of p38 MAPK in the regulation of synaptic plasticity, its implication in an animal model of neurodegeneration, and its potential as a neurotherapeutic target.

2. p38 MAPK and Synaptic Function

Several proteins have been found to be phosphorylated by MAPK, which implicates the role of this enzyme in a large range of cellular functions [5]. p38 MAPK is highly expressed in brain regions that are crucial for learning and memory, and is now emerging as a key player in the synaptic regulation and function. In recent decades, many reports have shown that the p38 MAPK signaling pathway plays important roles in synaptic plasticity (Figure 1).

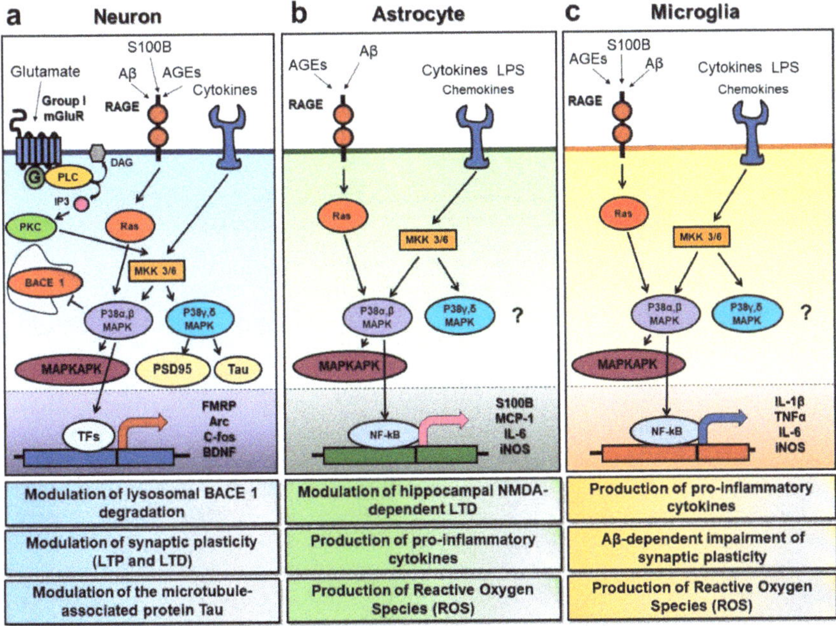

Figure 1. Overview of the p38 mitogen activated protein kinase (MAPK) signal transduction pathways in neurons, astrocytes, and microglia. p38 MAPK can be activated in response to various extracellular stimuli, such as glutamate, advanced glycation endproducts (AGEs), cytokines, and chemokines, leading to cell type-specific downstream effects. (**a**) In neuronal cells, the activation of the metabotropic glutamate receptors of group I (group I mGluRs) can turn on phospholipase C (PLC) and promote the phosphorylation of MAP kinase kinase 3/6 (MKK 3/6), with the subsequent activation of the p38 MAPKs. The α and β isoforms of p38 MAPK have been shown to inhibit beta-secretase 1 degradation, to promote the activation of MAP kinase activated protein kinase (MAPKAKP), and to act on specific transcription factors (TFs) to induce changes in the expression of the key proteins involved in synaptic plasticity, such as the fragile X mental retardation protein (FMRP), the activity-regulated cytoskeleton-associated protein (Arc), the c-fos protein, and the brain-derived neurotrophic factor (BDNF). On the other hand, the γ and δ isoforms seem to have a role in the modulation of synaptic proteins, such as postsynaptic density protein 95 (PSD95) and the microtubule-associated protein Tau. Furthermore, AGEs binding to the receptor for advanced glycation endproducts (RAGE) can turn on the Ras protein, and predominantly lead to the activation of p38 α and β isoforms, while cytokines and chemokines can trigger p38 MAPK by acting through their specific receptors and via MKKs. (**b**) In astrocytes, AGEs and pro inflammatory molecules such as lipopolysaccharide (LPS) can lead to p38 MAPK activation as well. Moreover, p38 activation in this specific cell type has been demonstrated to modulate hippocampal n-methyl-d-aspartate (NMDA)-dependent long-term depression (LTD), to induce the production of the S100 calcium binding protein (S100B), of pro-inflammatory molecules such as the monocyte chemoattractant proten-1 (MCP-1) and interleukin-6, and to increase the production of reactive oxygen species (ROS) via the expression of inducible nitric oxide synthase (iNOS). (**c**) In microglia, the signaling pathways upstream of p38 MAPK activation are very similar to those found in astrocytes. However, in addition, they lead to the production of pro-inflammatory cytokines such as interleukin-1β, the tumor necrosis factor α, and IL-6, and to increase ROS production, microglial activation of the p38 α isoform has been demonstrated to play a key role in the Aβ-dependent synaptic dysfunction.

For example, at the level of hippocampal formation, which is known to express long-term forms of synaptic plasticity, such as long-term potentiation (LTP) and long-term depression (LTD) [11], more information has been collected about the molecular pathways underlying these opposing forms of synaptic modifications. The mechanisms involved depend on the specific synapse and circuit, and the different type of stimulation pattern used to induce LTP and LTD [37]. p38 MAPK has been mostly implicated in synaptic depression, either the n-methyl-d-aspartate (NMDA)R-dependent or the mGluR-dependent form. The activation of NMDAR or metabotropic glutamate receptors (mGluRs) [11] triggers a diversity of signaling cascades, which results in a rapid and sustained decrease in synaptically evoked excitatory postsynaptic potentials (EPSPs). The activation of p38 MAP kinase in the hippocampus has been found to be necessary for the induction of mGluR-dependent LTD at the excitatory synapses between the CA3 and CA1 pyramidal neurons [38]. Indeed, p38 MAPK has been found in the hippocampus and has been demonstrated to be activated in response to a synaptic stimulation protocol that induces LTD [38]. In addition, it has been shown that a specific genetic ablation of the p38α isoform virtually abolishes NMDA-dependent LTD when targeting astrocytes, while producing no effect or slightly enhancing LTD when targeting neurons. These data indicate that astrocytic p38α is involved in activity-dependent glutamate release from astrocytes, contributing to astrocyte-to-neuron communication [39]. The authors concluded that the activity of p38α MAPK in the astrocyte contributes to hippocampal NMDA-dependent LTD, and is capable of modulating long-term memory in vivo [39]. Another study investigated the changes in the mRNA expression levels of p38 MAPK, demonstrating its implication in the induction of LTD in response to low-frequency stimulation (LFS) [40]. The requirement of p38 MAPK for the expression of NMDA-dependent LTD was also suggested in other brain circuits using entorhinal cortex slices, where the application of the SB203580 MAPK inhibitor completely suppressed LFS-induced LTD in superficial layer II [41]. Furthermore, in the mouse primary visual cortex, p38 MAPK has been demonstrated to mediate anisomycin-induced LTD by promoting α-amino-3-hydroxy-5-methyl-4-isoxazolepropionic acid (AMPA endocytosis). In fact, anisomycin administration produced a time-dependent decline in field excitatory post-synaptic potentials (fEPSPs) amplitude in acute brain slices of V1, and this decline could be rescued by the application of SB203580 [42]. However, SB203580 is proposed as a multi-target kinase inhibitor, which makes its use in support of a specific p38MAPK biological role more questionable [43].

Although, as reported above, p38 MAPK inhibition at the CA3-CA1 synapses does not affect LTP induction, p38 MAPK still appears to have a role in long-term potentiation, at least in pathological models. For example, reducing p38 MAPK activation by improved synaptic plasticity in angiotensin II-dependent hypertensive mice, either through genetic knock-down or pharmacological inhibition with SKF86002, as assessed by the LTP recording in the hippocampal slices [44]. Similar results were obtained in the entorhinal cortex (EC), where [45,46] the suppression of synaptic plasticity by the administration of Amyloid-β in slices could be prevented by a selective pharmacological inhibition of p38 MAPK using the MW 108 compound. In a more recent study, it was found that inhibition of the MAPK signaling pathway in an AD mouse model resulted in an improvement in hippocampal LTP [47]. Indeed, these data demonstrate that the impairment of LTP observed in APP/PS1 mice, was reversed by up-regulating mitogen-activated protein kinase (MAPK) phosphatase 1 (MKP-1), an essential negative regulator of MAPKs [47]. Finally, the importance of environmental factors in determining the role of p38 MAPK signaling cascade in LTP induction has been demonstrated [37]. In particular, the p38 MAPK was not required for hippocampal LTP in adolescent mice reared in standard conditions, but its activation was involved in LTP expression after exposure to an "enriched environment" [37]. Moreover, the NMDA glutamate receptor-dependent activation of p38 MAPK rescued the LTP in adolescent Ras-GRF knockout mice. This study revealed a new level of cell signaling control, whereby environmental factors influence the efficacy of a specific cascade to control LTP expression in adolescent animals [37].

3. p38 MAPK Neuroinflammation and Synaptic Dysfunction

The process of acute inflammation in mammalian tissue is one of extreme importance, as it is the immediate cellular response to injury and it is a defensive mechanism to prevent damage to the cells. The p38 module plays a critical role in normal immune and inflammatory responses; indeed, many studies have revealed its involvement in the production of inflammatory cytokines leading to chronic inflammation [11]. Thus, p38 is activated by numerous extracellular mediators of inflammation, including cytokines, chemoattractants, chemokines, and bacterial lipopolysaccharide (LPS). However, a major function of p38 isoforms is in turn the production of proinflammatory cytokines, and it has been proven that p38 can regulate cytokine expression by modulating transcription factors, such as nuclear factor-kB (NF-KB) [48], or at the mRNA level, by modulating their stability and translation through the regulation of MNK1 [49] and MNK2/3 [50]. It is also known that chronic inflammation occurs when there are persistent inflammatory stimuli that can have a damaging rather than protective effect. For example, chronic glial cell activation is present in neurodegenerative diseases [2]. The p38 MAPK pathway contributes to neuroinflammation mediated by glial cells, including microglia and astrocyte, and p38 α appears to be the main isoform involved in the inflammatory response [5]. In the brain, one of the physiological roles of microglia and astrocytes is to respond to stress and other cellular stimuli, defend the brain tissue, and take part in an inflammatory response by acting as mediators in inflammation and neuroprotection. Changes in morphology and transcriptional activation take place in the transition of microglial cells from a resting state, which exhibit a ramified morphology at the microscope, to an activated state with less extensive branching and processes. Activated microglia and reactive astrocytes are able to produce reactive oxygen species (ROS) and neurotoxic molecules that can induce molecular processes leading to neuronal death [51], but a prolonged and sustained activation of glial cells can result in an exaggerated inflammatory response and, as a result, cause neuronal cell death through the elevated release of proinflammatory cytokines, which have a potential neurotoxic effect, leading to increased neurodegeneration [2]. In addition to cellular damage, high concentrations of pro-inflammatory cytokines have been shown to affect neuronal synaptic functioning via p38 MAPK signaling in a variety of brain regions. For example, p38 MAPK mediates the inhibitory effect of the pro-inflammatory cytokine interleukin-1β (IL-1β) against LTP in the rodent dentate gyrus [52]. In the CNS, IL-1β levels increase in response to a number of different stimuli, such as the peripheral administration of lipopolysaccharide (LPS) [53], traumatic brain injury [54], acute stress [55], and β-adrenoceptor agonist administration [56]. IL-1β has been demonstrated to inhibit both NMDA-dependent and -independent forms of LTP in the hippocampus, and to progressively increase during aging in parallel with the age-related impairment of LTP in rodents [57], suggesting that it may represent one possible cause of cognitive decline. Indeed, two weeks of IL-1β overexpression in an inducible transgenic mouse was demonstrated to impair long-term contextual and spatial memory, but did not affect short-term and non-hippocampal memory [58]. Moreover, besides IL-1β, other pro-inflammatory cytokines have been shown to influence synaptic plasticity via p38 MAPK signaling, such as tumor necrosis factor alpha (TNF-α). TNF-α inhibits LTP in CA1 and DG in the rat hippocampus at pathophysiological levels [59], and the inhibition of p38 MAPK reverses the effect of TNF-α on early phase LTP without affecting late LTP [60]. Therefore, p38 MAPK appears to play an important role in cytokine-induced synaptic dysfunction, and there is evidence that it is also a key molecule in neurodegenerative diseases. Neurodegeneration represents a common pathological condition to several brain disorders, including Alzheimer's disease (AD), multiple sclerosis (MS), Parkinson's disease (PD), Huntington's disease (HD), and amyotrophic lateral sclerosis (ALS), and highlighting the functional role of specific p38 MAPK substrates will be of particular importance, as these could be potential signaling targets.

p38 MAPK, AD Neurodegeneration and Synaptic Dysfunction

Alzheimer's disease (AD) is the most prevalent age-related, progressive, and irreversible neurodegenerative disorder, characterized by memory dysfunction and cognitive impairment that are thought to result from the formation in the brain both of senile plaques containing amyloid-β (Aβ), as well as neurofibrillary tangles containing the microtubule-associated protein tau [36]. Aβ toxicity and tau hyperphosphorylation increase the activation of mitogen-activated protein kinase (MAPK) and MAPK signaling [61]. p38 MAPK is one of the key regulators of Aβ induced toxicity from this family [62]. In this regard, the Aβ-induced synaptic dysfunction has been well characterized. First, it has been demonstrated that either a synthetic form of Aβ 1-42, in the nanomolar range, or cell-derived naturally secreted Aβ oligomers, have a strong inhibitory effect on the induction of hippocampal LTP, both in vitro and in vivo in the CA1 area [63,64]. It has been shown that higher Aβ levels are also able to depress glutamatergic synaptic transmission and surface receptor number [65]. This effect was described as a partial occlusion of LTD, and suggests that Aβ-induced depression shares some mechanisms also necessary for LTD expression, including activation of the p38 MAPK pathway. However, it was first demonstrated that the activation of p38 MAPK is involved in the inhibition of hippocampal LTP by Aβ [63]. In subsequent studies, the differential activation of stress related kinases, p38mapk, and JNK, involved in the progressive Aβ-dependent synaptic dysfunction, has been investigated in entorhinal cortex slices. A concentration-dependent effect of Aβ was described, with a lower nM concentration that selectively impairs LTP through the neuronal activation of p38 MAPK [45], while increasing Aβ concentration up to 1 μM induces specific phosphorylation of both p38 MAPK and JNK that would consequently affect glutamatergic synaptic transmission and the expression of LTD [41]. In particular, it was reported that Aβ was able to phosphorylate p38MAPK in cultured cortical neurons at concentrations and incubation times comparable with those used for LTP. Moreover, a dual role emerged for p38 MAPK, as it was required for LTD expression, but also contributed to LTD impairment induced by higher Aβ levels. Notably, Aβ exposure increased the phosphorylated levels of p38 MAPK, which were further enhanced after low frequency stimulation (LFS), the protocol used to induce LTD and capable of phosphorylating p38 MAPK [41]. Concerning the possible cell surface targets that are able to bind Aβ and trigger p38 MAPK cascade, the receptor for advanced glycation end-products (RAGE) was identified as being capable of binding Aβ, in monomeric, fibrillized, and oligomeric forms, and to contribute to the progressive deleterious effects of the Aβ (1-42) peptide on EC synaptic function. In particular, Aβ-mediated the enhancement of p38 MAPK phosphorylation in cortical neurons, and was reduced by blocking antibodies to RAGE [45]. Moreover, increased phospho-p38 MAPK after Aβ exposure was reduced in EC slices from RAGE defective mice [41].

Indeed, the activation of p38 MAPK has been verified in the brain during the early stages of AD, both clinically [6,66] and in mouse models [11,62,67,68].

Different transgenic mouse models have been used to investigate the role of p38 MAPK. The link between the Aβ/RAGE axis and p38 MAPK over-activation has been confirmed in double transgenic mice (APPswe/Ind J20 expressing defective-RAGE in microglia), demonstrating that the activation of RAGE inflammatory signaling in vivo caused by an Aβ enriched-environment represents an important early event during progressive EC dysfunction. In the APPJ20 model, neuronal plasticity is progressively impaired in EC slices, while the inhibition of RAGE signaling in microglia ameliorates synaptic and behavioral impairment in APPJ20xDNMSR mice, reducing the neuronal activation of p38MAPK. This finding supports the hypothesis that microglial RAGE interaction with Aβ may therefore contribute to triggering p38 MAPK signaling involved in cognitive dysfunction in vulnerable brain areas, resulting in the spreading of AD pathology.

More recently, double transgenic mice, which express both the human APP mutation and endophilin A1(EP), demonstrated that the upregulation of the EP expression in Aβ-rich environments leads to changes in both hippocampal LTP and learning and memory. Specifically, EP, a synaptic protein elevated in AD patients and AD transgenic animal models, increases cerebral Aβ accumulation. The EP-mediated signal transduction involved reactive oxygen species (ROS) and p38 MAPK, contributing to Aβ-induced mitochondrial dysfunction, synaptic injury, and cognitive decline. The neurodegenerative phenotype could be rescued by blocking either the ROS or p38 MAP kinase activity [69,70].

Evidence also exists that genetically targeting the alpha-isoform of p38MAPK is sufficient to ameliorate synaptic dysfunction. With immunological and biochemical methods, it has been observed that the reduction of the p38α MAPK expression facilitates the lysosomal degradation of BACE1, a key enzyme in Aβ generation that is potentially up regulated by neuroinflammation. This led to an attenuation of Aβ protein generation in the brain of APP/PS1 double transgenic mice, suggesting that p38α MAPK plays a role in the process of Aβ deposition in vivo [71]. Moreover, the selective pharmacological inhibition of p38α MAPK was neuroprotective in either amyloid or tau models of AD [46,68]. However, the role of p38α in neurodegeration can be different depending on the model. A recent report demonstrated that neuron-specific p38α-knockout mice show increased levels of anxiety in behaviour tests, an effect that was mediated by increased JNK activity [72].

The novel isoform selective p38α MAPK inhibitor was tested in two different animal models characterized by early synaptic and behavioral dysfunctions, and was found to be effective in ameliorating hippocampal-dependent associative and spatial memory. The first evidence was that selective p38α MAPK inhibition was capable of reducing LTP and memory deficits in the APP/PS1 Tg model, whose main feature is represented by Aβ deposition [25]. In contrast, no overexpression of APP was present in the APP/PS KI mouse, which can be considered a model of aging with physiological levels of APP. This mouse shows a slower pathology progression and offers the opportunity to study either the early or the late stage of neurodegeneration. In 11-month-old KI mice, the overproduction of cytokine leads to a spatial memory deficit that can be reduced by pharmacological treatment targeting p38α MAPK [25]. In addition, several inhibitors of p38α MAPK were effective in ameliorating the behavioral deficit in a mouse model of tau-related neurodegeneration, the reversible transgenic (rTg4510) mouse model. These mice overexpressed a human, mutant tau form (P301L) and developed age-related cognitive impairment, neurofibrillary tangles, and neuronal loss [73]. Overall, this evidence favors the hypothesis that p38 MAPK and, in particular, the α isoform, contribute to different pathological process, and therefore represents a promising therapeutic target for the treatment of AD [11,44].

An opposite role in neurodegeneration emerged for p38γ MAPK. This isoform mediates phosphorylation on tau at Threonine-205 (T205), a site-specific phosphorylation that improved memory deficits in APP transgenic mice [43], in particular interfering with the synaptic action of Aβ on glutamatergic neurotransmission [74,75].

4. Small Molecules Targeting p38α MAPK

Serine–threonine (S/T) protein kinases are important for CNS function and have been implicated in pathophysiology, yet there is a dearth of highly selective, CNS-active kinase inhibitors for in vivo investigations.

A challenge for all of life sciences is that approved protein kinase inhibitor drugs and research molecular probe inhibitors are multi-kinase inhibitors. Results should be taken with caution when using small-molecule inhibitors of protein kinases to investigate the physiological roles of these enzymes [76–81]. Figure 2 presents a graphical presentation of the kinome off-target issue for the S/T protein kinase, p38α MAPK, which is the focus of this review. As a consequence, the outcome of experiments are confounded by the use of p38MAPK inhibitors that also inhibit other kinases such as ABL1, ABL2, p38β, p38γ, CK1δ, CK1ε, PDGFRβ, SRC, BLK, CDK5, CDK8, DDR1, DDR2, EPHA3, EPHA7, EPHA8, EPHB2, FLT1, FRK, NTRK1, JNK1, JNK2, JNK3, KIT, MAP4K4, MRCKβ, PTK2β, RET, SLK, STK10, TIE1, TIE2, TNIK, TRKB, TRKC, ZAK, BRAF, CIT, DMPK, GAK, JNK2, JNK3, NLK, RIPK2, STK36, or TNIK. For example, MW150 has no direct effect on amyloid plaques [25,82] and avoids ABL as a target [25]. The pharmacological difference from p38MAPK inhibitors that have off-target ABL inhibition might be explained by the known effects of ABL inhibitors (e.g., Imatinib, Nilotinib and Bosutinib) on amyloid plaques [83–85]. Similarly, the medicinal chemistry refinement of the original fragment expansion hit, MW069a, to yield the more selective MW181 and MW108 class of refined inhibitors involved removal of the previously identified liability of protein kinase CK1d inhibition and removal of cross-over to potential excitotoxic GPCR targets. There was a coincident improvement in pharmacological safety as the neurotoxicity seen at high doses of MW069a was removed. The coincident removal of both off-target activities and improvement in pharmacological safety does not allow conclusions about exactly what kinase or GPCR target might represent a risk for higher doses of p38MAPK inhibitors. However, clinical findings and associated animal model studies describe susceptibility to migraine and sleep disorders in the presence of reduced CK1d activity suggesting risk in this off-target kinase [86]. An additional challenge for neurosciences is the fact that most protein kinase inhibitors used in research studies, as well as kinase inhibitor approved drugs, lack sufficient CNS exposure to allow adequate molecular target exposure. The blood–brain barrier challenge is not limited to protein kinase inhibitor drugs, however, as it is estimated that >95% of approved drugs lack sufficient brain tissue exposure. The CNS exposure limitation is most often linked to the molecular properties of the small molecule drugs [78]. However, this is a barrier that can be addressed through medicinal chemistry refinement [73,79].

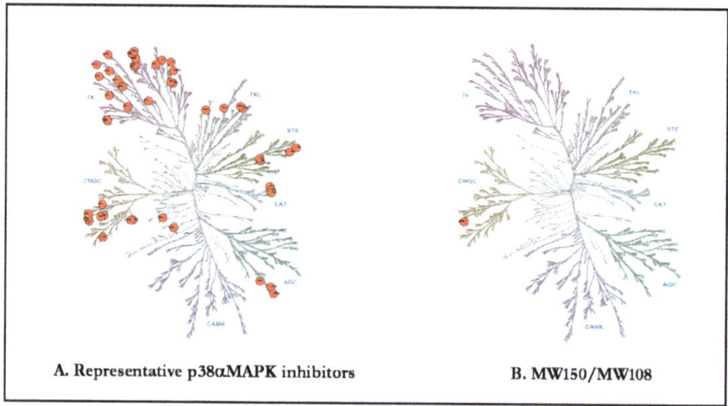

Figure 2. Kinome target selectivity of p38 MAPK Inhibitors. Differences in off-target kinase and GPCR liabilities provide potential explanations for pharmacological differences. Red circles denote kinase inhibition below the canonical IC50 < 1 μM. (**A**) Common off-target kinases among widely used p38α MAPK inhibitors: VX-745 (neflamapimod) includes ABL1, ABL2, p38β, PDGFRβ, and SRC; BIRB-796 includes BLK, CDK5, CDK8, DDR1, DDR2, EPHA3, EPHA7, EPHA8, EPHB2, p38β p38γ, FLT1, FRK, NTRK1, JNK1, JNK2, JNK3, KIT, MAP4K4, MRCKβ, PTK2β, RET, SLK, STK10, TIE1, TIE2, TNIK, TRKB, TRKC, and ZAK; and SB203580 includes BRAF, CIT, CK1δ, CK1ε, DMPK, GAK, JNK2, JNK3, NLK, p38β, RIPK2, STK36, and TNIK. (**B**) MW150 and MW108 have IC50 <1 μM for p38α MAPK in kinome-wide hierarchal screens.

Recent deliverables for p38α MAPK, MW108, and MW150 [25,73,78] avoid the molecular target selectivity and brain exposure challenges that plagued prior works. They provide precedents for CNS S/T protein inhibitors based on their high kinome selectivity, avoidance of high-risk off-target effects, and in vivo efficacy. The selectivity of MW108 and MW150 was demonstrated by large-scale kinome screens, functional GPCR agonist and antagonist analyses, and selected ion-channel and transporter screens. Furthermore, MW150 treatment at efficacious doses does not produce detectable pharmacodynamic effects in knock-in cells that have the endogenous kinase p38α MAPK replaced with p38α MAPK (T106M), an active p38α MAPK that is MW150 resistant. In vitro and in vivo assays demonstrated cellular target engagement, and dose dependent studies in diverse animal models documented their pharmacodynamic and efficacy functions. For example, the MW108/MW150 series ameliorated beta amyloid-induced and tau-induced synaptic and cognitive dysfunction in neurodegeneration [25,46,73–75,87], as well as attenuate behavioral symptoms and pathophysiological biomarkers in a genetic susceptibility model of autism spectrum disorders [88]. Clearly, the MW108/MW150 series allows for the pursuit of preclinical and clinical therapeutic hypotheses involving p38α MAPK that were not feasible previously.

The activation of p38α MAPK mediated signaling cascades are implicated in synaptic dysfunction in neurodegenerative disorders through clinical observations and preclinical investigations. Activation in both neurons and glia offers the unusual potential to generate enhanced phenotypic responses through targeting a single kinase in two distinct cell types involved in pathophysiology progression. The pathophysiology mechanism, plus a highly selective and bioavailable p38α MAPK inhibitor drug, could provide a novel form of pleiotropy (one drug, one target, multiple clinical effects), in contrast to the pleiotropic effects of drugs such as steroids, where the same drug engages multiple molecular targets. The high safety potential for MW150 in preclinical toxicology screens and first-in-human clinical trials suggests that such a pleiotropic mechanism could be highly desirable clinically [73]. However, it raises the need for caution if MW150 is used as a molecular probe reagent in basic research experimental design.

In summary, recent progress has directly addressed the scientific challenges in the development and use of p38α MAPK inhibitors as therapeutic candidates, and greatly improved the delivery of small molecule inhibitors as in vivo research tools. Overall, the paradigm shift and new deliverables allow for more robust testing of various hypotheses about S/T protein kinases, such as p38α MAPK in neuropathology progression and its potential for disease modification.

Author Contributions: C.F. and N.O. formulated the theme and outline of the review. All authors reviewed the literature, drafted and revised the manuscript. C.F., F.T., O.A., D.M.W. and N.O. reviewed, revised, and finalized the manuscript. All authors have read and agreed to the published version of the manuscript.

Funding: This research received no external funding.

Conflicts of Interest: The authors declare no conflicts of interest.

Abbreviations

MAPK	mitogen activated protein kinases
CNS	central nervous system
LTP	long-term potentiation
LTD	long-term depression
AD	Alzheimer's disease
ERK	extracellular signal-regulated kinases
EPSPs	synaptically evoked excitatory postsynaptic potentials
JNK	c-Jun amino-terminal kinases
ATP	adenosine triphosphate
NFAT	nuclear factor of activated T cells
ATF	activating transcription factor
STAT1	signal transducer and activator of transcription
LPS	lipopolysaccharide
IL-1β	interleukin-1β
TNFα	tumor necrosis factor
AGEs	advanced glycation endproducts
mGluRs	metabotropic glutamate receptors
TFs	transcription factors
FMRP	fragile X mental retardation protein
Arc	activity-regulated cytoskeleton
BDNF	brain derived neurotrophic factor
PSD95	postsynaptic density protein 95
RAGE	advanced glycation endproducts receptor
MCP-1	monocyte chemoattractant proten-1
iNOS	inducible nitric oxide synthase
NMDA	n-methyl-d-aspartate
LFS	low-frequency stimulation
AMPA	α-amino-3-hydroxy-5-methyl-4-isoxazolepropionic acid
Aβ	amyloid-β
EC	entorhinal cortex
Tg	transgenic
APP	amyloid precursor protein
PS1	mutant human presenilin 1
EP	endophilin

NF-KB	nuclear factor-kB
ROS	reactive oxygen species
MS	multiple sclerosis
PD	Parkinson's disease
HD	Huntington's disease
ALS	amyotrophic lateral sclerosis
KI	knock in
KO	knock out
S/T	serine-threonine
FPCR	G protein-coupled receptor

References

1. Origlia, N.; Arancio, O.; Domenici, L.; Yan, S.S. MAPK, beta-amyloid and synaptic dysfunction: The role of RAGE. *Expert Rev. Neurother.* **2009**, *9*, 1635–1645. [CrossRef] [PubMed]
2. Correa, S.A.; Eales, K.L. The Role of p38 MAPK and Its Substrates in Neuronal Plasticity and Neurodegenerative Disease. *J. Signal Transduct.* **2012**, *2012*, 649079. [CrossRef] [PubMed]
3. Johnson, G.L.; Lapadat, R. Mitogen-activated protein kinase pathways mediated by ERK, JNK, and p38 protein kinases. *Science* **2002**, *298*, 1911–1912. [CrossRef]
4. Jagodzik, P.; Tajdel-Zielinska, M.; Ciesla, A.; Marczak, M.; Ludwikow, A. Mitogen-Activated Protein Kinase Cascades in Plant Hormone Signaling. *Front. Plant Sci.* **2018**, *9*, 1387. [CrossRef] [PubMed]
5. Cargnello, M.; Roux, P.P. Activation and function of the MAPKs and their substrates, the MAPK-activated protein kinases. *Microbiol. Mol. Biol. Rev. MMBR* **2011**, *75*, 50–83. [CrossRef]
6. Pei, J.J.; Braak, E.; Braak, H.; Grundke-Iqbal, I.; Iqbal, K.; Winblad, B.; Cowburn, R.F. Localization of active forms of C-jun kinase (JNK) and p38 kinase in Alzheimer's disease brains at different stages of neurofibrillary degeneration. *J. Alzheimer's Dis. JAD* **2001**, *3*, 41–48. [CrossRef] [PubMed]
7. Troy, C.M.; Rabacchi, S.A.; Xu, Z.; Maroney, A.C.; Connors, T.J.; Shelanski, M.L.; Greene, L.A. Beta-Amyloid-induced neuronal apoptosis requires c-Jun N-terminal kinase activation. *J. Neurochem.* **2001**, *77*, 157–164. [CrossRef]
8. Zhu, X.; Mei, M.; Lee, H.G.; Wang, Y.; Han, J.; Perry, G.; Smith, M.A. P38 activation mediates amyloid-beta cytotoxicity. *Neurochem. Res.* **2005**, *30*, 791–796. [CrossRef]
9. Hasegawa, Y.; Toyama, K.; Uekawa, K.; Ichijo, H.; Kim-Mitsuyama, S. Role of ASK1/p38 Cascade in a Mouse Model of Alzheimer's Disease and Brain Aging. *J. Alzheimer's Dis. JAD* **2018**, *61*, 259–263. [CrossRef]
10. Ono, K.; Han, J. The p38 signal transduction pathway: Activation and function. *Cell. Signal.* **2000**, *12*, 1–13. [CrossRef]
11. Lee, J.K.; Kim, N.J. Recent Advances in the Inhibition of p38 MAPK as a Potential Strategy for the Treatment of Alzheimer's Disease. *Molecules* **2017**, *22*, 1287. [CrossRef]
12. Hasegawa, M.; Cuenda, A.; Spillantini, M.G.; Thomas, G.M.; Buee-Scherrer, V.; Cohen, P.; Goedert, M. Stress-activated protein kinase-3 interacts with the PDZ domain of alpha1-syntrophin. A mechanism for specific substrate recognition. *J. Biol. Chem.* **1999**, *274*, 12626–12631. [CrossRef]
13. Sabio, G.; Arthur, J.S.; Kuma, Y.; Peggie, M.; Carr, J.; Murray-Tait, V.; Centeno, F.; Goedert, M.; Morrice, N.A.; Cuenda, A. p38gamma regulates the localisation of SAP97 in the cytoskeleton by modulating its interaction with GKAP. *EMBO J.* **2005**, *24*, 1134–1145. [CrossRef] [PubMed]
14. Sabio, G.; Reuver, S.; Feijoo, C.; Hasegawa, M.; Thomas, G.M.; Centeno, F.; Kuhlendahl, S.; Leal-Ortiz, S.; Goedert, M.; Garner, C.; et al. Stress- and mitogen-induced phosphorylation of the synapse-associated protein SAP90/PSD-95 by activation of SAPK3/p38gamma and ERK1/ERK2. *Biochem. J.* **2004**, *380*, 19–30. [CrossRef] [PubMed]
15. Feijoo, C.; Campbell, D.G.; Jakes, R.; Goedert, M.; Cuenda, A. Evidence that phosphorylation of the microtubule-associated protein Tau by SAPK4/p38delta at Thr50 promotes microtubule assembly. *J. Cell Sci.* **2005**, *118*, 397–408. [CrossRef]
16. Lee, S.H.; Park, J.; Che, Y.; Han, P.L.; Lee, J.K. Constitutive activity and differential localization of p38alpha and p38beta MAPKs in adult mouse brain. *J. Neurosci. Res.* **2000**, *60*, 623–631. [CrossRef]

17. Menon, R.; Papaconstantinou, J. p38 Mitogen activated protein kinase (MAPK): A new therapeutic target for reducing the risk of adverse pregnancy outcomes. *Expert Opin. Ther. Targets* **2016**, *20*, 1397–1412. [CrossRef]
18. Schieven, G.L. The biology of p38 kinase: A central role in inflammation. *Curr. Top. Med. Chem.* **2005**, *5*, 921–928. [CrossRef]
19. Bachstetter, A.D.; Xing, B.; de Almeida, L.; Dimayuga, E.R.; Watterson, D.M.; Van Eldik, L.J. Microglial p38alpha MAPK is a key regulator of proinflammatory cytokine up-regulation induced by toll-like receptor (TLR) ligands or beta-amyloid (Abeta). *J. Neuroinflamm.* **2011**, *8*, 79. [CrossRef]
20. Li, Z.; Li, J.; Bu, X.; Liu, X.; Tankersley, C.G.; Wang, C.; Huang, K. Age-induced augmentation of p38 MAPK phosphorylation in mouse lung. *Exp. Gerontol.* **2011**, *46*, 694–702. [CrossRef]
21. Papaconstantinou, J.; Hsieh, C.C. Activation of senescence and aging characteristics by mitochondrially generated ROS: How are they linked? *Cell Cycle* **2010**, *9*, 3831–3833. [CrossRef] [PubMed]
22. Li, H.; Liu, Y.; Gu, Z.; Li, L.; Liu, Y.; Wang, L.; Su, L. p38 MAPK-MK2 pathway regulates the heat-stress-induced accumulation of reactive oxygen species that mediates apoptotic cell death in glial cells. *Oncol. Lett.* **2018**, *15*, 775–782. [CrossRef] [PubMed]
23. Yang, C.; Zhu, Z.; Tong, B.C.; Iyaswamy, A.; Xie, W.J.; Zhu, Y.; Sreenivasmurthy, S.G.; Senthilkumar, K.; Cheung, K.H.; Song, J.X.; et al. A stress response p38 MAP kinase inhibitor SB202190 promoted TFEB/TFE3-dependent autophagy and lysosomal biogenesis independent of p38. *Redox Biol.* **2020**, *32*, 101445. [CrossRef] [PubMed]
24. Munoz, L.; Ammit, A.J. Targeting p38 MAPK pathway for the treatment of Alzheimer's disease. *Neuropharmacology* **2010**, *58*, 561–568. [CrossRef] [PubMed]
25. Roy, S.M.; Grum-Tokars, V.L.; Schavocky, J.P.; Saeed, F.; Staniszewski, A.; Teich, A.F.; Arancio, O.; Bachstetter, A.D.; Webster, S.J.; Van Eldik, L.J.; et al. Targeting human central nervous system protein kinases: An isoform selective p38alphaMAPK inhibitor that attenuates disease progression in Alzheimer's disease mouse models. *Acs Chem. Neurosci.* **2015**, *6*, 666–680. [CrossRef] [PubMed]
26. Criscuolo, C.; Fontebasso, V.; Middei, S.; Stazi, M.; Ammassari-Teule, M.; Yan, S.S.; Origlia, N. Entorhinal Cortex dysfunction can be rescued by inhibition of microglial RAGE in an Alzheimer's disease mouse model. *Sci. Rep.* **2017**, *7*, 42370. [CrossRef] [PubMed]
27. Veglianese, P.; Lo Coco, D.; Bao Cutrona, M.; Magnoni, R.; Pennacchini, D.; Pozzi, B.; Gowing, G.; Julien, J.P.; Tortarolo, M.; Bendotti, C. Activation of the p38MAPK cascade is associated with upregulation of TNF alpha receptors in the spinal motor neurons of mouse models of familial ALS. *Mol. Cell. Neurosci.* **2006**, *31*, 218–231. [CrossRef]
28. Guo, W.; Vandoorne, T.; Steyaert, J.; Staats, K.A.; Van Den Bosch, L. The multifaceted role of kinases in amyotrophic lateral sclerosis: Genetic, pathological and therapeutic implications. *Brain J. Neurol.* **2020**, *143*, 1651–1673. [CrossRef] [PubMed]
29. Gui, C.; Ren, Y.; Chen, J.; Wu, X.; Mao, K.; Li, H.; Yu, H.; Zou, F.; Li, W. p38 MAPK-DRP1 signaling is involved in mitochondrial dysfunction and cell death in mutant A53T alpha-synuclein model of Parkinson's disease. *Toxicol. Appl. Pharmacol.* **2020**, *388*, 114874. [CrossRef]
30. Brobey, R.K.; German, D.; Sonsalla, P.K.; Gurnani, P.; Pastor, J.; Hsieh, C.C.; Papaconstantinou, J.; Foster, P.P.; Kuro-o, M.; Rosenblatt, K.P. Klotho Protects Dopaminergic Neuron Oxidant-Induced Degeneration by Modulating ASK1 and p38 MAPK Signaling Pathways. *PLoS ONE* **2015**, *10*, e0139914. [CrossRef]
31. Kompa, A.R. Do p38 mitogen-activated protein kinase inhibitors have a future for the treatment of cardiovascular disease? *J. Thorac. Dis.* **2016**, *8*, E1068–E1071. [CrossRef] [PubMed]
32. Jeong, H.J.; Lee, H.J.; Vuong, T.A.; Choi, K.S.; Choi, D.; Koo, S.H.; Cho, S.C.; Cho, H.; Kang, J.S. Prmt7 Deficiency Causes Reduced Skeletal Muscle Oxidative Metabolism and Age-Related Obesity. *Diabetes* **2016**, *65*, 1868–1882. [CrossRef] [PubMed]
33. Wang, H.; Tu, S.; Yang, S.; Shen, P.; Huang, Y.; Ba, X.; Lin, W.; Huang, Y.; Wang, Y.; Qin, K.; et al. Berberine Modulates LPA Function to Inhibit the Proliferation and Inflammation of FLS-RA via p38/ERK MAPK Pathway Mediated by LPA1. *Evid. Based Complementary Altern. Med. Ecam* **2019**, *2019*, 2580207. [CrossRef] [PubMed]
34. Bredeson, S.; Papaconstantinou, J.; Deford, J.H.; Kechichian, T.; Syed, T.A.; Saade, G.R.; Menon, R. HMGB1 promotes a p38MAPK associated non-infectious inflammatory response pathway in human fetal membranes. *PLoS ONE* **2014**, *9*, e113799. [CrossRef]

35. Bachstetter, A.D.; Van Eldik, L.J. The p38 MAP Kinase Family as Regulators of Proinflammatory Cytokine Production in Degenerative Diseases of the CNS. *Aging Dis.* **2010**, *1*, 199–211. [PubMed]
36. Kim, E.K.; Choi, E.J. Compromised MAPK signaling in human diseases: An update. *Arch. Toxicol.* **2015**, *89*, 867–882. [CrossRef]
37. Li, S.; Tian, X.; Hartley, D.M.; Feig, L.A. The environment versus genetics in controlling the contribution of MAP kinases to synaptic plasticity. *Curr. Biol. CB* **2006**, *16*, 2303–2313. [CrossRef]
38. Bolshakov, V.Y.; Carboni, L.; Cobb, M.H.; Siegelbaum, S.A.; Belardetti, F. Dual MAP kinase pathways mediate opposing forms of long-term plasticity at CA3-CA1 synapses. *Nat. Neurosci.* **2000**, *3*, 1107–1112. [CrossRef]
39. Navarrete, M.; Cuartero, M.I.; Palenzuela, R.; Draffin, J.E.; Konomi, A.; Serra, I.; Colie, S.; Castano-Castano, S.; Hasan, M.T.; Nebreda, A.R.; et al. Astrocytic p38alpha MAPK drives NMDA receptor-dependent long-term depression and modulates long-term memory. *Nat. Commun.* **2019**, *10*, 2968. [CrossRef]
40. Tan, B.; Bitiktas, S.; Kavraal, S.; Dursun, N.; Donmez Altuntas, H.; Suer, C. Low-frequency stimulation induces a durable long-term depression in young adult hyperthyroid rats: The role of p38 mitogen-activated protein kinase and protein phosphatase 1. *Neuroreport* **2016**, *27*, 640–646. [CrossRef]
41. Origlia, N.; Bonadonna, C.; Rosellini, A.; Leznik, E.; Arancio, O.; Yan, S.S.; Domenici, L. Microglial receptor for advanced glycation end product-dependent signal pathway drives beta-amyloid-induced synaptic depression and long-term depression impairment in entorhinal cortex. *J. Neurosci. Off. J. Soc. Neurosci.* **2010**, *30*, 11414–11425. [CrossRef] [PubMed]
42. Xiong, W.; Kojic, L.Z.; Zhang, L.; Prasad, S.S.; Douglas, R.; Wang, Y.; Cynader, M.S. Anisomycin activates p38 MAP kinase to induce LTD in mouse primary visual cortex. *Brain Res.* **2006**, *1085*, 68–76. [CrossRef] [PubMed]
43. Ittner, A.; Asih, P.R.; Tan, A.R.P.; Prikas, E.; Bertz, J.; Stefanoska, K.; Lin, Y.; Volkerling, A.M.; Ke, Y.D.; Delerue, F.; et al. Reduction of advanced tau-mediated memory deficits by the MAP kinase p38gamma. *Acta Neuropathol.* **2020**. [CrossRef]
44. Dai, H.L.; Hu, W.Y.; Jiang, L.H.; Li, L.; Gaung, X.F.; Xiao, Z.C. p38 MAPK Inhibition Improves Synaptic Plasticity and Memory in Angiotensin II-dependent Hypertensive Mice. *Sci. Rep.* **2016**, *6*, 27600. [CrossRef] [PubMed]
45. Origlia, N.; Righi, M.; Capsoni, S.; Cattaneo, A.; Fang, F.; Stern, D.M.; Chen, J.X.; Schmidt, A.M.; Arancio, O.; Yan, S.D.; et al. Receptor for advanced glycation end product-dependent activation of p38 mitogen-activated protein kinase contributes to amyloid-beta-mediated cortical synaptic dysfunction. *J. Neurosci. Off. J. Soc. Neurosci.* **2008**, *28*, 3521–3530. [CrossRef]
46. Origlia, N.; Criscuolo, C.; Arancio, O.; Yan, S.S.; Domenici, L. RAGE inhibition in microglia prevents ischemia-dependent synaptic dysfunction in an amyloid-enriched environment. *J. Neurosci. Off. J. Soc. Neurosci.* **2014**, *34*, 8749–8760. [CrossRef]
47. Du, Y.; Du, Y.; Zhang, Y.; Huang, Z.; Fu, M.; Li, J.; Pang, Y.; Lei, P.; Wang, Y.T.; Song, W.; et al. MKP-1 reduces Abeta generation and alleviates cognitive impairments in Alzheimer's disease models. *Signal Transduct. Target. Ther.* **2019**, *4*, 58. [CrossRef] [PubMed]
48. Saha, R.N.; Jana, M.; Pahan, K. MAPK p38 regulates transcriptional activity of NF-kappaB in primary human astrocytes via acetylation of p65. *J. Immunol.* **2007**, *179*, 7101–7109. [CrossRef]
49. Buxade, M.; Parra-Palau, J.L.; Proud, C.G. The Mnks: MAP kinase-interacting kinases (MAP kinase signal-integrating kinases). *Front. Biosci. J. Virtual Libr.* **2008**, *13*, 5359–5373. [CrossRef]
50. Mody, N.; Leitch, J.; Armstrong, C.; Dixon, J.; Cohen, P. Effects of MAP kinase cascade inhibitors on the MKK5/ERK5 pathway. *FEBS Lett.* **2001**, *502*, 21–24. [CrossRef]
51. Xu, L.; He, D.; Bai, Y. Microglia-Mediated Inflammation and Neurodegenerative Disease. *Mol. Neurobiol.* **2016**, *53*, 6709–6715. [CrossRef] [PubMed]
52. Kelly, A.; Vereker, E.; Nolan, Y.; Brady, M.; Barry, C.; Loscher, C.E.; Mills, K.H.; Lynch, M.A. Activation of p38 plays a pivotal role in the inhibitory effect of lipopolysaccharide and interleukin-1 beta on long term potentiation in rat dentate gyrus. *J. Biol. Chem.* **2003**, *278*, 19453–19462. [CrossRef] [PubMed]
53. Laye, S.; Parnet, P.; Goujon, E.; Dantzer, R. Peripheral administration of lipopolysaccharide induces the expression of cytokine transcripts in the brain and pituitary of mice. *Brain Res. Mol. Brain Res.* **1994**, *27*, 157–162. [CrossRef]

54. Fan, L.; Young, P.R.; Barone, F.C.; Feuerstein, G.Z.; Smith, D.H.; McIntosh, T.K. Experimental brain injury induces expression of interleukin-1 beta mRNA in the rat brain. *Brain Res. Mol. Brain Res.* **1995**, *30*, 125–130. [CrossRef]
55. Sonti, G.; Ilyin, S.E.; Plata-Salaman, C.R. Anorexia induced by cytokine interactions at pathophysiological concentrations. *Am. J. Physiol.* **1996**, *270*, R1394–R1402. [CrossRef] [PubMed]
56. Maruta, E.; Yabuuchi, K.; Nishiyori, A.; Takami, S.; Minami, M.; Satoh, M. Beta2-adrenoceptors on the glial cells mediate the induction of interleukin-1beta mRNA in the rat brain. *Brain Res. Mol. Brain Res.* **1997**, *49*, 291–294. [CrossRef]
57. Hoshino, K.; Hasegawa, K.; Kamiya, H.; Morimoto, Y. Synapse-specific effects of IL-1beta on long-term potentiation in the mouse hippocampus. *Biomed. Res.* **2017**, *38*, 183–188. [CrossRef]
58. Hein, A.M.; Stasko, M.R.; Matousek, S.B.; Scott-McKean, J.J.; Maier, S.F.; Olschowka, J.A.; Costa, A.C.; O'Banion, M.K. Sustained hippocampal IL-1beta overexpression impairs contextual and spatial memory in transgenic mice. *BrainBehav. Immun.* **2010**, *24*, 243–253. [CrossRef]
59. Tancredi, V.; D'Arcangelo, G.; Grassi, F.; Tarroni, P.; Palmieri, G.; Santoni, A.; Eusebi, F. Tumor necrosis factor alters synaptic transmission in rat hippocampal slices. *Neurosci. Lett.* **1992**, *146*, 176–178. [CrossRef]
60. Butler, M.P.; O'Connor, J.J.; Moynagh, P.N. Dissection of tumor-necrosis factor-alpha inhibition of long-term potentiation (LTP) reveals a p38 mitogen-activated protein kinase-dependent mechanism which maps to early-but not late-phase LTP. *Neuroscience* **2004**, *124*, 319–326. [CrossRef]
61. Fang, F.; Yu, Q.; Arancio, O.; Chen, D.; Gore, S.S.; Yan, S.S.; Yan, S.F. RAGE mediates Abeta accumulation in a mouse model of Alzheimer's disease via modulation of beta- and gamma-secretase activity. *Hum. Mol. Genet.* **2018**, *27*, 1002–1014. [CrossRef] [PubMed]
62. Kheiri, G.; Dolatshahi, M.; Rahmani, F.; Rezaei, N. Role of p38/MAPKs in Alzheimer's disease: Implications for amyloid beta toxicity targeted therapy. *Rev. Neurosci.* **2018**, *30*, 9–30. [CrossRef] [PubMed]
63. Walsh, D.M.; Klyubin, I.; Fadeeva, J.V.; Cullen, W.K.; Anwyl, R.; Wolfe, M.S.; Rowan, M.J.; Selkoe, D.J. Naturally secreted oligomers of amyloid beta protein potently inhibit hippocampal long-term potentiation in vivo. *Nature* **2002**, *416*, 535–539. [CrossRef] [PubMed]
64. Wang, Q.; Walsh, D.M.; Rowan, M.J.; Selkoe, D.J.; Anwyl, R. Block of long-term potentiation by naturally secreted and synthetic amyloid beta-peptide in hippocampal slices is mediated via activation of the kinases c-Jun N-terminal kinase, cyclin-dependent kinase 5, and p38 mitogen-activated protein kinase as well as metabotropic glutamate receptor type 5. *J. Neurosci. Off. J. Soc. Neurosci.* **2004**, *24*, 3370–3378. [CrossRef]
65. Snyder, E.M.; Nong, Y.; Almeida, C.G.; Paul, S.; Moran, T.; Choi, E.Y.; Nairn, A.C.; Salter, M.W.; Lombroso, P.J.; Gouras, G.K.; et al. Regulation of NMDA receptor trafficking by amyloid-beta. *Nat. Neurosci.* **2005**, *8*, 1051–1058. [CrossRef]
66. Zhu, X.; Rottkamp, C.A.; Hartzler, A.; Sun, Z.; Takeda, A.; Boux, H.; Shimohama, S.; Perry, G.; Smith, M.A. Activation of MKK6, an upstream activator of p38, in Alzheimer's disease. *J. Neurochem.* **2001**, *79*, 311–318. [CrossRef]
67. Ferrer, I. Stress kinases involved in tau phosphorylation in Alzheimer's disease, tauopathies and APP transgenic mice. *Neurotox. Res.* **2004**, *6*, 469–475. [CrossRef]
68. Maphis, N.; Jiang, S.; Xu, G.; Kokiko-Cochran, O.N.; Roy, S.M.; Van Eldik, L.J.; Watterson, D.M.; Lamb, B.T.; Bhaskar, K. Selective suppression of the alpha isoform of p38 MAPK rescues late-stage tau pathology. *Alzheimer's Res. Ther.* **2016**, *8*, 54. [CrossRef]
69. Yu, Q.; Wang, Y.; Du, F.; Yan, S.; Hu, G.; Origlia, N.; Rutigliano, G.; Sun, Q.; Yu, H.; Ainge, J.; et al. Overexpression of endophilin A1 exacerbates synaptic alterations in a mouse model of Alzheimer's disease. *Nat. Commun.* **2018**, *9*, 2968. [CrossRef]
70. Chen, K.; Lu, Y.; Liu, C.; Zhang, L.; Fang, Z.; Yu, G. Morroniside prevents H2O2 or Abeta1-42-induced apoptosis via attenuating JNK and p38 MAPK phosphorylation. *Eur. J. Pharmacol.* **2018**, *834*, 295–304. [CrossRef]
71. Schnoder, L.; Hao, W.; Qin, Y.; Liu, S.; Tomic, I.; Liu, X.; Fassbender, K.; Liu, Y. Deficiency of Neuronal p38alpha MAPK Attenuates Amyloid Pathology in Alzheimer Disease Mouse and Cell Models through Facilitating Lysosomal Degradation of BACE1. *J. Biol. Chem.* **2016**, *291*, 2067–2079. [CrossRef] [PubMed]
72. Stefanoska, K.; Bertz, J.; Volkerling, A.M.; van der Hoven, J.; Ittner, L.M.; Ittner, A. Neuronal MAP kinase p38alpha inhibits c-Jun N-terminal kinase to modulate anxiety-related behaviour. *Sci. Rep.* **2018**, *8*, 14296. [CrossRef] [PubMed]

73. Roy, S.M.; Minasov, G.; Arancio, O.; Chico, L.W.; Van Eldik, L.J.; Anderson, W.F.; Pelletier, J.C.; Watterson, D.M. A Selective and Brain Penetrant p38alphaMAPK Inhibitor Candidate for Neurologic and Neuropsychiatric Disorders That Attenuates Neuroinflammation and Cognitive Dysfunction. *J. Med. Chem.* **2019**, *62*, 5298–5311. [CrossRef]
74. Ittner, A.; Chua, S.W.; Bertz, J.; Volkerling, A.; van der Hoven, J.; Gladbach, A.; Przybyla, M.; Bi, M.; van Hummel, A.; Stevens, C.H.; et al. Site-specific phosphorylation of tau inhibits amyloid-beta toxicity in Alzheimer's mice. *Science* **2016**, *354*, 904–908. [CrossRef] [PubMed]
75. Ittner, A.A.; Gladbach, A.; Bertz, J.; Suh, L.S.; Ittner, L.M. p38 MAP kinase-mediated NMDA receptor-dependent suppression of hippocampal hypersynchronicity in a mouse model of Alzheimer's disease. *Acta Neuropathol Commun.* **2014**, *2*, 149. [CrossRef] [PubMed]
76. Bain, J.; Plater, L.; Elliott, M.; Shpiro, N.; Hastie, C.J.; McLauchlan, H.; Klevernic, I.; Arthur, J.S.; Alessi, D.R.; Cohen, P. The selectivity of protein kinase inhibitors: A further update. *Biochem. J.* **2007**, *408*, 297–315. [CrossRef]
77. Kuma, Y.; Sabio, G.; Bain, J.; Shpiro, N.; Marquez, R.; Cuenda, A. BIRB796 inhibits all p38 MAPK isoforms in vitro and in vivo. *J. Biol. Chem.* **2005**, *280*, 19472–19479. [CrossRef]
78. Chico, L.K.; Van Eldik, L.J.; Watterson, D.M. Targeting protein kinases in central nervous system disorders. *Nat. Rev. Drug Discov.* **2009**, *8*, 892–909. [CrossRef]
79. Watterson, D.M.; Grum-Tokars, V.L.; Roy, S.M.; Schavocky, J.P.; Bradaric, B.D.; Bachstetter, A.D.; Xing, B.; Dimayuga, E.; Saeed, F.; Zhang, H.; et al. Development of Novel In Vivo Chemical Probes to Address CNS Protein Kinase Involvement in Synaptic Dysfunction. *PLoS ONE* **2013**, *8*, e66226. [CrossRef]
80. Fabian, M.A.; Biggs, W.H.; Treiber, D.K.; Atteridge, C.E.; Azimioara, M.D.; Benedetti, M.G.; Carter, T.A.; Ciceri, P.; Edeen, P.T.; Floyd, M. A small molecule-kinase interaction map for clinical kinase inhibitors. *Nat. Biotechnol.* **2005**, *23*, 329–336. [CrossRef]
81. Karaman, M.W.; Herrgard, S.; Treiber, D.K.; Gallant, P.; Atteridge, C.E.; Campbell, B.T.; Chan, K.W.; Ciceri, P.; Davis, M.I.; Edeen, P.T. A quantitative analysis of kinase inhibitor selectivity. *Nat. Biotechnol.* **2008**, *26*, 127–132. [CrossRef]
82. Zhou, Z.; Bachstetter, A.D.; Spani, C.B.; Roy, S.M.; Watterson, D.M.; Van Eldik, L.J. Retention of normal glia function by an isoform-selective protein kinase inhibitor drug candidate that modulates cytokine production and cognitive outcomes. *J. Neuroinflammation* **2017**, *14*, 75. [CrossRef] [PubMed]
83. Cancino, G.I.; Toledo, E.M.; Leal, N.R.; Hernandez, D.E.; Yevenes, L.F.; Inestrosa, N.C.; Alvarez, A.R. STI571 prevents apoptosis, tau phosphorylation and behavioural impairments induced by Alzheimer's beta-amyloid deposits. *Brain: A J. Neurol.* **2008**, *131*, 2425–2442. [CrossRef] [PubMed]
84. Lonskaya, I.; Hebron, M.L.; Desforges, N.M.; Franjie, A.; Moussa, C.E. Tyrosine kinase inhibition increases functional parkin-Beclin-1 interaction and enhances amyloid clearance and cognitive performance. *EMBO Mol. Med.* **2013**, *5*, 1247–1262. [CrossRef]
85. Lonskaya, I.; Hebron, M.L.; Selby, S.T.; Turner, R.S.; Moussa, C.E. Nilotinib and bosutinib modulate pre-plaque alterations of blood immune markers and neuro-inflammation in Alzheimer's disease models. *Neuroscience* **2015**, *304*, 316–327. [CrossRef] [PubMed]
86. Brennan, K.C.; Bates, E.A.; Shapiro, R.E.; Zyuzin, J.; Hallows, W.C.; Huang, Y.; Lee, H.Y.; Jones, C.R.; Fu, Y.H.; Charles, A.C. Casein kinase idelta mutations in familial migraine and advanced sleep phase. *Sci. Transl. Med.* **2013**, *5*, 183. [CrossRef] [PubMed]
87. Rutigliano, G.; Stazi, M.; Arancio, O.; Watterson, D.M.; Origlia, N. An isoform-selective p38alpha mitogen-activated protein kinase inhibitor rescues early entorhinal cortex dysfunctions in a mouse model of Alzheimer's disease. *Neurobiol. Aging* **2018**, *70*, 86–91. [CrossRef]
88. Robson, M.J.; Quinlan, M.A.; Margolis, K.G.; Gajewski-Kurdziel, P.A.; Veenstra-VanderWeele, J.; Gershon, M.D.; Watterson, D.M.; Blakely, R.D. p38alpha MAPK signaling drives pharmacologically reversible brain and gastrointestinal phenotypes in the SERT Ala56 mouse. *Proc. Natl. Acad. Sci. USA* **2018**, *115*, E10245–E10254. [CrossRef]

International Journal of
Molecular Sciences

Review

Is p38 MAPK Associated to Drugs of Abuse-Induced Abnormal Behaviors?

Rana El Rawas *, Inês M. Amaral and Alex Hofer

Experimental Addiction Research, Department of Psychiatry, Psychotherapy and Psychosomatics, Division of Psychiatry I, Medical University Innsbruck, 6020 Innsbruck, Austria; ines.amaral@i-med.ac.at (I.M.A.); a.hofer@i-med.ac.at (A.H.)

* Correspondence: rana.el-rawas@i-med.ac.at; Tel.: +43-512-504-23711

Received: 16 June 2020; Accepted: 7 July 2020; Published: 8 July 2020

Abstract: The family members of the mitogen-activated protein kinases (MAPK) mediate a wide variety of cellular behaviors in response to extracellular stimuli. p38 MAPKs are key signaling molecules in cellular responses to external stresses and regulation of pro-inflammatory cytokines. Some studies have suggested that p38 MAPK in the region of the nucleus accumbens is involved in abnormal behavioral responses induced by drugs of abuse. In this review, we discuss the role of the p38 MAPK in the rewarding effects of drugs of abuse. We also summarize the implication of p38 MAPK in stress, anxiety, and depression. We opine that p38 MAPK activation is more closely associated to stress-induced aversive responses rather than drug effects per se, in particular cocaine. p38 MAPK is only involved in cocaine reward, predominantly when promoted by stress. Downstream substrates of p38 that may contribute to the p38 MAPK associated-behavioral responses are proposed. Finally, we suggest p38 MAPK inhibitors as possible therapeutic interventions against stress-related disorders by potentially increasing resilience against stress and addiction relapse induced by adverse experiences.

Keywords: p38 MAPK; cocaine; conditioned place preference; reward; stress; anxiety; depression; nucleus accumbens; social interaction; k opioid receptors

---

1. Introduction

The mitogen-activated protein kinases (MAPK) superfamily is made up of three majorsignaling pathways: The extracellular signal-regulated protein kinases (ERKs), the c-Jun N-terminal kinases or stress-activated protein kinases (JNK/SAPK), and the p38 family of kinases. The MAPK is a serine/threonine kinase that is activated through phosphorylation by a MAPK kinase (MKK), which is a "dual-specific" kinase that phosphorylates at both serine/threonine and tyrosine residues within a threonine/any amino acid/tyrosine (Thr X Tyr) motif, in which the middle amino acid is different for each MAPK [1]. Enzymes in the p38 MAPK module are subject to dual phosphorylation at the Thr-Gly-Tyr motif situated within the kinase activation loop and are primarily activated by various environmental stresses, including heat, osmotic and oxidative stresses, as well as inflammatory cytokines [2] (Figure 1).

p38 MAPK is inactive in the non-phosphorylated state. Dual phosphorylation at Thr-180 and Tyr-182 residues by either MKK3 or MKK6 induces global conformational reorganizations that modify the alignment of the C- and N-terminal domains of p38 MAPK, consequently permitting the binding of ATP and the desired substrate [3–5]. However, selective activation by distinct MKKs has been observed among p38 isoforms, as evidenced by the inability of MKK3 to effectively activate one isoform of p38, p38β, while MKK6 is a potent activator [5,6]. MKK4 can also activate p38 in vitro or when co-expressed with p38 in cultured cells, although reported as a non-physiologic activator of p38 [2]. Phosphorylated p38 MAPK (pp38) can activate a wide range of substrates. These include transcription factors such as ATF2, CCAAT/enhancer-binding protein-homologous protein (CHOP), and myocyte enhancer factor 2C (MEF2C), as well as protein kinases such as the MAPK-activated protein (MAPKAP)-2, and cytosolic and nuclear proteins such as Cdc25 and Glycogen synthase (GS) [2,5]. Duration of signaling is controlled by phosphatases, including protein phosphatase 1, protein phosphatase 2A, or MAPK phosphatases [5]. Phosphorylated substrates go on to elicit varied biological responses that include inflammation, apoptosis, proliferation, cell-cycle regulation, and differentiation [5,7]. The growing list of specific downstream targets and various activation modes of the p38 MAPK pathway raise a possibility of the functional diversity of each p38 MAPK isoform [7]. Four different genes encoding p38 MAPK isoforms (p38α, p38β, p38γ, p38δ) with > 60% overall sequence homology and > 90% identity within the kinase domains have been described in human tissues [2]. Of these, p38α and p38β are ubiquitously expressed while p38γ and p38δ are differentially expressed depending on tissue type [6]. Generally, p38α and p38β are highly expressed in the brain with p38α being mainly expressed in neuronal cells and p38β being highly expressed in both neuronal and glial cells [7]. At the subcellular level, p38α is distributed in dendrites, in cytoplasmic and nuclear regions of the cell body of neurons, in contrast to p38β that is preferentially expressed in the nucleus of neurons [7].

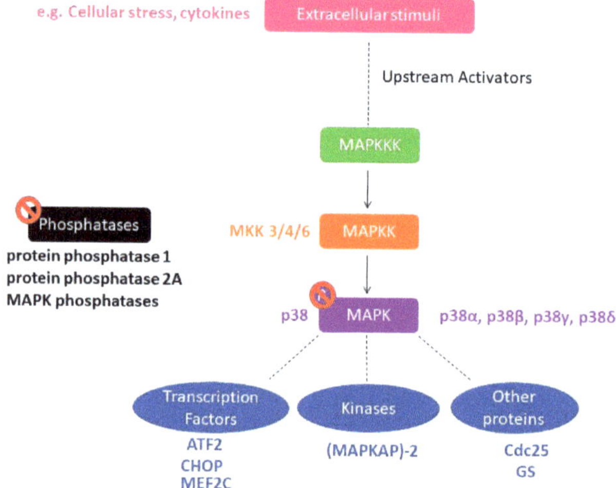

Figure 1. Schematic representation of the p38 mitogen-activated protein kinases (MAPK) signaling pathway. A variety of extracellular signals, such as cellular stress and cytokines, can activate the p38 MAPK pathway. This leads to the initiation of MAPK phosphorylation cascade, in which MAPKKK phosphorylate the p38 MAPK-specific MAPKKs MKK3, MKK4, or MKK6. Subsequently, the four isoforms of p38 MAPK (α, β, δ, and γ) are phosphorylated, thereby activating various p38 MAPK substrates. Substrates can be transcription factors such as ATF2, CHOP, and MEF2C, protein kinases such as (MAPKAP)-2 as well as other proteins such as Cdc25 and GS. Phosphorylated substrates go on to elicit varied biological responses. The duration of signaling is controlled by phosphatases such as protein phosphatase 1; protein phosphatase 2A or MAPK phosphatase. Abbreviations: CHOP, CCAAT/enhancer-binding protein-homologous protein; MEF2C, myocyte enhancer factor 2C; (MAPKAP)-2, MAPK-activated protein; GS, Glycogen synthase.

The pyridinylimidazole compounds, represented by SB203580, were initially seen as inflammatory cytokine synthesis inhibitors [8,9]. Subsequently, they were found to be selective inhibitors of p38 MAPK [8,9]. SB203580 inhibits the catalytic activity of p38 MAPK by competitive binding at the ATP binding site of the kinase [8,10]. MAPK inhibitors appear to be a therapeutic strategy in several disease models, particularly inflammatory disorders, due to their capability to reduce the synthesis and the signaling of pro-inflammatory cytokines [11]. This review discusses the role of p38 MAPK in the rewarding effects of drugs of abuse, as well as in stress-related behaviors.

2. Role of p38 MAPK in the Rewarding Effects of Drugs of Abuse

Environmental stimuli associated with drugs of abuse motivate animals to "prefer" the contexts associated with drug intake and to spend more time in these contexts. Such an approach, known as conditioned place preference (CPP), is widely used for evaluating the rewarding effects of drugs [12,13]. During CPP, the animal is "conditioned" in an experimental apparatus consisting of at least two compartments that have distinct visual and tactile cues. The animal can be injected with the drug in one compartment and with saline in the other compartment. The neutral environment associated with the drug acquires secondary motivational properties, such that it can act as a conditioned stimulus when the animal is subsequently exposed to this environment again [14]. During the test, the animal can "choose" the compartment in which it prefers to spend more time. If the drug has rewarding properties, the animal will "prefer" to spend more time in the compartment previously associated with the drug (Figure 2).

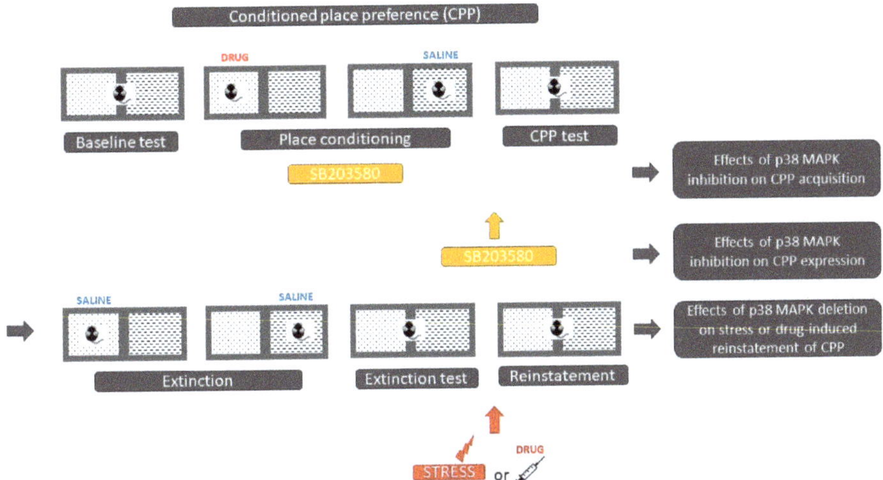

Figure 2. conditioned place preference (CPP). In the pre-conditioning day or baseline test, animals (such as mice) are allowed to freely explore the entire apparatus. The time that mice spend in each compartment is recorded to show the natural preference of animals. During conditioning, mice are randomly assigned to their drug and saline training compartments for the training sessions. Conditioned preference is assessed by allowing the mice to roam freely in all the compartments and recording the time spent in each. When the p38 MAPK inhibitor, SB203580, is injected before DRUG conditioning (the stimulus), the effects of p38 inhibition on CPP acquisition are investigated. Likewise, when SB203580 is injected before the CPP test, the effects of p38 inhibition on CPP expression are explored. After the CPP expression, mice can undergo extinction of CPP, during which they are conditioned repeatedly to saline in both compartments. By conditioning the animals with saline in their previous DRUG-associated compartment, mice progressively lose their preference to the compartment associated to the DRUG. When the animals reach the extinction criteria, which is checked by an extinction test, they can be tested for reinstatement of CPP. Reinstatement can be induced by an exposure to stress (stress-induced reinstatement of CPP) or also, by an exposure to a priming injection of the DRUG (drug-induced reinstatement). Place preference during reinstatement of CPP is again determined by allowing the mice to freely explore all the compartments of CPP.

The nucleus accumbens (NAc) is a critical element of the mesocorticolimbic system which has a well-established role in mediating the rewarding effects of drugs of abuse [15]. Gerdjikov et al. tested the hypothesis that the inhibition of MAPKs would inhibit the acquisition of NAc amphetamine CPP. They showed that NAc injections of the ERK inhibitor PD98059 or the p38 kinase inhibitor

SB203580 dose-dependently impaired CPP, but not the JNK inhibitor SP600125 [16]. These results suggest that ERK and p38, but not JNK, MAPKs may be necessary for the establishment of NAc amphetamine-produced CPP [16]. Moreover, Zhang et al. showed that repeated morphine treatment induced the acquisition of CPP and increased the phosphorylation of p38 in the NAc [17]. Consistently, the microinjection of the p38 inhibitor SB203580 into the NAc prior to the administration of morphine prevented the acquisition of CPP and inhibited the activation of p38, thereby indicating that the activation of p38 in the NAc may be necessary for morphine CPP [17]. The same group also found that, following 5 days of morphine treatment, p38 activation was induced in the NAc microglia but not in astrocytes or neurons [18]. They reported that the bilateral microinjection of minocycline, a putative inhibitor of microglia, or SB203580, a selective p38 MAPK inhibitor, into the NAc before each morphine treatment for five days impaired the acquisition of morphine CPP and suppressed the activation of p38 signaling in the microglia induced by morphine treatment [18]. Following the acquisition of morphine CPP, a single minocycline or SB203580 injection failed to block the expression of morphine CPP [18]. Thus, based on these studies, it appears that p38 signaling in the NAc microglia may play an important role in the acquisition but not the expression of morphine CPP [18]. Therefore, it has been proposed that p38 MAPK in the NAc is involved in abnormal behavioral responses induced by drugs of abuse [17]. Later, another study also showed that intraperitoneal (i.p.) injections of SB203580 dampened the acquisition of morphine-induced CPP in mice and reduced the phosphorylation of p38 MAPK in the NAc of morphine CPP mice [19].

The effect of p38 MAPK inhibition on cocaine-induced behaviors, such as the acquisition and expression of CPP, was also investigated. The expression of cocaine CPP was specifically blocked by i.p. SB203580 when injected on the post-conditioning test day [20]. By contrast, morphine CPP was not affected by i.p. SB203580 [20]. Thus, in line with previous studies, i.p. inhibition of p38 failed to block the expression of CPP to morphine. On the other hand, studies that performed intracerebroventricular (icv) injections of SB203580 before cocaine training found no effect on cocaine CPP [21,22]. Consistently, the deletion of p38α in the serotonergic neurons of the dorsal raphe nucleus (DRN) [23] or the conditional knock-out of p38α MAPK in the dopaminergic neurons of the ventral tegmental area (VTA) [24] does not affect cocaine CPP, suggesting that the deletion of p38α does not alter the associative learning required for place preference or the rewarding properties of cocaine. Interestingly, serotonergic p38α MAPK deletion blocked the reinstatement of cocaine preference induced by stress, but not by cocaine priming, thereby showing that serotonergic p38α MAPK deletion selectively alters only the stress-induced modulation of cocaine-seeking behaviors [23]. Table 1 summarizes the studies investigating p38 MAPK blockade on drug reward.

Similarly to drugs of abuse, rats acquire and express CPP for natural reward such as social interaction [25–31]. In general, social reward CPP is assessed by placing the rats during half the conditioning sessions into a compartment of the CPP with their assigned social partner, and during the other half of the sessions alone into the other compartment of the CPP [26–28,32]. We and others [33] perform social interaction reward CPP by pairing one compartment with a sex- and weight-matched male conspecific preceded by an i.p. injection of saline for 15 min and the other compartment with saline only [32]. Animals that spent more time in the social interaction-paired compartment than in the saline-paired compartment during the CPP test expressed preference for social interaction. When comparing social interaction CPP to cocaine CPP, almost the same brain regions were activated [31]. However, parts of the insular cortex, namely the granular insular cortex and the dorsal part of the agranular insular cortex, were more activated after cocaine CPP, whereas the prelimbic cortex and the core sub-region of the NAc were more activated after social interaction CPP [31]. Given that p38 MAPK activation was reported to be increased after morphine CPP, we assessed the expression of p38 MAPK activation in the NAc shell and core sub-regions at different time points after cocaine CPP and compared it to social interaction CPP [34]. It was expected that cocaine CPP would enhance the activation of p38 in the NAc as compared to social interaction CPP and to control rats that received saline in both compartments of the CPP. However, we found that control rats and cocaine CPP expressing rats showed similarly enhanced p38 activation compared to naïve untreated and social interaction CPP rats. Furthermore, 24 h after social interaction CPP, pp38 neuronal levels in the NAc shell decreased to the level of naïve untreated rats with pp38 expressed mainly in neurons (92%) (Figure 3).

Table 1. Summary of the findings describing the role of p38 MAPK in drug reward.

Drug	P38 Blockade	Results	References
Amphetamine	SB203580 into the NAc	impaired amphetamine conditioned place preference (CPP)	[16]
Morphine	SB203580 into the NAc	prevented the acquisition but not the expression of morphine CPP	[17,18]
Morphine	SB203580 i.p.	dampened the acquisition of morphine-induced CPP	[19]
Morphine	SB203580 i.p.	failed to block the expression of morphine CPP	[20]
Cocaine	SB203580 i.p.	blocked the expression of cocaine CPP	[20]
Cocaine	SB203580 icv	did not affect acquisition to cocaine CPP	[21,22]
Cocaine	deletion of p38α MAPK in the serotonergic neurons	did not affect cocaine CPP	[23]
Cocaine	conditional knock-out of p38α MAPK in the dopaminergic neurons	did not affect cocaine CPP	[24]
Cocaine	serotonergic p38α MAPK deletion	blocked the reinstatement of cocaine preference induced by stress	[23]
Cocaine	serotonergic p38α MAPK deletion	did not affect the reinstatement of cocaine preference induced by cocaine priming	[23]

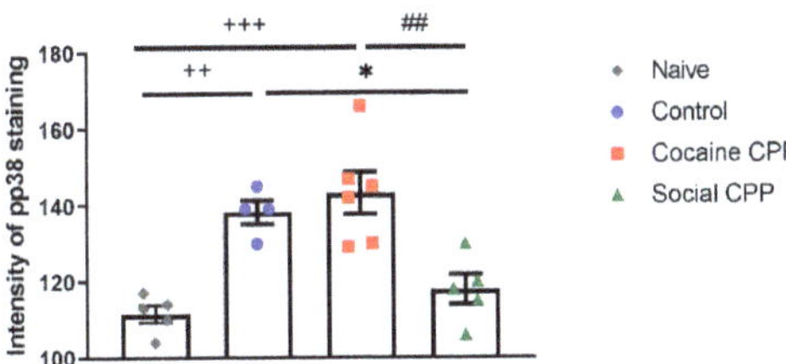

Figure 3. pp38 expression in the NAc shell 24 h after the cocaine CPP test [34]. Social interaction CPP decrease p38 MAPK activation to the levels of naïve untreated rats. Control rats receiving saline in both compartment of CPP and cocaine CPP expressing rats show similar increased levels of pp38 as compared to naïve untreated and social CPP expressing rats. Statistical test: one-way analysis of variance followed by Tukey's post hoc test. * $p < 0.05$, different from saline control; ++ $p < 0.01$, +++ $p < 0.001$ different from naïve; ## $p < 0.01$ different from cocaine CPP.

As p38 plays a role in stress and anxiety behaviors that will be detailed in the coming section, those results suggest that: First, cocaine treatment per se does not induce p38 activation as control rats and cocaine CPP-expressing rats show the same levels of pp38 in the NAc shell. Second, marginal stress such as injecting animals with saline and placing them into the CPP apparatus is sufficient to induce p38 activation in the NAc shell. In fact, compared to naïve untreated animals, control rats receiving saline injections in both compartments of the CPP, expressed significantly higher levels of activated p38 in the NAc shell. Finally, social interaction reward has anti-stress effects as social interaction-expressing CPP rats show levels of pp38 similarly to naïve untreated rats [34]. Indeed, rats receiving corticotropin-releasing factor (CRF) icv injections before cocaine conditioning showed an increase in cocaine CPP, whereas those receiving icv injections of the non-selective antagonist alpha-helical CRF (α CRF) showed a decrease in cocaine CPP [22]. Remarkably, when social interaction was made available in the alternative compartment, CRF-induced increase of cocaine preference was reversed completely to the level of rats receiving cocaine paired with α CRF. This reversal of cocaine preference was also paralleled by a reversal in altered behavioral sequencing of grooming, considered as a marker of stress [35] and by a CRF-induced increase of p38 MAPK expression in the NAc shell [22] (Figure 4). These results show that the modulation of the CRF system has a direct impact on p38

MAPK expression in the NAc shell, as p38 MAPK expression was increased after icv CRF injections prior to each cocaine conditioning and was decreased after icv injections of α CRF prior to each cocaine conditioning [22]- Figure 4. Accordingly, p38 is considered to be more closely related to stress modulation than to cocaine treatment. Indeed, CRF was reported to induce dynorphin-dependent k opioid receptor (KOR) activation in the NAc [36] with p38 being an important mediator of the KOR-dependent aversive properties of stress [21,37].

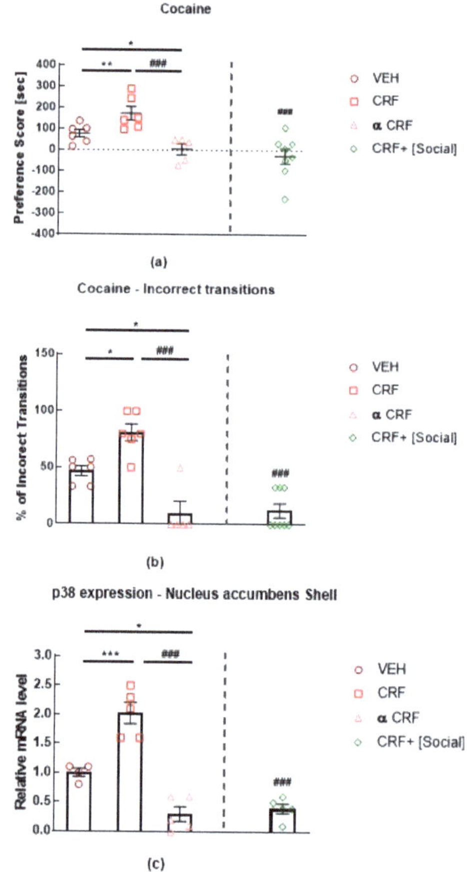

Figure 4. Effects of intracerebroventricular (icv) injections of vehicle, corticotropin-releasing factor (CRF), and alpha-helical CRF (α CRF) on (**a**) cocaine preference; on (**b**) associated percentage of incorrect transitions and on (**c**) associated p38 expression in the NAc shell [22]. Preference score is the time that the rat spent in the stimulus-associated compartment during the test-pretest. CRF+ [social] is a group of rats conditioned with cocaine that received icv injections of CRF prior to cocaine conditioning but also had the opportunity to social interaction in the alternative compartment of the CPP. The incorrect transitions of cephalocaudal grooming progression were evaluated during the test session of CPP for each treatment condition. Statistical test: one-way analysis of variance followed by Tukey's post hoc test. α CRF, alpha-helical CRF; VEH, vehicle * $p < 0.05$, ** $p < 0.01$, *** $p < 0.001$ different from VEH; ### $p < 0.001$, different from CRF.

We propose that the anti-stress effects of social interaction might be mediated by the KOR system in the ventral NAc shell, which was previously shown to drive aversion via KOR activation [38] through a decrease in the activation of p38 MAPK [34]. In parallel, we additionally propose that the rewarding effects of social interaction might be mediated by the KOR system in the dorsal NAc shell, which has also previously been shown to drive preference/reward via KOR activation [38] through an increase in ERK.

3. Role of p38 MAPK in Stress, Anxiety, and Depression

Sustained stressful experience can lead to maladaptive responses, including clinical depression, anxiety, and an increased risk for drug addiction [39,40]. After stress exposure, p38 MAPK is generally activated in different brain regions (Table 2). Repeated forced swim stress activated p38 MAPK in the cortex, the hippocampus, and the NAc [21], which decreased significantly following a pretreatment with SB203580 before bouts of forced swimming [21]. pp38 levels were also increased in the prefrontal cortex (PFC) after cold exposure [41], in the hippocampus after enhanced single prolonged stress [42] and in the DRN after social defeat stress [23]. Additionally, a significant positive correlation was found between early life stress and the percentage of monocytes staining positive for pp38 [43]. Neuro-inflammation, in response to bacterial endotoxin lipopolysaccharide (LPS)-induced depressive-like behaviors, has been reported to be accompanied by increased levels of pp38 in the habenula [44]. Both the p38 inhibitor SB203580 and the anti-depressant fluoxetine normalized the changes in p38 phosphorylation and reversed the depressive-like behaviors [44]. Interestingly, the depletion of neuronal p38α in mice resulted specifically in increased anxiety-related behaviors without affecting learning and memory processes or motor coordination and muscle function [45].

Table 2. Summary of the findings reporting the activation p38 MAPK after CPP to drugs and exposure to stress.

Stimuli/Protocol	Regions	Animals	References
morphine CPP	NAc	rats	[17,18]
		mice	[19]
repeated forced swim stress	cortex, hippocampus, NAc	mice	[21]
cold exposure	PFC	rats	[41]
enhanced single prolonged stress	hippocampus	rats	[42]
social defeat stress	dorsal raphe nucleus (DRN)	mice	[23]
early life stress	monocytes staining for pp38	monkeys	[43]
bacterial endotoxin LPS	habenula	rats	[44]

p38 activation appears to be an important mediator of KOR-induced aversive stress effects through G-protein-coupled receptor kinase 3 (GRK3)/β-arrestin, a KOR-associated protein, dependent mechanisms [21,37]. Inhibition of p38 MAPK was found to block stress-induced behavioral responses, including aversive responses to the KOR agonist U50, 488 [21,46]. Importantly, cell-specific deletion of p38α MAPK in serotonergic neurons blocked stress-induced aversion [23]. These effects seem to be regulated by KOR as KOR knockout (KO) mice did not develop conditioned place aversion (CPA) to U50,488; however, re-expression of KOR in the serotonergic neurons of the DRN or in the dopaminergic neurons of the VTA of KOR KO mice activated p38 and restored place aversion [24,47].

One possible mechanism encoding the behavior responses to stress is a change in gene expression downstream to p38 [21]. One candidate is zif268, whose induction is p38-dependent [21] and has previously shown to be a direct downstream target of p38 [48]. Indeed, multiple swim stress exposure has been reported to cause a significant up-regulation of zif268 in the striatum only in wild type but not in KOR$^{(-/-)}$ mice [21]. SB203580 had an inhibitory effect on zif268 induced by stress, suggesting that this immediate early gene may be involved in the aversive responses to KOR activation [21]. Another possible substrate of pp38 that may contribute to the behavioral responses is the serotonin transporter (SERT). In fact, p38 MAPK activation has been reported to regulate the activity of SERT in vitro [49,50]. Evidence of an in vivo relationship between activation of p38 MAPK signaling pathways and central serotonin function/metabolism was described by [43]. They found a significant negative correlation between the percentage of monocytes staining positive for intracellular pp38 and CSF concentrations of the serotonin metabolite 5-HIAA in non-human primates [43]. Thus, the activation of the p38 pathway would be expected to decrease synaptic availability of serotonin and to reduce the serotonin metabolites by increasing SERT expression/activity [43]. Further evidence was provided in mice demonstrating that SERT activity in nerve terminals of serotonergic neurons is positively modulated in a p38α dependent manner [23]. Stress-induced p38α MAPK caused translocation of SERT to the plasma membrane in the brain, thereby increasing the rate of transmitter uptake at serotonergic nerve terminals and inducing a hypo-serotonergic state that underlies depression-like and drug-seeking behaviors [23]. In line with these findings, cytokine induction by LPS produced SERT activation and behavioral despair, both requiring p38 MAPK pathway activation [51]. In mice exhibiting a selective elimination of p38α MAPK in serotonergic neurons, LPS failed to elevate brain SERT activity despite normal peripheral stress responses [52]. Moreover, p38α MAPK excision in serotonergic neurons resulted in behavioral resilience to anxiety and depression-like behaviors [52]. Thus, p38 activation can stimulate the expression of SERT, which is used as a major pharmacological target for depression treatment [44]. A third possible candidate could involve the modulation of proteins related to synaptic plasticity, such as AMPA receptors. p38 MAPK signaling has been shown to be an important mediator of AMPA receptor surface trafficking during synaptic plasticity in which activation of p38 MAPK may lead to synaptic removal of surface AMPA receptors [53,54]. Generally, compounds which augment signaling through AMPA receptors exhibit antidepressant-like behavioral effects in animal models [55]. For example, the antidepressant fluoxetine has been found to alter AMPA receptor phosphorylation in a manner that is expected to increase AMPA receptor signaling [55]. Therefore, it is possible that a decrease in the surface expression of AMPA receptors contributes to the behavioral responses associated to p38 MAPK. One additional target of p38 relevant to depression could involve the glucocorticoid receptor (GR). It was reported that GR function is reduced in patients with major depression [56]. p38 signaling pathways have been shown to be implicated in the inhibition of GR function. Indeed, activation of p38 MAPK has been demonstrated to disrupt transactivation of the GR [57], leading potentially to glucocorticoid resistance or decreased responsiveness to glucocorticoids, a primary feature of major depression [58]. Therefore, SB203580 through inhibition of p38 MAPK could recover the normal functioning of GR and alleviate the glucocorticoid resistance underlying depression [44]. Figure 5 summarizes the possible substrates that may contribute to the behavioral responses associated to p38 MAPK.

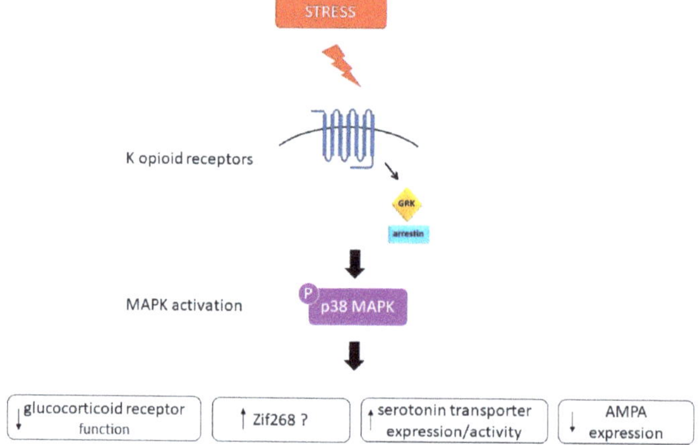

Figure 5. Arrestin-dependent signaling events result in p38 MAPK activation and subsequent dysphoric behavioral responses. Possible substrates encoding the behavior responses to stress downstream to p38 comprises changes in gene expression such as zif268, in serotonin transporter expression/activity, in glucocorticoid receptor function and modulation of proteins related to synaptic plasticity, such as AMPA receptors.

4. Conclusions

It is unclear why p38 MAPK is merely involved in the expression of cocaine CPP when the inhibitor is administered before the post-conditioning test in a drug-free state [20], i.e., when the animals encounter drug-associated cues, but not in the learning required for the rewarding properties of cocaine. It was previously observed that KOR activation before the presentation of cocaine-associated cues enhances approach behaviors to those cues [59], possibly via activation of p38 signaling pathway. This potentiation of cocaine CPP by KOR activation does not result from an enhancement of associative learning mechanisms, as KOR activation only occurred before the final preference test after the associative learning phases were already complete. Conversely, the inhibition of p38 signaling only after the post-conditioning test might reduce the rewarding value of cocaine-associated contexts. More studies are needed to emphasize this possibility, in particular because expression of cocaine CPP did not increase the levels of pp38 in the regions of NAc core or NAc shell [34]. Yet, the study by [20] suggested that p38 MAPK-mediated norepinephrine transporter (NET) up-regulation is linked to cocaine-induced CPP.

It appears that p38 MAPK activation is more closely associated to stress-induced aversive responses rather than drug effects per se. Mostly, studies show that p38 MAPK activation is only involved in cocaine reward, predominantly when promoted by stress. However, it remains open to discussion how p38 MAPK is implicated in CPP morphine acquisition. The first explanation could be that morphine might activate KOR as well as μ opioid receptors (MOR). Indeed, it has been reported that morphine is weakly selective to the MOR and possesses affinity to δ opioid receptors (DOR) and KORs [60,61]. This explanation is further supported by the fact that naloxone, a non-selective opioid antagonist, could block the acquisition of morphine CPP [62]. However, the rewarding effects of morphine are abolished in MOR-deficient animals [63], thereby showing that MOR gene product is the molecular target of morphine in vivo. In addition, the k-opioid antagonist nor-binaltorphimine did not affect morphine CPP [64]. Remarkably, it appears that DORs, rather than KORs, are implicated in the acquisition of morphine reward; as the administration of the selective delta-2-opioid receptor antagonist naltriben prior to morphine was able to block morphine-induced CPP [65], suggesting that this first explanation is unlikely to occur. The second explanation might be that opioid receptor-mediated p38 phosphorylation has also been demonstrated for MORs [66]. MOR opioids could to some extent induce activation of p38 [67,68]. It is therefore plausible that inhibition of p38 signaling during morphine training could abolish acquisition but not before the post-conditioning test, after that morphine acquisition was already established.

In conclusion, understanding the molecular and cellular mechanisms that control stress-induced behaviors could explain the neurobiological mechanisms involved in depression and addiction-like behaviors and provides insight to potential therapeutic targets. Emerging evidence demonstrates a role for p38 MAPK in depression, anxiety, and addiction relapse induced by stress. Targeting the p38 MAPK pathway for therapeutic advantage might appear standard, given the broad range of pathologies in which this pathway is implicated. However, the pathology-specific functions and targets of p38 MAPK together with its interaction with other intracellular regulatory pathways initiates many challenges to exploiting this pathway for therapeutic benefit [5]. Indeed, p38 MAPK inhibitors have been studied extensively in both preclinical experiments and clinical trials for inflammatory diseases. Here, we opine that p38 MAPK inhibitors are of growing interest as possible therapeutic interventions against stress-related disorders by potentially increasing resilience against stress and addiction relapse induced by adverse experiences.

Author Contributions: Writing—original draft preparation, R.E.R.; writing—review and editing, I.M.A. and A.H. All authors have read and agreed to the published version of the manuscript.

Funding: The work by our group described in this article was funded by Austrian Science Fund (FWF). Grant Numbers: P23824-B18; P27852-B21 and T758-BBL.

Acknowledgments: The authors wish to thank Ingie Zakaria for her assistance in English editing.

Conflicts of Interest: The authors declare no conflict of interest.

Abbreviations

ERK	Extracellular signal-regulated protein kinases
JNK	c-Jun N-terminal kinases
MAPK	Mitogen-activated protein kinases
MKK	MAP kinase kinase
NET	Norepinephrine transporter
pp38	Phosphorylated p38
CHOP	CCAAT/enhancer-binding protein-homologous protein
MAPKAP	MAPK-activated protein
MEF2C	Myocyte enhancer factor 2C
GS	Glycogen synthase
GRK3	G-protein-coupled receptor kinase 3
CPP	Conditioned place preference
KO	Knock out
NAc	Nucleus accumbens
i.p.	Intraperitoneal
icv	Intracerebroventricular
DRN	Dorsal raphe nucleus
VTA	Ventral tegmental area
CRF	Corticotropin-releasing factor
KOR	κ opioid receptor
αCRF	Alpha helical CRF
VEH	Vehicle
PFC	Prefrontal cortex
LPS	Lipopolysaccharide
CPA	Conditioned place aversion
SERT	Serotonin transporter
GR	Glucocorticoid receptor
MOR	μ opioid receptor
DOR	δ opioid receptor

References

1. Cowan, K.J.; Storey, K.B. Mitogen-activated Protein Kinases: New Signaling Pathways Functioning in Cellular Responses to Environmental Stress. *J. Exp. Biol.* **2003**, *206*, 1107–1115. [CrossRef] [PubMed]
2. New, L.; Han, J. The p38 MAP kinase pathway and its biological function. *Trends Cardiovasc. Med.* **1998**, *8*, 220–228. [CrossRef]
3. Wang, Z.; Harkins, P.C.; Ulevitch, R.J.; Han, J.; Cobb, M.H.; Goldsmith, E.J. The Structure of Mitogen-Activated Protein Kinase p38 at 2.1-A Resolution. *Proc. Natl. Acad. Sci. USA* **1997**, *94*, 2327–2332. [CrossRef] [PubMed]
4. Corrêa, S.A.; Eales, K.L. The Role of p38 MAPK and Its Substrates in Neuronal Plasticity and Neurodegenerative Disease. *J. Signal Transduct.* **2012**, *2012*, 1–12. [CrossRef] [PubMed]
5. Coulthard, L.R.; White, D.E.; Jones, D.L.; McDermott, M.F.; Burchill, S.A. p38(MAPK): Stress responses from molecular mechanisms to therapeutics. *Trends Mol. Med.* **2009**, *15*, 369–379. [CrossRef]
6. Zarubin, T.; Han, J. Activation and signaling of the p38 MAP kinase pathway. *Cell Res.* **2005**, *15*, 11–18. [CrossRef]
7. Lee, S.H.; Park, J.; Che, Y.; Han, P.L.; Lee, J.K. Constitutive activity and differential localization of p38alpha and p38beta MAPKs in adult mouse brain. *J. Neurosci. Res.* **2000**, *60*, 623–631. [CrossRef]
8. Lee, J.C.; Kumar, S.; Griswold, D.E.; Underwood, D.C.; Votta, B.J.; Adams, J.L. Inhibition of p38 MAP kinase as a therapeutic strategy. *Immunopharmacology* **2000**, *47*, 185–201. [CrossRef]
9. Lee, J.C.; Kassis, S.; Kumar, S.; Badger, A.; Adams, J.L. p38 mitogen-activated protein kinase inhibitors–mechanisms and therapeutic potentials. *Pharmacol. Ther.* **1999**, *82*, 389–397. [CrossRef]

10. Fisk, M.; Gajendragadkar, P.R.; Mäki-Petäjä, K.M.; Wilkinson, I.B.; Cheriyan, J. Therapeutic Potential of p38 MAP Kinase Inhibition in the Management of Cardiovascular Disease. *Am. J. Cardiovasc. Drugs* **2014**, *14*, 155–165. [CrossRef]
11. Kaminska, B. MAPK Signalling Pathways as Molecular Targets for Anti-Inflammatory Therapy–From Molecular Mechanisms to Therapeutic Benefits. *Biochim. Biophys. Acta* **2005**, *1754*, 253–262. [CrossRef] [PubMed]
12. Bardo, M.T.; Bevins, R.A. Conditioned place preference: What does it add to our preclinical understanding of drug reward? *Psychopharmacology* **2000**, *153*, 31–43. [CrossRef]
13. Tzschentke, T.M. Measuring reward with the conditioned place preference (CPP) paradigm: Update of the last decade. *Addict. Biol.* **2007**, *12*, 227–462. [CrossRef] [PubMed]
14. Tzschentke, T.M. Measuring reward with the conditioned place preference paradigm: A comprehensive review of drug effects, recent progress and new issues. *Prog. Neurobiol.* **1998**, *56*, 613–672. [CrossRef]
15. Carlezon, W.A.; Thomas, M.J. Biological Substrates of Reward and Aversion: A Nucleus Accumbens Activity Hypothesis. *Neuropharmacology* **2009**, *56* (Suppl. 1), 122–132. [CrossRef]
16. Gerdjikov, T.V.; Ross, G.M.; Beninger, R.J. Place preference induced by nucleus accumbens amphetamine is impaired by antagonists of ERK or p38 MAP kinases in rats. *Behav. Neurosci.* **2004**, *118*, 740–750. [CrossRef]
17. Zhang, X.; Cui, Y.; Jing, J.; Cui, Y.; Xin, W.; Liu, X. Involvement of p38/NF-κB signaling pathway in the nucleus accumbens in the rewarding effects of morphine in rats. *Behav. Brain Res.* **2011**, *218*, 184–189. [CrossRef]
18. Zhang, X.-Q.; Cui, Y.; Cui, Y.; Chen, Y.; Na, X.-D.; Chen, F.-Y.; Wei, X.-H.; Li, Y.-Y.; Liu, X.-G.; Xin, W.-J. Activation of p38 signaling in the microglia in the nucleus accumbens contributes to the acquisition and maintenance of morphine-induced conditioned place preference. *Brain. Behav. Immun.* **2012**, *26*, 318–325. [CrossRef]
19. Hong, S.-I.; Nguyen, T.-L.; Ma, S.-X.; Kim, H.-C.; Lee, S.-Y.; Jang, C.-G. TRPV1 modulates morphine-induced conditioned place preference via p38 MAPK in the nucleus accumbens. *Behav. Brain Res.* **2017**, *334*, 26–33. [CrossRef]
20. Mannangatti, P.; NarasimhaNaidu, K.; Damaj, M.I.; Ramamoorthy, S.; Jayanthi, L.D. A Role for p38 Mitogen-activated Protein Kinase-mediated Threonine 30-dependent Norepinephrine Transporter Regulation in Cocaine Sensitization and Conditioned Place Preference. *J. Biol. Chem.* **2015**, *290*, 10814–10827. [CrossRef]
21. Bruchas, M.R.; Land, B.B.; Aita, M.; Xu, M.; Barot, S.K.; Li, S.; Chavkin, C. Stress-Induced p38 Mitogen-Activated Protein Kinase Activation Mediates -Opioid-Dependent Dysphoria. *J. Neurosci.* **2007**, *27*, 11614–11623. [CrossRef] [PubMed]
22. Lemos, C.; Salti, A.; Amaral, I.M.; Fontebasso, V.; Singewald, N.; Dechant, G.; Hofer, A.; El Rawas, R. Social interaction reward in rats has anti-stress effects. *Addict. Biol.* **2020**, e12878. [CrossRef] [PubMed]
23. Bruchas, M.R.; Schindler, A.G.; Shankar, H.; Messinger, D.I.; Miyatake, M.; Land, B.B.; Lemos, J.C.; Hagan, C.E.; Neumaier, J.F.; Quintana, A.; et al. Selective p38α MAPK deletion in serotonergic neurons produces stress resilience in models of depression and addiction. *Neuron* **2011**, *71*, 498–511. [CrossRef]
24. Ehrich, J.M.; Messinger, D.I.; Knakal, C.R.; Kuhar, J.R.; Schattauer, S.S.; Bruchas, M.R.; Zweifel, L.; Kieffer, B.L.; Phillips, P.E.M.; Chavkin, C. Kappa Opioid Receptor-Induced Aversion Requires p38 MAPK Activation in VTA Dopamine Neurons. *J. Neurosci.* **2015**, *35*, 12917–12931. [CrossRef]
25. Calcagnetti, D.J.; Schechter, M.D. Place conditioning reveals the rewarding aspect of social interaction in juvenile rats. *Physiol. Behav.* **1992**, *51*, 667–672. [CrossRef]
26. Douglas, L.A.; Varlinskaya, E.I.; Spear, L.P. Rewarding properties of social interactions in adolescent and adult male and female rats: Impact of social versus isolate housing of subjects and partners. *Dev. Psychobiol.* **2004**, *45*, 153–162. [CrossRef]
27. Thiel, K.J.; Okun, A.C.; Neisewander, J.L. Social reward-conditioned place preference: A model revealing an interaction between cocaine and social context rewards in rats. *Drug Alcohol Depend.* **2008**, *96*, 202–212. [CrossRef]
28. Trezza, V.; Damsteegt, R.; Vanderschuren, L.J.M.J. Conditioned place preference induced by social play behavior: Parametrics, extinction, reinstatement and disruption by methylphenidate. *Eur. Neuropsychopharmacol.* **2009**, *19*, 659–669. [CrossRef] [PubMed]
29. Fritz, M.; El Rawas, R.; Salti, A.; Klement, S.; Bardo, M.T.; Kemmler, G.; Dechant, G.; Saria, A.; Zernig, G. Reversal of cocaine-conditioned place preference and mesocorticolimbic Zif268 expression by social interaction in rats. *Addict. Biol.* **2011**, *16*, 273–284. [CrossRef]

30. Kummer, K.; Klement, S.; Eggart, V.; Mayr, M.J.; Saria, A.; Zernig, G. Conditioned place preference for social interaction in rats: Contribution of sensory components. *Front. Behav. Neurosci.* **2011**, *5*, 80. [CrossRef]
31. El Rawas, R.; Klement, S.; Kummer, K.K.; Fritz, M.; Dechant, G.; Saria, A.; Zernig, G. Brain regions associated with the acquisition of conditioned place preference for cocaine vs. social interaction. *Front. Behav. Neurosci.* **2012**, *6*, 6. [CrossRef] [PubMed]
32. El Rawas, R.; Saria, A. The Two Faces of Social Interaction Reward in Animal Models of Drug Dependence. *Neurochem. Res.* **2016**, *41*, 492–499. [CrossRef] [PubMed]
33. Yates, J.R.; Beckmann, J.S.; Meyer, A.C.; Bardo, M.T. Concurrent choice for social interaction and amphetamine using conditioned place preference in rats: Effects of age and housing condition. *Drug Alcohol Depend.* **2013**, *129*, 240–246. [CrossRef] [PubMed]
34. Salti, A.; Kummer, K.K.; Sadangi, C.; Dechant, G.; Saria, A.; El Rawas, R. Social interaction reward decreases p38 activation in the nucleus accumbens shell of rats. *Neuropharmacology* **2015**, *99*, 510–516. [CrossRef] [PubMed]
35. Kalueff, A.V.; Tuohimaa, P. The grooming analysis algorithm discriminates between different levels of anxiety in rats: Potential utility for neurobehavioural stress research. *J. Neurosci. Methods* **2005**, *143*, 169–177. [CrossRef]
36. Land, B.B.; Bruchas, M.R.; Lemos, J.C.; Xu, M.; Melief, E.J.; Chavkin, C. The dysphoric component of stress is encoded by activation of the dynorphin kappa-opioid system. *J. Neurosci.* **2008**, *28*, 407–414. [CrossRef]
37. Bruchas, M.R.; Macey, T.A.; Lowe, J.D.; Chavkin, C. Kappa Opioid Receptor Activation of p38 MAPK Is GRK3- and Arrestin-dependent in Neurons and Astrocytes. *J. Biol. Chem.* **2006**, *281*, 18081–18089. [CrossRef]
38. Al-Hasani, R.; McCall, J.G.; Shin, G.; Gomez, A.M.; Schmitz, G.P.; Bernardi, J.M.; Pyo, C.-O.; Park, S.; Marcinkiewcz, C.M.; Crowley, N.A.; et al. Distinct Subpopulations of Nucleus Accumbens Dynorphin Neurons Drive Aversion and Reward. *Neuron* **2015**, *87*, 1063–1077. [CrossRef]
39. Sinha, R. Chronic stress, drug use, and vulnerability to addiction. *Ann. N. Y. Acad. Sci.* **2008**, *1141*, 105–130. [CrossRef]
40. Sousa, N. The Dynamics of the Stress Neuromatrix. *Mol. Psychiatry* **2016**, *21*, 302–312. [CrossRef]
41. Zheng, G.; Chen, Y.; Zhang, X.; Cai, T.; Liu, M.; Zhao, F.; Luo, W.; Chen, J. Acute cold exposure and rewarming enhanced spatial memory and activated the MAPK cascades in the rat brain. *Brain Res.* **2008**, *1239*, 171–180. [CrossRef] [PubMed]
42. Peng, Z.; Wang, H.; Zhang, R.; Chen, Y.; Xue, F.; Nie, H.; Wu, D.; Wang, Y.; Tan, Q. Gastrodin ameliorates anxiety-like behaviors and inhibits IL-1beta level and p38 MAPK phosphorylation of hippocampus in the rat model of posttraumatic stress disorder. *Physiol. Res.* **2013**, *62*, 537–545.
43. Sanchez, M.M.; Alagbe, O.; Felger, J.C.; Zhang, J.; Graff, A.E.; Grand, A.P.; Maestripieri, D.; Miller, A.H. Activated p38 MAPK is associated with decreased CSF 5-HIAA and increased maternal rejection during infancy in rhesus monkeys. *Mol. Psychiatry* **2007**, *12*, 895–897. [CrossRef]
44. Zhao, Y.-W.; Pan, Y.-Q.; Tang, M.-M.; Lin, W.-J. Blocking p38 Signaling Reduces the Activation of Pro-inflammatory Cytokines and the Phosphorylation of p38 in the Habenula and Reverses Depressive-Like Behaviors Induced by Neuroinflammation. *Front. Pharmacol.* **2018**, *9*, 511. [CrossRef]
45. Stefanoska, K.; Bertz, J.; Volkerling, A.M.; Van der Hoven, J.; Ittner, L.M.; Ittner, A. Neuronal MAP kinase p38α inhibits c-Jun N-terminal kinase to modulate anxiety-related behaviour. *Sci. Rep.* **2018**, *8*, 14296. [CrossRef]
46. Zan, G.-Y.; Wang, Q.; Wang, Y.-J.; Chen, J.-C.; Wu, X.; Yang, C.-H.; Chai, J.-R.; Li, M.; Liu, Y.; Hu, X.-W.; et al. p38 Mitogen-Activated Protein Kinase Activation in Amygdala Mediates κ Opioid Receptor Agonist U50,488H-induced Conditioned Place Aversion. *Neuroscience* **2016**, *320*, 122–128. [CrossRef]
47. Land, B.B.; Bruchas, M.R.; Schattauer, S.; Giardino, W.J.; Aita, M.; Messinger, D.; Hnasko, T.S.; Palmiter, R.D.; Chavkin, C. Activation of the kappa opioid receptor in the dorsal raphe nucleus mediates the aversive effects of stress and reinstates drug seeking. *Proc. Natl. Acad. Sci. USA* **2009**, *106*, 19168–19173. [CrossRef]
48. Rolli, M.; Kotlyarov, A.; Sakamoto, K.M.; Gaestel, M.; Neininger, A. Stress-induced Stimulation of Early Growth Response gene-1 by p38/stress-activated Protein Kinase 2 Is Mediated by a cAMP-responsive Promoter Element in a MAPKAP Kinase 2-independent Manner. *J. Biol. Chem.* **1999**, *274*, 19559–19564. [CrossRef]

49. Zhu, C.-B.; Carneiro, A.M.; Dostmann, W.R.; Hewlett, W.A.; Blakely, R.D. p38 MAPK activation elevates serotonin transport activity via a trafficking-independent, protein phosphatase 2A-dependent process. *J. Biol. Chem.* **2005**, *280*, 15649–15658. [CrossRef]
50. Samuvel, D.J.; Jayanthi, L.D.; Bhat, N.R.; Ramamoorthy, S. A role for p38 mitogen-activated protein kinase in the regulation of the serotonin transporter: Evidence for distinct cellular mechanisms involved in transporter surface expression. *J. Neurosci.* **2005**, *25*, 29–41. [CrossRef]
51. Zhu, C.-B.; Lindler, K.M.; Owens, A.W.; Daws, L.C.; Blakely, R.D.; Hewlett, W.A. Interleukin-1 Receptor Activation by Systemic Lipopolysaccharide Induces Behavioral Despair Linked to MAPK Regulation of CNS Serotonin Transporters. *Neuropsychopharmacology* **2010**, *35*, 2510–2520. [CrossRef] [PubMed]
52. Baganz, N.L.; Lindler, K.M.; Zhu, C.B.; Smith, J.T.; Robson, M.J.; Iwamoto, H.; Deneris, E.S.; Hewlett, W.A.; Blakely, R.D. A requirement of serotonergic p38α mitogen-activated protein kinase for peripheral immune system activation of CNS serotonin uptake and serotonin-linked behaviors. *Transl. Psychiatry* **2015**, *5*, e671. [CrossRef] [PubMed]
53. Huang, C.-C.; You, J.-L.; Wu, M.-Y.; Hsu, K.-S. Rap1-induced p38 mitogen-activated protein kinase activation facilitates AMPA receptor trafficking via the GDI.Rab5 complex. Potential role in (S)-3,5-dihydroxyphenylglycene-induced long term depression. *J. Biol. Chem.* **2004**, *279*, 12286–12292. [CrossRef] [PubMed]
54. Liang, Y.-C.; Huang, C.-C.; Hsu, K.-S. A Role of p38 Mitogen-Activated Protein Kinase in Adenosine A_1 Receptor-Mediated Synaptic Depotentiation in Area CA1 of the Rat Hippocampus. *Mol. Brain* **2008**, *1*, 13. [CrossRef]
55. Bleakman, D.; Alt, A.; Witkin, J.M. AMPA Receptors in the Therapeutic Management of Depression. *CNS Neurol. Disord. Drug Targets* **2007**, *6*, 117–126. [CrossRef] [PubMed]
56. Pariante, C.M.; Miller, A.H. Glucocorticoid receptors in major depression: Relevance to pathophysiology and treatment. *Biol. Psychiatry* **2001**, *49*, 391–404. [CrossRef]
57. Wang, X.; Wu, H.; Miller, A.H. Interleukin 1alpha (IL-1alpha) Induced Activation of p38 Mitogen-Activated Protein Kinase Inhibits Glucocorticoid Receptor Function. *Mol. Psychiatry* **2004**, *9*, 65–75. [CrossRef]
58. Pace, T.W.W.; Hu, F.; Miller, A.H. Cytokine-effects on glucocorticoid receptor function: Relevance to glucocorticoid resistance and the pathophysiology and treatment of major depression. *Brain. Behav. Immun.* **2007**, *21*, 9–19. [CrossRef]
59. Schindler, A.G.; Li, S.; Chavkin, C. Behavioral stress may increase the rewarding valence of cocaine-associated cues through a dynorphin/kappa-opioid receptor-mediated mechanism without affecting associative learning or memory retrieval mechanisms. *Neuropsychopharmacology* **2010**, *35*, 1932–1942. [CrossRef]
60. Yamada, H.; Shimoyama, N.; Sora, I.; Uhl, G.R.; Fukuda, Y.; Moriya, H.; Shimoyama, M. Morphine can produce analgesia via spinal kappa opioid receptors in the absence of mu opioid receptors. *Brain Res.* **2006**, *1083*, 61–69. [CrossRef]
61. Pan, Z.Z.; Tershner, S.A.; Fields, H.L. Cellular mechanism for anti-analgesic action of agonists of the kappa-opioid receptor. *Nature* **1997**, *389*, 382–385. [CrossRef]
62. Neisewander, J.L.; Pierce, R.C.; Bardo, M.T. Naloxone enhances the expression of morphine-induced conditioned place preference. *Psychopharmacology* **1990**, *100*, 201–205. [CrossRef] [PubMed]
63. Matthes, H.W.; Maldonado, R.; Simonin, F.; Valverde, O.; Slowe, S.; Kitchen, I.; Befort, K.; Dierich, A.; Le Meur, M.; Dollé, P.; et al. Loss of morphine-induced analgesia, reward effect and withdrawal symptoms in mice lacking the mu-opioid-receptor gene. *Nature* **1996**, *383*, 819–823. [CrossRef]
64. Wu, G.; Huang, W.; Zhang, H.; Li, Q.; Zhou, J.; Shu, H. Inhibitory Effects of Processed Aconiti Tuber on Morphine-Induced Conditioned Place Preference in Rats. *J. Ethnopharmacol.* **2011**, *136*, 254–259. [CrossRef]
65. Billa, S.K.; Xia, Y.; Morón, J.A. Disruption of morphine-conditioned place preference by a delta2-opioid receptor antagonist: Study of mu-opioid and delta-opioid receptor expression at the synapse. *Eur. J. Neurosci.* **2010**, *32*, 625–631. [CrossRef]
66. Al-Hasani, R.; Bruchas, M.R. Molecular mechanisms of opioid receptor-dependent signaling and behavior. *Anesthesiology* **2011**, *115*, 1363–1381. [CrossRef]

67. Li, Z.-H.; Chu, N.; Shan, L.-D.; Gong, S.; Yin, Q.-Z.; Jiang, X. Inducible Expression of Functional Mu Opioid Receptors in Murine Dendritic Cells. *J. Neuroimmune Pharmacol.* **2009**, *4*, 359–367. [CrossRef]
68. Rozenfeld-Granot, G.; Toren, A.; Amariglio, N.; Nagler, A.; Rosenthal, E.; Biniaminov, M.; Brok-Simoni, F.; Rechavi, G. MAP kinase activation by mu opioid receptor in cord blood CD34+CD38- cells. *Exp. Hematol.* **2002**, *30*, 473–480. [CrossRef]

MDPI
St. Alban-Anlage 66
4052 Basel
Switzerland
Tel. +41 61 683 77 34
Fax +41 61 302 89 18
www.mdpi.com

International Journal of Molecular Sciences Editorial Office
E-mail: ijms@mdpi.com
www.mdpi.com/journal/ijms

www.ingramcontent.com/pod-product-compliance
Lightning Source LLC
LaVergne TN
LVHW070949170726
843515LV00005B/947